MAPPERLEY HOSPITAL
MEDICAL LIBRARY

Author................................Class Mark........................

Title...

16 MAY 1995	01 DEC 2008	
-5 OCT 1999	18 DEC 2008	
-9 NOV 2001		
28 MAR 2003	30 OCT 2009	
25 JUL 2003	22 DEC 2015	
12 JUN 2006		
19 DEC 2006		
16 JAN 2007		
07 MAR 2008		
07 ... 2008		

Alzheimer disease,
Down syndrome, and
their relationship

John Langdon Haydon Down

Alois Alzheimer

Photographs provided by the National Library of Medicine

Alzheimer disease, Down syndrome, and their relationship

Edited by

J. M. BERG
*Surrey Place Centre and
University of Toronto*

H. KARLINSKY
*Riverview Hospital and
University of British Columbia*

and

A. J. HOLLAND
*Ida Darwin Hospital and
University of Cambridge*

OXFORD · NEW YORK · TOKYO · TORONTO

OXFORD UNIVERSITY PRESS

1993

Oxford University Press, Walton Street, Oxford OX2 6DP
Oxford   New York   Toronto
Delhi   Bombay   Calcutta   Madras   Karachi
Kuala Lumpur   Singapore   Hong Kong   Tokyo
Nairobi   Dar es Salaam   Cape Town
Melbourne   Auckland   Madrid
and associated companies in
Berlin   Ibadan

Oxford is a trade mark of Oxford University Press

Published in the United States
by Oxford University Press Inc., New York

A catalogue record for this book is available from the British Library

Library of Congress Cataloging in Publication Data
Alzheimer disease, Down syndrome, and their relationship/edited by
J.M. Berg, H. Karlinsky, and A.J. Holland. — 1st ed.
Includes bibliographical references and index.
1. Alzheimer's disease—Risk factors. 2. Down's syndrome.
3 Alzheimer's disease—Etiology. I. Berg, J.M. (Joseph Maurice)
II. Karlinsky, Harry, 1954– . III. Holland, A.J., Dr.
[DNLM: 1. Alzheimer's Disease. 2. Down Syndrome. WM 220 A477823 1993]
RC523.A379   1993   616.8'31–dc20   93–28474
ISBN 0–19–262382–6 (hb.)

Typeset by Selwood Systems,
Midsomer Norton, Avon
Printed in Great Britain by
Biddles Ltd, Guildford & King's Lynn

Foreword

by Sir Martin Roth

Alzheimer disease, Down syndrome, and their relationship is the most wide-ranging and comprehensive work to have been devoted to the nature of the association between these two clinical and biological phenomena. Its appearance is timely. Until a few decades ago the high mortality of those with Down syndrome (DS) during the first two decades of life, owing to concomitant congenital cardiac malformation, respiratory infection, and other intercurrent diseases, resulted in few affected individuals surviving into adult life. Following the advent of antibiotics and advances in cardiac surgery a growing proportion have survived until the fifth decade and beyond.

Having reached middle age they have proved to be at high risk of developing features consistent with Alzheimer disease (AD) as observed in the general population. As it was relatively common for strong emotional bonds to be established between those with DS and their parents and family, they were often brought up at home and continued to be cared for during adult life by their kindred. With the onset of dementia, persons with DS are liable to manifest disturbed behaviour, aggressiveness, incontinence, epilepsy, and other neurological complications and general helplessness. Families have found themselves increasingly unable to cope and other provisions for nursing care and supervision have become imperative. There is need for local and health authorities and the community at large to be made aware of the onerous stresses to which carers are being exposed by the changing clinical profile of DS to the detriment of their own mental health. The chapters devoted to the management of DS and the familial and social aspects in this volume will make a valuable contribution to this end.

The authoritative and comprehensive chapters devoted to the description and diagnostic aspects convey in a clear manner the exceptional clinical and scientific interest of AD as it manifests itself in DS. The paradoxical character of some aspects of the phenomenon adds to its fascination. AD of late life in the general population is widely regarded as having some kinship with the phenomenon of ageing. The pathological changes in the brain are indeed similar in a qualitative sense to those commonly found in the brains of well-preserved elderly persons. However, outside the central nervous system there is often very little in the way of ageing to observe in AD.

Many persons with well-established AD look healthy and relatively youthful and at post-mortem examination organs outside the brain may exhibit little or no pathology. This disorder differs from relatively convincing models of ageing such as Werner syndrome in which early greying of hair, atrophy of the skin, cataracts, peripheral vascular disease, maturity onset diabetes, osteoporosis, and testicular atrophy are associated with a prematurely aged appearance and in women an early menopause. Precocious ageing is also manifest in a proportion of persons with DS. Chronic cardio-respiratory disease and other disorders and a life spent in institutions may, however, have contributed to this in the past. But ageing in this sense is not a feature of AD.

The great majority of persons with AD do, however, appear in old age with a 10% prevalence among those aged 70 years and over and 20–25% in those aged more than 80 years. Cases of the early onset type may commence before the age of 50 but are exceedingly rare. In contrast with the high prevalence of AD, DS has an incidence of 1 in about 800 live births. Severe mental impairment is manifest in a high proportion from the time of birth onwards. The first pathological changes in the brain in the form of amorphous amyloid deposits with or without some neurofibrillary tangles (NFT) appear as early as in the second decade in some cases. They rise steeply in prevalence and the distribution and severity of the changes increases thereafter. The generalisation that AD neuropathology is peculiar to old age is thus refuted.

The fact that the incidence of dementia in DS is on the average delayed by about 20 to 30 years is yet another paradox, for those with low brain weight and a relatively small population of neurones might, on general grounds, have been expected to suffer a more rapid progression than AD in the general population which generally runs its course within a period of 7–8 years.

The transitional period in DS of some decades before the commencement of dementia and the appearance for the first time of a range of pathological changes comparable with those in AD has some far-reaching implications for the pathogenesis of the Alzheimer form of cerebral degeneration. To take up just one point here, the earliest pathological changes are confined in a high proportion of cases entirely to amyloid deposits, with neuritic plaques and NFT being scarce until the commencement of dementia 20–30 years later. But, as no dementia is manifest in this early phase, the findings are difficult to reconcile with the claim that amyloid is by itself damaging to neurones.

Only in recent years has the closeness of the resemblance and the character of the pathological neurotransmitters and certain immunochemical changes been recognized in the brains of those with AD whether or not they also have DS. The contributions of Mann fully summarized in his chapter have been important for the establishment of this detailed correspondence between the two phenomena. The density of neuritic plaques, NFT, granulovacuolar degeneration and the fallout of pyramidal neurones typical of AD in the general population are closely replicated in DS cases as age advances.

The markers for cholinergic, noradrenergic, and serotonergic neurotransmitters in cortical pyramidal cells and also in GABAergic cells in the cortex are reduced in aged persons with DS as in AD in general. Neuronal fallout of the subcortical nuclei which form the starting point of the projections of the pathways for these transmitters to the cerebral cortex are identical in affected individuals with and without DS. The topography of the cortical areas that are the sites of neuronal loss are also closely similar.

In respect of immunochemical reactivity of the NFT in DS to antibodies against tau and ubiquitin and other proteins, the identity has not yet been shown to be complete. The total body of evidence for the present suggests that the pathological degenerative process of the dementia of DS is likely to prove homologous in the main with that of AD in the general population. However, there may be significant clinical disparities between the two phenomena. The symptoms of the early stages of the process appear similar in dementia of DS cases and AD of later life in persons without DS, but the final clinical stages may differ in those with DS by increased presence of more florid neurological symptoms, such as myoclonus and seizures, disturbances of gait and severe disorganization of speech (Evenhuis 1990).

There are also discrepancies between the published reports as regards the prevalence of dementia in DS. Older series of subjects studied in institutional cohorts of cases have reported 90 to 100% of individuals aged 40 and over as developing a clinical syndrome closely similar to AD in the general population. Considerably lower figures have been reported from recent enquiries (Franchesci *et al.* 1990). In a clinico-pathological study Wisniewski *et al.* (1985) reported that 56 out of 100 DS brains exhibited Alzheimer-like changes. They noted considerable disparity between the pathological findings and their clinical manifestations, which were less frequent than they had been led to expect from the widespread cerebral pathology they found in a high proportion of the subjects (Wisniewski and Rabe 1986). This is an important claim and if confirmed another departure from the associations observed in AD in the general population where those who exhibit more than 'threshold' levels of tangles and neuritic plaques in hippocampus and cerebral cortex are almost invariably demented while those with sub-threshold levels of pathology are mentally intact (Roth 1986). Differences in the diagnostic criteria employed and variations in methods of quantifying pathological change may be responsible for some of these disparities. Standardization of methods of measurement of cognitive function and quantification of cerebral pathology might reduce or eliminate them.

The dementia of typical AD and that of DS are separated by two other differences that merit attention. The first is that at a conservative estimate about half or more of those affected with DS who develop amorphous amyloid deposits around the third decade will have an AD-like syndrome by the time they reach the age of 50 years and exhibit cerebral pathology identical with that of AD in the general population. In the latter case, only a small fraction of those who have cerebral amyloid deposits at 70 years proceed to develop dementia. The second is that a transitional period of some 20–25 years exists between the appearance of the first pathological changes in DS and the commencement of dementia. In contrast, AD in the general population frequently runs its course within less than a decade. The disparities create research opportunities not available for investigating the aetiology and pathogenesis of AD in the general population.

Another question is posed by the different forms of chromosomal abnormality associated with DS. In the majority of cases this is a trisomy of the whole of chromosome 21. But in some 5% of cases, the condition arises from unbalanced translocation of the long arm of chromosome 21. There is evidence to suggest that the part of chromosome 21 which is responsible for the genesis of the DS phenotype is the segment of the chromosome 21 distal to the gene that codes for superoxide dismutase (Whalley *et al.* 1982). If the segment is so limited, it excludes the gene for the β-amyloid precursor protein (APP). It would therefore be important to determine whether those with this translocation do or do not develop a typical clinical syndrome of AD.

The answer to this question also has bearing upon certain problems of aetiology that have arisen in relation to AD in the general population. When the locus of the APP gene was first discovered to be on chromosome 21 in families in which the disease was transmitted as a major autosomal dominant (St George-Hyslop *et al.* 1987) it was suspected that it might be identical with the gene that was responsible for the transmission of the disease. But, early investigations demonstrated failure of the APP gene to segregate with the disease in several families (van Broeckhoven *et al.* 1987).

However a point mutation has recently been located at residue 717 of APP which entails a substitution of isoleucine for valine (Goate *et al.* 1991). Two other mutations have been more recently identified at the same locus. Only a very small proportion of the rare familial forms are due to a single point mutation and no mutations have been found in the common late life sporadic forms of AD. The point mutations formed in each case suggested that the deposition of the β-amyloid portion of APP might be the first event in the development of certain rare forms of AD. It also raised the possibility that it might constitute an early step in the final common pathway of causation of all forms of AD. The fact that amyloid deposition is the first observable change in the evolution of the AD of DS to be followed after an interval of years by the emergence of the full range of pathological and other cerebral changes in AD have provided some support for the β-amyloid hypotheses. Recently, Schellenberg *et al.* (1992) and subsequently other groups of investigators have adduced genetic linkage evidence for a familial AD locus on chromosome 14, while Corder *et al.* (1993) have identified apolipoprotein E type 4 allele on chromosome 19 as a risk factor for both sporadic and familial late-onset AD. Whether these additional genetic loci are involved with the processing of APP needs to be clarified.

Moreover, several lines of evidence conflict with the theory that the deposition of amyloid cleaved and abnormally processed from APP is the specific cause of AD. Deposits of amyloid are common among normal non-demented persons aged 75 and over, but only a very small proportion of such persons proceed to develop the full clinical changes and cerebral pathology of AD. The presence of amyloid deposits alone is not correlated with significant cognitive impairment, nor is there evidence of neuronal destruction in those in whom the cerebral cortex contains deposits of amyloid alone (Braak and Braak 1991). Amyloid deposits in the cerebellum and striatum of patients with AD are not correlated with neuronal damage nor have relevant neurological signs been found in patients with AD during life.

It is also noteworthy that all the single point mutations on the APP gene found in a small proportion of familial cases have been located near to but outside the region of the gene which codes for the β-amyloid protein. There is a point mutation at residue 693 that is within the β-amyloid portion of the sequence. It is associated with amyloid deposition in the walls of cerebral arteries. But there are neither neuritic plaques nor NFT (Levy *et al.* 1990) and patients die of cerebral haemorrhage without symptoms of dementia during life.

The findings in hereditary cerebral haemorrhage of the Dutch type, and other rare syndromes with similar genetic and pathological characteristics are consistent with the hypothesis that it is the neuritic plaques and NFT and the paired helical filaments common to both that are most closely related to the destruction of neurones which are responsible for the mental deterioration of AD.

They are also consistent with the results of recent pathological studies in persons with DS ranging in age from 13 to 71 years (Mann *et al.* 1990). The question at issue is whether the β-amyloid is both the first event and the specific cause of dementia or, as Mann has put it in his chapter, '. . . whether it simply represents a relatively innocuous marker of a wider ranging process that carries in its wake other changes some of which might lead to neurofibrillary alterations and cell death'.

Deposition of NFT in the cerebral cortex is highly correlated with measures of dementia during life (Wilcock and Esiri 1982; Braak and Braak 1991). Structural molecular and immunochemical studies in AD in the general population have shown the paired helical filaments of which tangles are constituted to be intimately associated with dismemberment of neurones; their shape expands and becomes deformed, the cell membrane and nucleus disappears and the tangles are extruded into extra-cellular space. These tangles cease to interact with antibodies prepared against certain segments of tau protein but interact strongly with antibodies prepared against the middle repeat section of this protein (Bondareff *et al.* 1990).

The lesions detected by antibodies prepared against tau protein correspond with those that discriminate sharply between normal ageing and dementia (Duyckaerts *et al.* 1990). The sites of tau immunoreactivity are located in extra-cellular tangles, dystrophic neurites and neuritic plaques. The main missing links in the process of development towards dementia consist of the changes that intervene between deposition of β-amyloid protein as in the early stages of DS and in normal elderly subjects and the development of paired helical filaments and tangles which contain no amyloid protein but are closely associated with the process of destruction of neurones and the development of mental deterioration.

Further exploration needs to be undertaken to define the course of events during the transitional period. Intracellular tangles are immuno-labelled by antibodies which differ from those that interact with extracellular tangles: the former are prepared against the N-terminal of tau protein, the latter against its central repeat region and, as is recently shown, with antibodies against its C-terminal (Bondareff *et al.* 1993). Tau protein formed in PHF's is differentiated from normal tau protein by its immunochemical features (Novak *et al.* 1991). It is also redistributed from the axonal pool which is its normal location to the somato-dendritic part of the neurone where it is found in a highly insoluble and abnormal form (Novak *et al.* 1991; Mukaetove-Ladinska *et al.* 1992). It is found in this insoluble form in the PHF of NFT, neuritic plaques and dystrophic neurites of AD in the general population.

It is not known whether these immunochemical phenomena and the molecular changes underlying them are replicated when AD occurs in DS. Application of the methods employed in the enquiries described above into the AD syndrome of DS during the long transitional period of AD development in individuals with DS might shed light on the steps in the process of evolution from the stage of plaque deposition to that of tau bearing NFT and neurite formation and thence to neuronal destruction and the causal factors underlying this process. Such enquiries could not be undertaken in the typical forms of AD in the general population. The ultimate objective would be to arrest the degenerative process at the stage reached when only amorphous amyloid plaques and very few NFT are manifest. Whether or not such an objective can be attained only a large amount of effort and the passage of time can tell.

The editors and the authors of the chapters have performed a valuable service with the publication of this lucid, coherent, well documented and wide ranging volume which covers the clinical, psychological, social and neurobiological aspects of a hybrid phenomenon of exceptional interest. The book defines some areas of ignorance where further observations might shed fresh light both on Alzheimer disease and on some aspects of Down syndrome and is likely to prove a reference source for some years to come.

References

Bondareff, W., Wischik, C.M., Novak, M., Amos, W.B., Klug, A., and Roth, M. (1990). Molecular analysis of neurofibrillary degeneration in Alzheimer's disease: An immunochemical study. *American Journal of Pathology*, 37, 711–23.

Bondareff, W., Harrington, C., Wischik, C.M., Novak, M., Amos, W.B., Hauser, D.L., and Roth, M. (1993). Confocal microscopic localisation of phosphorylated and nonphosphorylated Tau epitopes in neurofibrillary tangles in Alzheimer's disease. (Submitted for publication).

Braak, H., and Braak, E. (1991). Neuropathological stageing of Alzheimer-related changes. *Acta Neuropathologica*, 82, 239–59.

Corder, E.H., Saunders, A.M., Strittmatter, W.J., Schmechel, D.E., Gaskell, P.C., Small, G.W., *et al.* (1993). Gene dose of apolipoprotein E type 4 allele and the risk of Alzheimer's disease in late onset families *Science*, 261, 921–3.

Duyckaerts, C., Delaere, P., Hauw, J.J., Abbamondi-Pinto, A.L., Sorbi, S., Allen, I., *et al.* (1990). Rating of the lesions in senile dementia of the Alzheimer type: concordance between laboratories. A European multicenter study under the auspices of EURAGE. *Journal of the Neurological Sciences*, 97, 295–323.

Evenhuis, H.M. (1990). The natural history of dementia in Down's syndrome. *Archives of Neurology*, 47, 263–7.

Franchesci, C., Monti, D., Cossarizza, A., Tomasi, A., Sola, P., and Zannotti, M. (1990). Oxidative stress, poly (ADP) ribolysation and ageing: in vitro studies in lymphocytes from normal and Down's syndrome subjects of different age and from patients with Alzheimer's dementia. *Advances in Experimental Medicine and Biology*, 264, 499–502.

Goate, A., Chartier-Harlin, M.-C., Mullan, M., Brown, J., Crawford, F., Fidani, L., *et al.* (1991). Segregation of a missense mutation in the amyloid precursor protein gene with familial Alzheimer's disease. *Nature*, 349, 704–6.

Levy, E., Carman, M.D., Fernandez-Madrid, I.J., Power, M.D., Lieberburg, I., Van Duinen, S.G., *et al.* (1990). Mutation of the Alzheimer's disease amyloid gene in hereditary cerebral hemorrhage, Dutch type. *Science*, 248, 1124–6.

Mann, D.M.A., Jones, D., Prinja, D., and Purkiss, M.S. (1990). The prevalence of amyloid (A4) protein deposits within the cerebral and cerebellar cortex in Alzheimer's disease and Down's syndrome. *Acta Neuropathologica*, 80, 318–27.

Mukaetova-Ladinska, E.B., Harrington, C.R., Hills, R., O'Sullivan, A., Roth, M., and Wischik, C.M. (1992). Regional distribution of paired helical filaments and normal tau proteins in ageing and in Alzheimer's disease with and without occipital lobe involvement. *Dementia*, 3, 61–9.

Novak, M., Jakes, R., Edwards, P.C., Milstein, C., and Wischik, C.M. (1991). Difference between the tau protein of Alzheimer paired helical filament core and normal tau revealed by epitope analysis of monoclonal antibodies 423 and 7.51. *Proceedings of the National Academy of Sciences of the United States of America*. 88, 5837–41.

Roth, M. (1986). The association of clinical and neurobiological findings and its bearing on the classification and aetiology of Alzheimer's disease. In *Alzheimer's Disease and Related Disorders*, British Medical Bulletin, Vol. 42, (ed. M. Roth and L.L. Iversen), pp. 42–50.

Schellenberg, G.D., Bird, T.D., Wijsman, E.M., Orr, H.T., Anderson, L., Nemens, E., *et al.* (1992). Genetic linkage evidence for a familial Alzheimer's disease locus on chromosome 14. *Science*, 258, 668–71.

St. George-Hyslop, P.H., Tanzi, R., Polinsky, R.J., Haines, J.L., Nee, L., Watkins, P.C., *et al.* (1987). The genetic defect causing familial Alzheimer's disease maps on chromosome 21. *Science,* 235, 885–90.

Van Broeckhoven, C., Genthe, A.M., Vandenberghe, A., Hortsthemke, B., Backhovens, H., Raeymaekers, P., *et al.* (1987). Failure of familial Alzheimer's disease to segregate with the A4-amyloid gene in several European families. *Nature*, 329, 153–5.

Whalley, L.J., Carothers, S.D., Collyer, S., De May, R., and Frackiewicz, A. (1982). A study of familial factors in Alzheimer's disease. *British Journal of Psychiatry,* 140, 249–56.

Wilcock, G.K., and Esiri, M.M. (1982). Plaques, tangles and dementia. *Journal of the Neurological Sciences*, 56, 343–56.

Wisniewski, H.M., and Rabe, A. (1986). Discrepancy between Alzheimer-type neuropathology and dementia in persons with Down's syndrome. *Annals of the New York Academy of Sciences*, 477, 247–60.

Wisniewski, K.E., Wisniewski, H.M., and Wen, G.Y. (1985). Occurrence of neuropathological changes and dementia of Alzheimer's disease in Down's syndrome. *Annals of Neurology*, 17, 278–82.

Preface

At first glance, there may appear to be little, if anything, in common between Alzheimer disease and Down syndrome. Alzheimer disease is a neurodegenerative disorder, in which the karyotype is apparently normal, with onset in relatively late adulthood, and with inexorably progressive dementia as its major clinical manifestation. Down syndrome, on the other hand, is a disorder in which a characteristic chromosomal aberration is associated with distinctive morphological phenotypic anomalies, clinically recognizable initially at, or soon after, birth, and with mental deficit of variable degree as a persistent feature. However, in the last few decades it has become increasingly clear that important links between the two conditions exist. This is most apparent in the presence of typical neuropathological changes of Alzheimer disease in persons with Down syndrome some 20–30 years before these changes usually occur in individuals with Alzheimer disease in the general population. In recent years, intensive study of this phenomenon and its ramifications and implications in clinical and biological terms has yielded significant data that are leading to greater understanding of both disorders.

The present volume, beginning with a general overview of each of these disorders, addresses from various perspectives the characteristics and nature of the Alzheimer disease/Down syndrome relationship, with contributions reflecting the current status of both research and practice. It is hoped that the book will be of interest and useful both to researchers on the subject and to health professionals in general who work with adults with Down syndrome and their families.

An explanatory observation on terminology used throughout the text seems appropriate here. In particular, the term Alzheimer disease has historically had a number of different meanings. Currently, it is generally utilized to denote the insidious, progressive dementia that accompanies proven or presumed Alzheimer-type neuropathological brain changes, irrespective of the age of onset in affected individuals. Perhaps rather arbitrarily, the editors have therefore uniformly applied the term Alzheimer disease to persons with Down syndrome who demonstrate essentially the same characteristic clinical and neuropathological features of the disease as those observed in the general population. As the aetiological and clinical heterogeneity of Alzheimer disease is now unequivocally established, such usage does not preclude subsequent definition of unique aetiological or clinical features of Alzheimer disease in Down syndrome as compared to other subtypes of the disease.

Finally, we gladly record our thanks to the invited contributing authors whose expertise and up-to-date knowledge have resulted in highly informative, scholarly and insightful chapters. It is a pleasure to acknowledge the support of the Queen Elizabeth Hospital, the Toronto Hospital, Riverview Hospital, Surrey Place Centre, the University of Toronto's Division of Geriatric Psychiatry and the Alzheimer Society For Metropolitan Toronto. Appreciation is also expressed, no less enthusiastically, to Ms Marika Korossy, Ms Joan Edwards, Ms Anne Lennox, Ms Ginny Strain, and Ms Anita Thibert for their diligent technical and secretarial assistance which was provided with much good grace and beyond the call of duty.

Toronto
October 1992

J. M. B.
H. K.
A. J. H.

Contents

Contributors

M. S. Adlin, M.D.
Assistant Professor,
Department of Medicine,
Section on Geriatrics and Gerontology,
University of Wisconsin-Madison,
425 Henry Mall, Madison, Wisconsin 53706, USA

T. G. Beach, M.D., Ph.D.
Resident in Neuropathology,
University of British Columbia,
Vancouver General Hospital,
Department of Pathology,
855 West 12th Avenue, Vancouver, British Columbia V5Z 1M9,
Canada

J. M. Berg, M.B., B.Ch., M.Sc., F.R.C.Psych., F.C.C.M.G.
Professor Emeritus,
Faculty of Medicine, University of Toronto;
Director of Biomedical Services and Research,
Surrey Place Centre, 2 Surrey Place, Toronto, Ontario M5S 2C2,
Canada

I. Campbell-Taylor, Ph.D.
Assistant Professor,
Faculty of Medicine,
University of Toronto;
Consultant, Communication Disorders,
Surrey Place Centre, 2 Surrey Place, Toronto, Ontario M5S 2C2,
Canada

L. Crayton, Ph.D.
Research Psychologist,
Institute of Psychiatry,
University of London,
London, SE5 8AF, UK

A. J. Dalton, Ph.D.
Deputy Director Grants/Research Development and Head,
Aging Processes Laboratory,
New York State Institute for Basic Research,
1050 Forest Hill Road, Staten Island, New York 10314, USA

S. B. Dunnett, M.A., Ph.D.
University Lecturer,
Department of Experimental Psychology,
University of Cambridge,
Downing Street, Cambridge CB2 3EB, UK

C. J. Epstein, M. D.
Professor of Pediatrics,
University of California,
San Francisco, California 94143, USA

J. B. Fotheringham, M.D., F.R.C.P. (C)
Associate Professor,
Department of Psychiatry,
Queen's University,
Kingston, Ontario K7L 3N6, Canada

J. A. Hardy, Ph.D.
Pfeiffer Chair for Alzheimer Research,
Department of Psychiatry,
University of South Florida,
3515 East Fletcher Avenue,
Tampa, Florida 33613, USA

J. J. A. Holden, Ph.D.
Professor,
Department of Psychiatry,
Queen's University,
c/o Ongwanada Resource Centre,
191 Portsmouth Avenue, Kingston, Ontario K7M 8A6, Canada

A. J. Holland, B.Sc., M.B., B.S., M.R.C.P., M.R.C.Psych.
University Lecturer in Developmental Psychiatry (Learning Disability),
Department of Psychiatry,
University of Cambridge,
Ida Darwin Hospital,
Fulbourn, Cambridge, CB1 5EE, UK

D. M. Holtzman, M.D.
Adjunct Assistant Professor,
Department of Neurology,
University of California,
San Francisco, California 94143, USA

H. Karlinsky, M.D., M.Sc., F.R.C.P.(C)
Clinical Associate Professor,
Department of Psychiatry,
University of British Columbia;
Director, Alzheimer Disease Research Program,
Riverview Hospital,
500 Lougheed Highway, Port Coquitlam, British Columbia V3C 4J2,
Canada

D. M. A. Mann, Ph.D., M.R.C.Path.
Senior Lecturer in Neuropathology,
Department of Pathological Sciences,
Division of Molecular Pathology,
University of Manchester,
Oxford Road, Manchester M13 9PT, UK

B. D. McCreary, M.D., F.R.C.P. (C)
Associate Professor and Chairman,
Division of Developmental Disabilities,
Department of Psychiatry, Queen's University,
Kingston, Ontario K7L 3N6, Canada

C. Oliver, Ph.D.
Senior Lecturer in Psychology,
Institute of Psychiatry,
University of London,
London, SE5 8AF, UK

H. Ouellette-Kuntz, R.N., M.Sc.
Lecturer,
Department of Psychiatry,
Queen's University,
c/o Ongwanada Resource Centre,
191 Portsmouth Avenue, Kingston, Ontario K7M 8A6, Canada

M. E. Percy, Ph.D.
Associate Professor,
Department of Obstetrics and Gynaecology,
University of Toronto;
Neurogeneticist, Division of Biomedical Services and Research, Surrey Place Centre,
2 Surrey Place, Toronto, Ontario M5S 2C2, Canada

S-J. Richards, B.Sc., Ph.D.
Senior Research Fellow,
Department of Medicine,
University of Cambridge,
Addenbrooke's Hospital,
Hills Road, Cambridge CB2 2QQ, UK

D. M. Robertson, M.D., F.R.C.P. (C)
Professor,
Department of Pathology,
Queen's University,
c/o Kingston General Hospital,
76 Stuart Street, Kingston, Ontario K7L 2V7, Canada

M. N. Rossor, M.A., M.D., F.R.C.P.
Consultant Neurologist,
The National Hospital for Neurology and Neurosurgery,
Queen Square, London WC1 3BN;
Department of Neurology,
St. Mary's Hospital,
Praed Street London W2 1NY, UK

Sir Martin Roth, M.D. (Cantab. and Lond.),
F.R.C.P., F.R.C.Psych, Sc.D. (Hon.)
Professor Emeritus of Psychiatry,
University of Cambridge, and Fellow of Trinity College.
Addenbrooke's Hospital,
Hills Road, Cambridge CB2 2QQ, UK

M. B. Schapiro, M.D.
Chief,
Section on Brain Aging and Dementia,
Laboratory of Neurosciences,
National Institute on Aging,
Clinical Center,
Bethesda, Maryland 20892, USA

G. B. Seltzer, Ph.D.
Clinic Director,
Program on Aging and Developmental Disabilities,
The Waisman Center,
University of Wisconsin-Madison,
1500 Highland Avenue, Madison, Wisconsin 53705, USA

L. J. Whalley, M.D., F.R.C.Psych.
Professor of Mental Health,
Department of Mental Health,
Aberdeen University Medical School,
Foresterhill, Aberdeen AB9 2ZD, UK

H. M. Wisniewski, M.D., Ph.D.
Director,
New York State Institute for Basic Research,
1050 Forest Hill Road, Staten Island, New York 10314, USA

F. Yamaguchi, M.D., Ph.D.
Research Assistant,
Department of Medicine,
University of Cambridge,
Addenbrooke's Hospital,
Hills Road, Cambridge CB2 2QQ, UK

I

Alzheimer disease and Down syndrome: an overview of each disorder

Alzheimer disease

H. Karlinsky, J. A. Hardy, and M. N. Rossor

Summary

Persons with Down syndrome are at increased risk for the development of Alzheimer disease. This remarkable association constitutes the theme of the present book. As a preamble to detailed discussions regarding the link between these two disorders, this chapter presents a general overview of Alzheimer disease, including its evolving conceptualization, clinical characteristics, neuropathological findings, diagnosis, aetiology and treatment. Despite considerable advances, definitive treatment for Alzheimer disease is not yet available. Nevertheless, recent dramatic molecular genetic findings, fuelled in part by knowledge of the above-mentioned association with Down syndrome, bodes well for future preventive and curative interventions.

> *'I fear I am not in my perfect mind.*
> *Methinks I should know you and know this man;*
> *Yet I am doubtful: for I am mainly ignorant*
> *What place this is, and all the skill I have*
> *Remembers not these garments; nor I know not*
> *Where I did lodge last night.' (King Lear IV vii 63)*

Introduction

As Shakespeare's depiction of King Lear poignantly demonstrates, recognition of memory loss and disorientation as a potential accompaniment of ageing occurred long before the twentieth century (Howells 1991). However it was not until 1910 that the term Alzheimer disease (AD), now considered to be the most common cause of intellectual deterioration in the elderly, was first applied by Kraepelin (1910) to describe a presenile dementia characterized by what were then believed to be unique histopathological findings. The condition was named in honour of Alois Alzheimer (see Frontispiece), a German neuropathologist, who in 1907 published a report of the clinicopathological findings in a

56-year-old woman whose progressive intellectual deterioration followed the onset of paranoia at the age of 51. At autopsy, Alzheimer identified the presence of neuritic plaques and intensely staining fibrils which 'stained differently from normal fibrils'. Alzheimer concluded that he had described 'a unique entity' (Alzheimer 1907; English translation by Hochberg and Hochberg 1977).

Not surprisingly, the scientific community's conceptualization of AD has significantly evolved during the approximately 85 years since Alzheimer's initial case report. As Fox (1989) summarized, 'the disease has emerged from an obscure, rarely applied medical diagnosis to its characterization as the fourth or fifth leading cause of death in the United States...' Of fundamental importance to this transition has been the reconsideration of AD as an age-specific illness. Although the diagnosis of AD is now generally applied to all individuals with the typical clinical and neuropathological findings, regardless of age of onset, the diagnosis had initially been limited to individuals in the presenile age range (i.e. under the age of 65 years). The turning point was a series of significant studies beginning in the late 1950s which convincingly established that many persons over the age of 65 years with what was then usually termed 'senile dementia', had the same clinical and pathological findings as those with presenile AD (Beach 1987). Redesignating these elderly individuals as suffering from AD dramatically increased the public profile of this disorder, particularly because of the large number of people calculated to be affected (Katzman 1976). Although epidemiological studies have been fraught with methodological limitations, the prevalence of AD clearly increases markedly with age (Jorm *et al.* 1987) with some evidence suggesting that almost 50 per cent of individuals over the age of 85 are affected (Evans *et al.* 1989).

There has also been a dramatic acceleration of research on AD during the past one to two decades, with significant contributions from diverse disciplines (Khachaturian 1985). Recently, innovative molecular genetic techniques have begun to uncover further crucial characteristics of the disease (Hardy *et al.* 1991). A major stimulant to this research progress has been the accumulating clinical and neuropathological evidence that persons with Down syndrome (DS) are at increased risk for development of AD. Aspects of this remarkable association, including hypotheses that attempt to account for its occurrence, are considered in chapters throughout this book. As a preamble, the present chapter provides a general overview of AD. The clinical, neuropathological, diagnostic, aetiological and therapeutic aspects of AD are summarized, with a highlighting of those issues particularly relevant to its association with DS.

Clinical characteristics

Progressive intellectual deterioration is the defining clinical feature of AD. Cognitive deficits begin insidiously and caregivers typically have difficulty in dating the precise onset of manifestations. By the late stages of the disease, however, affected individuals are profoundly impaired and completely dependent on others for their activities of daily living. Although the duration of illness can vary remarkably, with average survival having increased recently due

Stage I (duration of disease 1–3 years)
    Memory: new learning defective, remote recall mildly impaired
    Visuospatial skills: topographic disorientation, poor complex constructions
    Language: poor wordlist generation, anomia
    Personality: indifference, occasional irritability
    Psychiatric features: sadness or delusions in some
    Motor system: normal

Stage II (duration of disease 2–10 years)
    Memory: recent and remote recall more severely impaired
    Visuospatial skills: poor constructions, spatial disorientation
    Language: fluent aphasia
    Calculation: acalculia
    Praxis: ideomotor apraxia
    Personality: indifference or irritability
    Psychiatric features: delusions in some
    Motor system: restlessness, pacing

Stage III (duration of disease 8–12 years)
    Intellectual functions: severely deteriorated
    Motor system: limb rigidity and flexion posture
    Sphincter control: urinary and fecal incontinence

Table 1.1. Principal clinical findings in each stage of AD (abridged from Cummings and Benson 1992)

to improved general nursing and medical care, mean survival is approximately eight years from onset of clinical manifestations (Barclay *et al.* 1985). Aspiration pneumonia, urinary tract infection or infection of decubitus ulcers is often the immediate cause of death (Cummings and Benson 1992, p. 57).

Early attempts to characterize the clinical course of AD utilized three broad stages of increasing severity of illness, often represented by the terms forgetfulness, confusion and dementia, respectively (Reisberg 1983). However, the sequence in which cognitive changes occur has become much clearer with recent detailed neuropsychological investigations of the disease. It is now recognized, for example, that although recent memory loss is a consistent and prominent early feature of AD (Haxby *et al.* 1986), visuospatial skills and wordfinding abilities (Filley *et al.* 1986; Whitworth and Larson 1989) are also impaired in the early stages. Each of these spheres of cognitive functioning—memory, visuospatial skills and language—undergoes characteristic changes as the illness advances (see Table 1.1). Other intellectual deficits, such as the onset of ideomotor apraxia (Della Sala *et al.* 1987), also become evident as the disease progresses. Ultimately, in the final stage of AD, virtually all intellectual capacities are lost.

In addition to the inevitable progression of cognitive deficits, AD is also characterized by the frequent co-existence of a variety of psychiatric and neurological symptoms and signs. Affective and psychotic symptomatology are particularly common. Co-existing depressive illnesses have been diagnosed with a surprisingly wide range of reported frequencies (Wragg and Jeste 1989), probably reflecting the overlap of symptomatology that occurs in both depression and dementia. The subjective symptoms of depression may be especially difficult to elicit in severe dementia (Reifler *et al.* 1982), an observation that may equally apply to a proportion of individuals with DS. Delusions, most often persecutory in nature, have been reported in 50 per cent of individuals with presumed AD at some point during their illness (Cummings *et al.* 1987). Hallucinations can also occur, though relatively less frequently (Cummings *et al.* 1987).

Although personality and social graces are often stated to remain relatively intact in AD, changes in personality frequently take place as the disease progresses, particularly the almost invariable development of passive behaviour (Rubin *et al.* 1987). Individuals with AD can also exhibit intermittent angry outbursts and assaultive behaviour (Swearer *et al.* 1988) but, in general, behavioural disturbances have been poorly studied.

From a neurological perspective, extrapyramidal abnormalities have been reported in as many as 60 per cent of affected individuals, particularly akinesia and increased muscle tone in the later stages of the disease (Sulkava 1982). Neurological examination also frequently reveals the presence of primitive developmental reflexes (Huff and Growdon 1986). Generalized seizures (Sulkava 1982) and myoclonus (Faden and Townsend 1976) occur in some affected individuals, but generally not until the later stages of illness; if observed early, it should raise the possibility of other potential diagnoses. Similarly, bowel and bladder incontinence also tends to occur relatively late in the course of AD.

Although persons with AD generally show patterns of intellectual decline consistent with those described above, prominent clinical features in any one affected individual at a particular stage of the disease, as well as that individual's rate of deterioration, is variable. Studies of clinical heterogeneity in persons with AD have tentatively identified at least four subgroups with differing rates and profiles of cognitive decline (e.g. Chui *et al.* 1985; Mayeux *et al.* 1985). Although this heterogeneity may reflect different aetiologies of AD, the variable clinical features may also be due to the interaction of a single disease process with distinct genotypes or brain cytoarchitecture. This latter explanation may account for the discrepancy between the clinical and pathological features of AD in individuals with DS, as well as possible differences in the clinical presentation of AD in individuals with and without DS (see Chapter 4).

Neuropathological findings

The key features described by Alzheimer using silver staining were the argyrophilic neuritic or senile plaques (SPs) and intraneuronal neurofibrillary tangles (NFTs). Also apparent by microscopy, but more difficult to demonstrate other than by quantitative neuronal counts, is cell loss from the cerebral cortex and selective subcortical nuclei. Cerebral cortical cell loss occurs particularly from the deeper layers of the cortex and preferentially involves the large pyramidal cells. Cells from a number of subcortical nuclei are also lost, particularly from the locus ceruleus and nucleus basalis of Meynert, although cell shrinkage suggests that the cell loss may not be as dramatic as originally supposed.

SPs are found predominantly in the association areas of the neocortex. They vary in diameter between 15 and 100 μm. The classical SP contains a central amyloid core, which may have additional aluminosilicate deposition (Candy *et al.* 1986), and is surrounded by abnormal nerve processes (often referred to as dystrophic neurites) as well as by glial processes. The major component of the amyloid core is a 39–42 amino acid peptide referred to variously as βA4, A4 protein, amyloid β-peptide or β-amyloid, which is derived from a large amyloid precursor protein (APP) (see later). The β-amyloid deposition is also substantial in cerebral blood vessels in AD, but to a far lesser extent than in the hereditary cerebral amyloid angiopathies (Levy *et al.* 1990; Hendricks *et al.* 1992).

Some authors also refer to 'burnt out' plaques in which there is a prominent amyloid core, but only a very thin rim of dystrophic neurites. More importantly, with regard to the formation and progression of SPs, β-amyloid deposition may occur in the absence of dystrophic changes and NFT formation, for example in the cerebellum (Joachim *et al.* 1989). Using anti-β-amyloid immunostaining, pre-amyloid deposits can be demonstrated quite widely (Bugiani *et al.* 1989).

At the ultrastructural level, NFTs consist of paired helical filaments with an individual filament diameter of 10 nm wound in a double helix (Kidd 1963; Terry *et al.* 1964). The central core of the paired helical filaments consists of the microtubule-associated protein tau, which exists in an abnormally hyper-phosphorylated form (Goedert *et al.* 1992). Immunostaining and direct protein analysis has demonstrated that ubiquitin also forms an important component of the NFT and may represent attempts by the cell to degrade the abnormal protein deposits. NFTs are found predominantly in the pyramidal cells of layer 3 in the archicortex and the temporoparietal neocortex. Early in the course of AD there is relative sparing of layer 4, but as the disease progresses NFTs become more widespread (Pearson *et al.* 1985). NFTs are not exclusive to AD and can be found, generally to a lesser degree, in other conditions (e.g. dementia pugilistica, progressive supranuclear palsy). As reviewed by Mann (Chapter 5), the presence in adults with DS of NFTs, as well as SPs and β-amyloid deposition, is now well-known.

Other histological features of AD include granulovacuolar degeneration of neurones, characterized by clear, round cytoplasmic zones of approximately 5 μm diameter, particularly within the hippocampus. Hirano bodies are eosinophilic inclusions also predominantly noted within the pyramidal layer of the hippocampus.

Neurotransmitter changes present in AD are considered to be a direct result both of neuronal loss and of axonal and dendritic attrition consequent upon neurite involvement in plaque formation. A wide array of neurotransmitter systems are involved (for a review see Rossor 1991). Loss of choline acetyltransferase is a consistent feature and reflects disruption to the ascending cholinergic projection from basal forebrain to the cerebral cortex. The noradrenergic and serotinergic subcortical ascending systems are also involved. The neuropeptide somatostatin is consistently reduced in the cerebral cortex. Uptake of a glutamate analogue is also reduced (Hardy *et al.* 1987). Many receptor binding sites are lost, but most studies have indicated preservation of the muscarinic receptor pool, which has encouraged trials of cholinergic replacement therapy.

Diagnosis

Consistent with Alzheimer's initial case description, the diagnosis of AD remains dependent on the histological verification of SPs and NFTs in the brain of an individual with progressive dementia. Over the past decade, however, both the pathological and clinical diagnostic criteria for AD have been significantly refined. A panel of neuropathology experts has now recommended microscopic criteria necessary for a post-mortem histological diagnosis (Khachaturian 1985). These quantitative criteria are age-specific and specify the minimum number of SPs and NFTs required to establish the presence of AD. Greater minimum counts are required for older individuals, as both SPs and NFTs can be present in limited numbers with 'normal ageing' in the absence of associated cognitive impairment (Tomlinson 1977). Of significance, SPs and NFTs can also be found in individuals with DS without apparent clinical consequences (see Chapter 5).

The molecular and cellular constituents of SPs and NFTs are also now being utilized for diagnostic purposes. Antibodies directed against β-amyloid can sensitively detect amyloid deposits, while immunohistochemical studies can accurately identify the presence of hyperphosphorylated tau (see Chapter 5). Tests for 'AD-associated protein', which includes an epitope of tau, are currently being suggested as an adjunct means of diagnosing AD in autopsy brain (Ghanbari *et al.* 1990). In the future, reductions in neuronal cells and/or neurochemical markers may be incorporated into pathological diagnostic criteria.

The clinical diagnosis of AD has also undergone significant change over the past decade, particularly with the introduction of explicit diagnostic criteria. For example, criteria proposed by the National Institute of Neurological and Communicative Disorders and Stroke (NINCDS) and the Alzheimer's Disease and Related Disorders Association (ADRDA) work group distinguish between probable, possible and definite AD (McKhann *et al.* 1984). A clinical diagnosis of probable AD requires the presence of a progressive dementia confirmed by neuropsychological testing in the absence of other systemic or brain diseases that could account for those deficits. Of note, the criteria for probable AD specify onset between the ages of 40 and 90 years, which arbitrarily excludes a proportion of affected individuals with DS (Dalton and Wisniewski 1990). A clinical diagnosis of possible AD applies if either there are atypical variations in the onset, presentation or clinical course of the dementia, or if a progressive dementia is felt to be due to AD despite the presence of other disorders that may affect cognitive functioning. A diagnosis of definite AD is reserved for those cases of probable AD with histopathological confirmation of the diagnosis by biopsy or autopsy.

The NINCDS–ADRDA criteria emphasize that individuals suspected of having AD must undergo a detailed evaluation that should encompass: a medical history, neurological, psychiatric and clinical examinations, neuropsychological tests and laboratory tests. The purpose is twofold. Traditionally this evaluation has focused on excluding other causes of dementia that may mimic AD, particularly those that are reversible. An example in the context of this book is routine exclusion of thyroid disease, a condition particularly prevalent in DS, which can be associated with cognitive impairment (see Chapter 7). Ongoing debate revolves around which laboratory investigations should be regularly undertaken, particularly the issue of selected versus routine use of neuroimaging procedures (Clarfield 1990; Katzman 1990).

More recently, diagnostic evaluations are attempting to identify clinical features or biological markers specific for AD. For example, the medical history should detail the precise sequence of an affected individual's cognitive deficits to ensure that the pattern is consistent with AD. An intensive search for specific biological markers for AD is also being undertaken and includes cerebrospinal fluid analysis, screening for extraneural and genetic markers, as well as the utilization of sophisticated electrophysiological and neuroimaging techniques (for a review see Verdon and Beattie 1991; see also Chapter 11). Apart from the possible presence of rare mutations in the APP gene (see below), a sensitive and specific biological marker for AD has not yet been found. One promising approach may be to combine several investigative results, such as the combination of magnetic resonance imaging (MRI) and single photon emission computed tomography (SPECT) measures recently recommended by Pearlson *et al.* (1992).

Another commonly used diagnostic guideline, the revised version of the Diagnostic and Statistical Manual of Mental Disorders (DSM-III-R) (American Psychiatric Association 1987) specifies that individuals affected with 'Primary Degenerative Dementia of the Alzheimer Type' must demonstrate a dementia characterized by an insidious onset with a generally progressive deteriorating course. The criteria also require exclusion of other specific causes of dementia, again highlighting that different illnesses may result in progressive intellectual deterioration similar to that seen in AD.

With the introduction of specific diagnostic criteria and detailed clinical evaluations, the accurate clinical diagnosis of AD (as confirmed by neuropathological findings) has significantly improved, although, in general, a 10–15 per cent misdiagnosis rate still occurs (Joachim *et al.* 1988). Differentiating individuals with ischaemic vascular dementia alone from those with both this dementia and AD remains particularly difficult (Katzman and Jackson 1991). Other diagnostic challenges include: distinguishing individuals with AD who have significant extrapyramidal features from those with other illnesses in which both extrapyramidal features and dementia may be present (e.g. Parkinson disease, progressive supranuclear palsy, Lewy Body disease), distinguishing individuals with AD who have prominent personality changes from those with frontal lobe dementias such as Pick disease and distinguishing individuals with early symptoms of AD from those with age-associated memory impairment. The last-mentioned distinction is also contained within the unique set of challenges associated with diagnosing AD in individuals with pre-existing mental retardation secondary to DS (see Chapter 6).

Aetiology

Although advanced age and a positive family history of AD have been known for some time as risk factors for development of AD, there was little progress in research focused on the aetiology of AD until a series of recent and dramatic molecular genetic studies. A significant impetus to this work was the recognition that persons with DS are also at increased risk for development of AD. Knowledge of this relationship, details of which are documented throughout this book, led investigators to query whether genetic material present on chromosome 21 either directly or indirectly led to the development of AD. St George-Hyslop *et al.* (1987), using data derived from the genetic analysis of four large kindreds in whom AD segregated as an autosomal-dominant disease, reported that a genetic defect causing early-onset familial AD (FAD) was located on chromosome 21. Contemporaneously, four groups (Goldgaber *et al.* 1987; Kang *et al.* 1987; Robakis *et al.* 1987; Tanzi *et al.* 1987a) reported that the cloning of the gene encoding APP from which β-amyloid was derived, was also localized to chromosome 21 and was thus a candidate gene for AD. However, genetic analysis of the segregation of the APP gene in several families with AD failed to show co-inheritance of the APP gene with the disease (Tanzi *et al.* 1987b; Van Broeckhoven *et al.* 1987) and further analysis showed that the APP was apparently of normal sequence in cases with AD (Vitek *et al.* 1988).

In subsequent studies, genetic linkage to chromosome 21 probes was also found in three different sets of FAD pedigrees (Van Broeckhoven *et al.* 1988; Goate *et al.* 1989; Heston *et al.* 1991). However, two other studies (Pericak-Vance *et al.* 1988; Schellenberg *et al.* 1988) failed to show linkage to chromosome 21 probes. A large collaborative genetic analysis confirmed that FAD was causally heterogeneous, with some cases resulting from genetic defects on chromosome 21 and others due to factors as yet unidentified (St George-Hyslop *et al.* 1990).

With the realization that AD was heterogeneous, Hardy and his colleagues switched from group analysis of many families to analysis of individual families in whom only a single aetiology was presumed to be present. Detailed analysis of one such family revealed that the APP gene segregated with the disease. Furthermore, sequencing revealed a mutation at codon 717, just outside the C-terminal of the β-amyloid sequence (β-amyloid's amino acids correspond to residues 672–714 of the APP_{770} transcript), changing a valine residue to isoleucine in the transmembrane domain of APP (Goate *et al.* 1991). Subsequent screening has revealed the presence of this mutation in a further nine families with early-onset AD (Hardy *et al.* 1991; Naruse *et al.* 1991; Yoshioka *et al.* 1991; Karlinsky *et al.* 1992b and unpublished findings). Two additional mutations of the same amino acid have also now been identified, changing it to glycine and phenylalanine in two other families with early-onset AD (Chartier-Harlin *et al.* 1991b; Murrell *et al.* 1991).

The mechanism by which the mutations could cause AD is unclear. However, more recently, two other mutations have been found in the β-amyloid region of the APP gene. The first changes alanine to glycine at codon 692 of the APP gene (Hendricks *et al.* 1992). Perhaps not surprisingly, since the mutation is adjacent to that causing hereditary cerebral haemorrhage with amyloidosis of Dutch type (HCHAA-D) (Levy *et al.* 1990), some individuals with this mutation have a clinical course similar to HCHAA-D. However, others with the mutation have a clinical picture similar to, if not identical with, AD. The second mutation is a double one at codons 670/1 of the APP gene, just outside the N-terminal of the β-amyloid sequence (Mullan *et al.* 1992). Thus, a tentative hypothesis is that mutations framing β-amyloid can cause AD, presumably by altering the cleavage of APP, which thereby leads to more β-amyloid formation (Hardy 1992). This, in turn, is suggested to lead to NFT formation and neuronal death (Hardy and Higgins 1992). This hypothesis directly impinges on the design of treatment strategies for AD (Hardy and Allsop 1991).

There are two routes currently known for APP breakdown: the secretase site, which cleaves the molecule just outside the membrane within the β-amyloid domain (Esch *et al.* 1990), and the recently proposed endosomal–lysosomal pathway, which results in a number of C-terminal fragments, including intact β-amyloid (Estus *et al.* 1992; Golde *et al.* 1992; Haass *et al.* 1992). The mechanism by which the deposited β-amyloid might result in neuronal loss and NFT formation is unclear, but is supported by the presence of the entire panoply of AD cytoskeletal pathology in cases of FAD with APP mutations (Lantos *et al.* 1992; Mann *et al.* 1992). There have been reports both of the neurotrophic and neurotoxic effect of β-amyloid and there is now evidence that, in cell culture, the addition of β-amyloid enhances the neurotoxic effect of glutamate-containing pyramidal cells and hyperphosphorylation of tau, which occurs in NFT formation and is calcium-dependent (Mattson *et al.* 1992).

Despite these dramatic molecular genetic findings, the basic cause(s) of the vast majority of AD cases is (are) still obscure. Screening surveys have shown that the APP mutations are extremely rare and, thus far, confined to a small proportion of early-onset FAD cases (Chartier-Harlin *et al.* 1991a; van Duijn *et al.* 1991; Fidani *et al.* 1992). Recently, genetic linkage evidence for an early-onset FAD locus on chromosome 14 has been provided by Schellenberg *et al.* (1992) and several subsequent investigators. Whether the chromosome 14 locus is involved in APP processing remains to be determined. Furthermore, AD is more frequently a late-onset disease (Evans *et al.* 1989), the aetiology of which is complex and probably involves various genetic (e.g. apoliprotein E type 4 allele on chromosome 19) and environmental risk factors (Farrer *et al.* 1991; McLachlan *et al.* 1991; Pericak-Vance *et al.* 1991; Strittmatter *et al.* 1993), quite possibly in combination. The significant discordance for AD in reported monozygotic twin pairs also does not support an exclusively genetic aetiology of AD (see Karlinsky *et al.* 1992a).

A variety of hypotheses concerning environmental causation has therefore been proposed for AD. One hypothesis relates to the possible role of aluminium as an environmental toxin (for a review see Crapper-McLachlan *et al.* 1991b). Other proposed risk factors have emerged from case-control studies, albeit with contradictory results. In a recent collaborative reanalysis of 11 case-control studies, the relative risk of developing AD increased with a positive family history of Parkinson disease, late maternal age at time of birth and previous history of head trauma, hypothyroidism and depression (van Duijn and Hofman 1992). The analysis also suggested that AD is associated with a positive family history of DS (see Chapter 12).

Conceptually, Katzman and Jackson (1991) have suggested that dual mechanisms may underlie how risk factors are operative. First, the onset of the clinical manifestations of AD appears to be quantitatively related to a loss of synapses (Terry *et al.* 1991). Events or processes that independently reduce the number of synapses, such as ageing, could therefore accelerate the onset of AD. Secondly, conditions that lead to the deposition of diffuse plaques, such as head trauma, could also lead to the evolution of AD. The amyloid cascade hypothesis of Hardy and Higgins (1992) also suggests that non-APP gene mutation causes of AD act by initially triggering β-amyloid protein deposition. In DS, this deposition may arise as a consequence of gene dosage effects of APP (see Chapter 12). The validity of these proposals remains to be determined. Nevertheless, as an increasing array of potential risk factors for the development of AD emerge, their placement within pathogenic frameworks such as APP mismetabolism will be useful organizing tools.

Treatment

To date, treatment research into AD has largely been focused on efforts to enhance cognitive functioning, either by providing immediate symptomatic therapy or by reducing the rate of anticipated cognitive deterioration. For these purposes, a wide range of pharmacological agents have been utilized, including stimulants, vasodilators, metabolic enhancers, neurotransmitters, neuroendocrine drugs, neurotrophic factors, aluminium-chelating agents and phospholipid derivatives (for a detailed review see Patterson 1991). Currently selected compounds tend to be theoretically based most often on enhancing one or more of the identified neurotransmitter deficits of AD, particularly cholinergic augmentation. In view of the recent molecular genetic findings, future treatments may focus on the potential aetiological role of β-amyloid deposition. Individuals at risk of developing AD, such as asymptomatic carriers of an APP mutation or relatively young adults with DS, may eventually benefit from preventive therapeutic strategies.

Despite intermittent optimistic reports, definitive treatment for the cognitive changes of AD has not yet been established. However, some innovations have emerged in the context of clinical drug trials, including videotaped behavioural observations to monitor drug effects (Crapper-McLachlan *et al.* 1991a) and novel drug delivery systems, such as continuous intracranial infusion

(Harbaugh *et al.* 1984). There is also the possibility of neural tissue transplantation, although ethical issues related to this potential treatment for AD remain extremely contentious (Dunnett 1991).

While specific therapies are lacking for the cognitive changes of AD, there have been important gains in the recognition and management, both pharmacological and non-pharmacological, of many non-cognitive behavioural manifestations that occur in a proportion of AD-affected individuals. It is often the presence of these manifestations that leads to institutionalization of these individuals. Medications can be appropriately introduced for co-existing depression, psychotic symptoms, anxiety, insomnia and poor aggressive impulse control (for a review see Katzman and Jackson 1991). However, investigation of the potential aetiology of the symptom and consideration of possible therapeutic environmental manipulations should precede pharmacological intervention (see Chapter 6).

Treatable neurological conditions may also occur in the course of AD. Generalized seizures and myoclonus can usually be relatively easily controlled with anticonvulsants although extrapyramidal signs tend to respond poorly to anti-parkinsonian medication. Physicians must have a high index of suspicion for intercurrent illnesses, particularly when cognitive functioning appears to abruptly deteriorate. In the end stages of the disease, urinary tract infections, bed sores and pneumonia often require medical intervention.

A major treatment advance has been the growing recognition of the need to support the affected individual's caregivers, who are at increased risk for stress-induced psychological and physical illnesses (Zarit *et al.* 1985). An appropriate and comprehensive treatment plan must include education of caregivers about AD and about available support services in their community. Preferably, such support services should include home care programmes, specialized day and respite programmes, caregiver support groups and a local AD society. Safety issues, such as fitness to drive, must be continuously addressed. Finally, in the absence of definitive treatment for this devastating disease, institutional care for severely impaired individuals may be required.

Conclusion

Although definitive treatment for the cognitive deficits of AD is not yet available, significant therapeutic interventions now exist for both affected individuals and their caregivers. Diagnostic accuracy has greatly improved as well over the past one to two decades. These gains in diagnosis and management are also of benefit to individuals with DS who develop AD. In turn, the recognition that the presence of DS is a risk factor for the development of AD has provided insight into the aetiology and pathophysiology of AD in the general population, particularly via the exciting recent molecular genetic findings.

References

Alzheimer, A. (1907). Über eine eigenartige Erkrankung der Hirnrinde. *Allgemeine Zeitschrift für Psychiatrie und Psychisch-Gerichtliche Medizin*, **64**, 146-8.

American Psychiatric Association. (1987). *Diagnostic and statistical manual of mental disorders*, (3rd edn revised). Washington, DC.

Barclay, L. L., Zemcov, A., Blass, J. P., and Sansone, J. (1985). Survival in Alzheimer's disease and vascular dementias. *Neurology*, **35**, 834-40.

Beach, T. G. (1987). The history of Alzheimer's disease: three debates. *Journal of the History of Medicine and Allied Sciences*, **42**, 327-49.

Bugiani, D., Giaccone, G., Frangione, B., Ghetti, B., and Tagliavini, F. (1989). Alzheimer patients: preamyloid deposits are more widely distributed than senile plaques throughout the central nervous system. *Neuroscience Letters*, **103**, 263-8.

Candy, J. M., Oakley, A. E., Klinowski, J., Carpenter, T. A., Perry, R. H., Atack, J.R., *et al.* (1986). Aluminosilicates and senile plaque formation in Alzheimer's disease. *Lancet*, **1**, 354-7.

Chartier-Harlin, M.-C., Crawford, F., Hamandi, K., Mullan, M., Goate, A., Hardy, J., *et al.* (1991a). Screening for the β-amyloid precursor mutation (APP717 Val→Ile) in extended pedigrees with early onset Alzheimer's disease. *Neuroscience Letters*, **129**, 134-5.

Chartier-Harlin M.-C., Crawford, F., Houlden H., Warren A., Hughes, D., Fidani, L., *et al.* (1991b). Early-onset Alzheimer's disease caused by mutations at codon 717 of the β-amyloid precursor protein gene. *Nature*, **353**, 844-6.

Chui, H. C., Teng, E. L., Henderson, V. W., and Moy, A. C. (1985). Clinical subtypes of dementia of the Alzheimer type. *Neurology*, **35**, 1544-50.

Clarfield, A. M. (1990). Should a major imaging procedure (CT or MRI) be required in the workup of dementia? An opposing view. *The Journal of Family Practice*, **31**, 405-10.

Crapper-McLachlan , D. R., Dalton, A. J., Kruck, T. P. A., Bell, M. Y., Smith, W. L., Kalow, W., and Andrews, D. F. (1991a). Intramuscular desferrioxamine in patients with Alzheimer's disease. *Lancet*, **337**, 1304-8.

Crapper-McLachlan , D. R., Kruck, T. P., Lukiw, W. J., and Krishnan, S. S. (1991b). Would decreased aluminum ingestion reduce the incidence of Alzheimer's disease? *Canadian Medical Association Journal*, **145**, 793-804.

Cummings, J. L., and Benson, D. F. (1992). *Dementia: a clinical approach*. Butterworth-Heinemann, Boston.

Cummings, J. L., Miller, B., Hill, M. A., and Neshkes, R. (1987). Neuropsychiatric aspects of multi-infarct dementia and dementia of the Alzheimer type. *Archives of Neurology*, **44**, 389-93.

Dalton, A. J., and Wisniewski, H. M. (1990). Down's syndrome and the dementia of Alzheimer disease. *International Review of Psychiatry*, **2**, 43-52.

Della Sala, S., Lucchelli, F., and Spinnler, H. (1987). Ideomotor apraxia in patients with dementia of Alzheimer type. *Journal of Neurology*, **234**, 91-3.

Dunnett, S. B. (1991). Neural transplants as a treatment for Alzheimer's disease? *Psychological Medicine*, **21**, 825-30.

Esch, F. S., Keim, P. S., Beattie, E. C., Blacher, R. W., Culwell, A. R., Oltersdorf, T., *et al.* (1990). Cleavage of amyloid β peptide during constitutive processing of its precursor. *Science*, **248**, 1122-4.

Estus, S., Golde, T. E., Kunishita, T., Blades, D., Lowery, D., Eisen, M., *et al.* (1992). Potentially amyloidogenic, carboxyl-terminal derivatives of the amyloid protein precursor. *Science*, **255**, 726-8.

Evans, D. A., Funkenstein, H., Albert, M. S., Scherr, P. A., Cook, N. R., Chown, M. J., *et al.* (1989). Prevalence of Alzheimer's disease in a community population of older persons: higher than previously reported. *Journal of the American Medical Association*, **262**, 2551-6.

Faden, A. I., and Townsend, J. J. (1976). Myoclonus in Alzheimer disease: a confusing sign. *Archives of Neurology*, **33**, 278-80.

Farrer, L. A., Myers, R. H., Connor, L., Cupples, L. A., and Growdon, J. H. (1991). Segregation analysis reveals evidence of a major gene for Alzheimer's disease. *American Journal of Human Genetics*, **48**, 1026-33.

Fidani, L., Rooke, K., Chartier-Harlin, M.-C., Hughes, D., Tanzi, R., Mullan, M., *et al.* (1992). Screening for mutations in the open reading frame and promoter of the β-amyloid precursor protein gene in familial Alzheimer's disease: identification of a further family with APP717 Val→Ile. *Human Molecular Genetics*, 1, 165-8.

Filley, C. M., Kelly, J., and Heaton, R. K. (1986). Neuropsychologic features of early- and late-onset Alzheimer's disease. *Archives of Neurology*, 43, 574-6.

Fox, P. (1989). From senility to Alzheimer's disease: the rise of the Alzheimer's disease movement. *Millbank Quarterly*, 67, 58-102.

Ghanbari, H. A., Miller, B. E., Haigler, H. J., Arato, M., Bissette, G., Davies, P., *et al.* (1990). Biochemical assay of Alzheimer's disease-associated protein(s) in human brain tissue. *Journal of the American Medical Association*, 263, 2907-10.

Goate, A. M., Haynes, A. R., Owen, M. J., Farrall, M., James, L. A., Lai, L. Y., *et al.* (1989). Predisposing locus for Alzheimer's disease on chromosome 21. *Lancet*, 1, 352-5.

Goate, A., Chartier-Harlin, M.-C., Mullan, M., Brown, J., Crawford, F., Fidani, L., *et al.* (1991). Segregation of a missense mutation in the amyloid precursor protein gene with familial Alzheimer's disease. *Nature*, 349, 704-6.

Goedert, M., Spillantini, M. G., Cairns, N. J., and Crowther, R. A. (1992). Tau proteins of Alzheimer paired helical filaments: abnormal phosphorylation of all six brain isoforms. *Neuron*, 8, 159-68.

Golde, T. E., Estus, S., Younkin, L. H., Selkoe, D. J., and Younkin, S. G. (1992). Processing of the amyloid protein precursor to potentially amyloidogenic derivatives. *Science*, 255, 728-30.

Goldgaber, D., Lerman, M. I., McBride, O. W., Saffiotti, U., and Gajdusek, D. C. (1987). Characterization and chromosomal localization of a cDNA encoding brain amyloid of Alzheimer's disease. *Science*, 235, 877-80.

Haass, C., Koo, E. H., Mellon, A., Hung, A. Y., and Selkoe, D. J. (1992). Targeting of cell surface β-amyloid precursor protein to lysosomes: alternative processing into amyloid bearing fragments. *Nature*, 357, 500-3.

Harbaugh, R. E., Roberts, D. W., Coombs, D. W., Saunders, R. L., and Reeder, T. M. (1984). Preliminary report: intracranial cholinergic drug infusion in patients with Alzheimer's disease. *Neurosurgery*, 15, 514-8.

Hardy, J. (1992). Framing β-amyloid. *Nature Genetics*, 1, 233-4.

Hardy, J., and Allsop, D. (1991). Amyloid deposition as the central event in the aetiology of Alzheimer's disease. *Trends in Pharmacological Sciences*, 12, 383-8.

Hardy, J. A., and Higgins, G. A. (1992). Alzheimer's disease: the amyloid cascade hypothesis. *Science*, 256, 184-5.

Hardy, J., Cowburn, R., Barton, A., Reynolds, G., Lofdahl, E., O'Carroll, A. -M., *et al.* (1987). Region-specific loss of glutamate innervation in Alzheimer's disease. *Neuroscience Letters*, 73, 77-80.

Hardy, J., Mullan, M., Chartier-Harlin, M.-C., Brown, J., Goate, A., Rossor, M., *et al.* (1991). Molecular classification of Alzheimer's disease. *Lancet*, 337, 1342-3.

Haxby, J. V., Grady, C. L., Duara, R., Schlageter, N., Berg, G., and Rapoport, S. I. (1986). Neocortical metabolic abnormalities precede nonmemory cognitive deficits in early Alzheimer's-type dementia. *Archives of Neurology*, 43, 882-5.

Hendricks, L., van Duijn, C. M., Cras, P., Cruts, M., van Hul, W., van Harskamp, F., *et al.* (1992). Presenile dementia and cerebral haemorrhage caused by a mutation at codon 692 of the beta-amyloid precursor protein gene. *Nature Genetics*, 1, 218-21.

Heston, L. L., Orr, H. T., Rich, S. S., and White, J. A. (1991). Linkage of an Alzheimer disease susceptibility locus to markers on human chromosome 21. *American Journal of Human Genetics*, 40, 449-53.

Hochberg, C. N., and Hochberg, F. H. (1977). A unique illness involving the cerebral cortex. In *Neurological classics in modern translation,* (ed. D.A. Rottenberg and F.H. Hochberg), pp. 41-3. Hafner Press, New York.

Howells, J. G. (1991). Dementia in Shakespeare's King Lear. In *Alzheimer and the dementias,* (ed. G. E. Berrios and H. L. Freeman), pp. 101-9. Royal Society of Medicine Services Limited, London.

Huff, F. J., and Growdon, J. H. (1986). Neurological abnormalities associated with severity of dementia in Alzheimer's disease. *Canadian Journal of Neurological Sciences*, 13, 403-5.

Joachim, C. L., Morris, J. H., and Selkoe, D. J. (1988). Clinically diagnosed Alzheimer's disease: autopsy results in 150 cases. *Annals of Neurology*, 24, 50-6.

Joachim, C. L., Morris, J. H., and Selkoe, D. J. (1989). Diffuse amyloid plaques occur commonly in the cerebellum in Alzheimer's disease. *American Journal of Pathology*, 135, 309-19.

Jorm, A. F., Korten, A. E, and Henderson, A. S. (1987). The prevalence of dementia: a quantitative integration of the literature. *Acta Psychiatrica Scandinavica*, 76, 465-79.

Kang, J., Lemaire, H. -G., Unterbeck, A., Salbaum, J. M., Masters, C. L., Grzeschik, K.H., *et al.* (1987). The precursor of Alzheimer's disease amyloid A4 protein resembles a cell-surface receptor. *Nature*, 325, 733-6.

Karlinsky, H., Macdonald, A. M., and Berg, J. M. (1992a). Primary degenerative dementia of the Alzheimer type in twins: initial findings from the Maudsley Hospital twin register. *International Journal of Geriatric Psychiatry*, 7, 603-10.

Karlinsky, H., Vaula, G., Haines, J. L., Ridgley, J., Bergeron, C., Mortilla, M., *et al.* (1992b). Molecular and prospective phenotypic characterization of a pedigree with familial Alzheimer's disease and a missense mutation in codon 717 of the β-amyloid precursor protein gene. *Neurology*, 42, 1445-53.

Katzman, R. (1976). The prevalence and malignancy of Alzheimer disease: a major killer. *Archives of Neurology*, 33, 217-8.

Katzman, R. (1990). Should a major imaging procedure (CT or MRI) be required in the workup of dementia? An affirmative view. *The Journal of Family Practice*, 31, 401-5.

Katzman, R., and Jackson, J. E. (1991). Alzheimer disease: basic and clinical advances. *Journal of the American Geriatrics Society*, 39, 516-25.

Khachaturian, Z. S. (1985). Diagnosis of Alzheimer's disease. *Archives of Neurology*, 42, 1097-105.

Kidd, M. (1963). Paired helical filaments in electron microscopy of Alzheimer's disease. *Nature*, 197, 192-3.

Kraepelin, E. (1910). *Psychiatrie. Ein Lehrbuch für Studierende und Ärzte*, Vol. 1. (8th edn), pp. 616-32. Barth, Leipzig.

Lantos, P. L., Luthert, P. J., Hanger, D., Anderton, B. H., Mullan, M., and Rossor, M. (1992). Familial Alzheimer's disease with the amyloid precursor protein position 717 mutation and sporadic Alzheimer's disease have the same cytoskeletal pathology. *Neuroscience Letters*, 137, 221-4.

Levy, E., Carman, M. D., Fernandez-Madrid, I. J., Power, M. D., Lieberburg, I., van Duinen, S. G., *et al.* (1990). Mutation of the Alzheimer's disease amyloid gene in hereditary cerebral hemorrhage, Dutch type. *Science*, 248, 1124-6.

McKhann, G., Drachman, D., Folstein, M., Katzman, R., Price, D., and Stadlan, E. M. (1984). Clinical diagnosis of Alzheimer's disease: report of the NINCDS-ADRDA Work Group under the auspices of the Department of Health and Human Services Task Force on Alzheimer's disease. *Neurology*, 34, 939-44.

McLachlan, D. R. C., Rupert, J. L., Kung-Sutherland, M., and Grima, E. A. (1991). Etiology of Alzheimer's disease. In *Alzheimer's disease research: ethical and legal issues,* (ed. J. M. Berg, H. Karlinsky and F. Lowy), pp. 93-119. Carswell, Toronto.

Mann, D. M. A., Jones, D., Snowden, J. S., Neary, D., and Hardy, J. (1992). Pathological changes in the brain of a patient with familial Alzheimer's disease having a missense mutation at codon 717 in the amyloid precursor protein gene. *Neuroscience Letters*, 137, 225-8.

Mattson, M. P., Cheng, B., Davis, D., Bryant, K., Lieberburg, I., and Rydel, R. E. (1992). β-amyloid peptides destabilize calcium homeostasis and render human cortical neurons vulnerable to excitotoxicity. *Journal of Neuroscience*, 12, 376-89.

Mayeux, R., Stern, Y., and Sparton, S. (1985). Heterogeneity in dementia of the Alzheimer type: evidence of subgroups. *Neurology*, 35, 453-6.

Mullan, M., Crawford, F., Axelman, K., Houlden, H., Lilius, L., Winblad, B., and Lannfelt, L. (1992). A new mutation in APP demonstrates that pathogenic mutations for probable Alzheimer's disease frame the beta-amyloid sequence. *Nature Genetics*, 1, 345-7.

Murrell, J., Farlow, M., Ghetti, B., and Benson, M. D. (1991). A mutation in the amyloid precursor protein associated with hereditary Alzheimer's disease. *Science*, **254**, 97-9.

Naruse, S., Igarashi, S., Kobayachi, H., Aoki, K., Inuzuka, T., Kaneko, K., *et al.* (1991). Mis-sense mutation Val→Ile in exon 17 of amyloid precursor protein gene in Japanese familial Alzheimer's disease. *Lancet*, **337**, 978-9.

Patterson, C. (1991). Treatment research in Alzheimer's disease: a review. In *Alzheimer's disease research: ethical and legal issues,* (ed. J.M. Berg, H. Karlinsky and F. Lowy), pp. 273-95. Carswell, Toronto.

Pearlson, G. D., Harris, G. J., Powers, R. E., Barta, P. E., Camargo E. E., Chase, G. A., *et al.* (1992). Quantitative changes in mesial temporal volume, regional cerebral blood flow, and cognition in Alzheimer's disease. *Archives of General Psychiatry*, **49**, 402-8.

Pearson, R. C. A., Esiri, M. M., Hiorns, R. W., Wilcock, G. K., and Powell, T. P. S. (1985). Anatomical correlates of the distribution of the pathological changes in the neocortex in Alzheimer's disease. *Proceedings of the National Academy of Sciences of the United States of America*, **82**, 4531-4.

Pericak-Vance, M. A., Yamaoka, L. H., Haynes, C. S., Gaskell, P. C., Hung, W. -Y., Clark, C. M., *et al.* (1988). Genetic linkage studies in Alzheimer's disease families. *Experimental Neurology*, **102**, 271-9.

Pericak-Vance, M. A., Bebout, J. H., Gaskell, P. C. Jr, Yamaoka, L. H., Hung, W. Y., Alberts, M. J., *et al.* (1991). Linkage studies in familial Alzheimer's disease: evidence for chromosome 19 linkage. *American Journal of Human Genetics*, **48**, 1034-50.

Reifler, B. V., Larson, E., and Hanley, R. (1982). Coexistence of cognitive impairment and depression in geriatric out-patients. *American Journal of Psychiatry*, **139**, 623-6.

Reisberg, B. (1983). An overview of current concepts of Alzheimer's disease, senile dementia, and age-associated cognitive decline. In *Alzheimer's disease. The standard reference,* (ed. B. Reisberg), pp. 3-20. The Free Press, New York.

Robakis, N. K., Ramakrishna, N., Wolfe, G., and Wisniewski, H. M. (1987). Molecular cloning and characterization of a cDNA encoding the cerebrovascular and the neuritic plaque amyloid peptides. *Proceedings of the National Academy of Sciences of the United States of America*, **84**, 4190-4.

Rossor, M. (1991). Primary degenerative dementia. In *Neurology in clinical practice,* (ed. W. G. Bradley, R. B. Doroff, G. M. Fenichel and C. D. Marsden), pp. 1409-22. Butterworth-Heinemann, Boston.

Rubin, E. H., Morris, J. C., and Berg, L. (1987). The progression of personality changes in senile dementia of the Alzheimer's type. *Journal of the American Geriatrics Society*, **35**, 721-5.

Schellenberg, G. D., Bird, T. D., Wijsman, E.M., Moore, D. K., Boehnke, M., Bryant, E. M., *et al.* (1988). Absence of linkage of chromosome 21q21 markers to familial Alzheimer's disease. *Science*, **241**, 1507-10.

Schellenberg, G. D., Bird, T. D., Wijsman, E. M., Orr, H. T., Anderson, L., Nemens, E., *et al.* (1992). Genetic linkage evidence for a familial Alzheimer's disease locus on chromosome 14. *Science*, **258**, 668–71.

St George-Hyslop, P. H., Tanzi, R. E., Polinsky, R. J., Haines, J. L., Nee, L., Watkins, P. C., *et al.* (1987). The genetic defect causing familial Alzheimer's disease maps on chromosome 21. *Science*, **235**, 885-90.

St George-Hyslop, P. H., Haines, J. L., Farrer, L. A., Polinsky, R., Van Broeckhoven, C., Goate, A., *et al.* (1990). Genetic linkage studies suggest that Alzheimer's disease is not a single homogeneous disorder. FAD Colaborative Study Group. *Nature*, **347**, 194-7.

Strittmatter, W. J., Saunders, A. M., Schmechel, D., Pericak-Vance, M., Enghild, J., Salvesen, G.S., and Roses, A. D. (1993). Apolipoprotein E: High-avidity binding to β-amyloid and increased frequency of type 4 allele in late-onset familial Alzheimer disease. *Proceedings of the National Academy of Sciences of the United States of America*, **90**, 1977–81.

Sulkava, R. (1982). Alzheimer's disease and senile dementia of Alzheimer type: a comparative study. *Acta Neurologica Scandinavica*, **65**, 636-50.

Swearer, J. M., Drachman, D. A., O'Donnell, B. F., and Mitchell, A. L. (1988). Troublesome and disruptive behaviors in dementia. Relationships to diagnosis and disease severity. *Journal of the American Geriatrics Society*, **36**, 784-90.

Tanzi, R. E., Gusella, J. F., Watkins, P. C., Bruns, G. A., St George-Hyslop, P., Van Keuren, M. L., *et al.* (1987a). Amyloid beta–protein gene: cDNA, mRNA distribution, and genetic linkage near the Alzheimer locus. *Science*, **235**, 880-4.

Tanzi, R. E., St George-Hyslop, P. H., Haines, J. L., Polinsky, R.J., Nee, L., Foncin, J. F., *et al.* (1987b). The genetic defect in familial Alzheimer's disease is not tightly linked to the amyloid beta–protein gene. *Nature*, **329**, 156-7.

Terry, R. D., Gonatas, N. K., and Weiss, M. (1964). Ultrastructural studies in Alzheimer's presenile dementia. *American Journal of Pathology*, **44**, 269-97.

Terry, R. D., Masliah, E., Salmon, D. P., Butters, N., DeTeresa, R., Hill, R., *et al.* (1991). Physical basis of cognitive alterations in Alzheimer's disease: synapse loss is the major correlate of cognitive imparment. *Annals of Neurology,* **30**, 572-80.

Tomlinson, B. E. (1977). The pathology of dementia. In *Dementia,* (2nd edn), (ed. C. E. Wells), pp. 113-53. F.A. Davis, Philadelphia.

Van Broeckhoven, C., Genthe, A. M., Vandenberghe, A., Horsthemke, B., Backhovens, H., Raeymaekers, P., *et al.* (1987). Failure of familial Alzheimer's disease to segregate with the A4-amyloid gene in several European families. *Nature*, **329**, 153-5.

Van Broeckhoven, C., Van Hul, W., Backhovens, H., Van Camp, G., Wehnert, A., Stinissen, P., *et al.* (1988). The familial Alzheimer's disease gene is located close to the centromere of chromosome 21. *American Journal of Human Genetics*, **43** (Supplement), A205.

van Duijn, C. M., and Hofman, A. (1992). Risk factors for Alzheimer's disease: The EURODEM collaborative re-analysis of case-control studies. *Neuroepidemiology*, **11** (Supplement 1), 106-13.

van Duijn, C. M., Hendriks L., Cruits, M., Hardy, J., Hofman, A., and Van Broeckhoven, C. (1991). Amyloid precursor protein gene mutation in early-onset Alzheimer's disease. *Lancet*, **337**, 978.

Verdon, J., and Beattie, B. L. (1991). Diagnostic research in Alzheimer's disease: An overview. In *Alzheimer's disease research: ethical and legal issues,* (ed. J.M. Berg, H. Karlinsky and F. Lowy), pp. 242-62. Carswell, Toronto.

Vitek, M. P., Rasool, C. G., de Sauvage, F., Vitek, S. M., Bartus, R. T., Beer, B., *et al.* (1988). Absence of mutation in the β-amyloid cDNAs cloned from the brains of three patients with sporadic Alzheimer's disease. *Molecular Brain Research*, **4**, 121-31.

Whitworth, R. H., and Larson, C. M. (1989). Differential diagnosis and staging of Alzheimer's disease with an aphasia battery. *Neuropsychiatry, Neuropsychology and Behavioral Neurology*, **1**, 255-65.

Wragg, R. E., and Jeste, D. V. (1989). Overview of depression and psychosis in Alzheimer's disease. *American Journal of Psychiatry*, **146**, 577-87.

Yoshioka, K., Miki, T., Katsuya, T., Ogihara, T., and Sakaki, Y. (1991). The [717]Val→Ile substitution in amyloid precursor protein is associated with familial Alzheimer's disease regardless of ethnic groups. *Biochemical and Biophysical Research Communications*, **178**, 1141-6.

Zarit, S. H., Orr, N. K., and Zarit, J. M. (1985). *The hidden victims of Alzheimer's disease: families under stress.* New York University Press, New York.

2

Down syndrome

J. M. Berg

Summary

As a background to detailed accounts and analyses of the relationships between Down syndrome and Alzheimer disease throughout the present book, a general overview of the nature and major characteristics of Down syndrome is presented in this chapter. Following a brief historical note on the syndrome, consideration is given to reproduction and life expectancy, to clinical manifestations (both physical and mental) at various ages, and to cytogenetic and molecular genetic data that have emerged in recent years. With the markedly improved survival of persons with Down syndrome into adulthood, including middle age and beyond, clinical, pathological and biological links between the syndrome and Alzheimer disease have become increasingly apparent, with each condition contributing to greater understanding of the other.

> Looking at persons with Down syndrome *'is like looking at the stars; the more one looks, the more one sees. New things are always appearing, and the suggestion is aroused that much more remains hidden from view.'*
>
> Bleyer (1934)

Introduction

Down syndrome (DS), the commonest known specific entity associated with mental retardation, has been the subject of intensive study since the initial definitive description of the condition in 1866 (*vide infra*). A vast array of biomedical and psychosocial data on the syndrome has now accumulated, documented in over 3,000 publications up to 1973 (Koch 1973) and in several thousand more since then, with a considerable proportion having a bearing also on other types of mental and related disabilities. Until recent years most of these data concerned children and teenagers with the syndrome, but dramatic improvements in life expectancy have led to an increasing focus on affected adults, including recognition and investigation of the remarkable relationship between DS and Alzheimer disease (AD). An attempt is made in this chapter to provide a general synopsis of major characteristics of DS, with particular attention to aspects of relevance in understanding the relationship mentioned above.

Historical note

It is something of a surprise that a condition as clinically distinctive and as common as DS, with occurrence in all peoples and countries, was first unequivocally described in phenotypic respects (despite an untenable ethnic classification approach) only about 125 years ago by Down (1866). Such surprise should be tempered by recognition of what must have been very high infant mortality of affected individuals in the past and by the tendency, well into the nineteenth century (Kanner 1964), to regard persons with mental deficit as being more or less in the same category, with little attention to them in terms of specific phenotypic or aetiological heterogeneity. It is of historic interest, however, that evidence has accumulated of the existence of DS long before Down wrote about it. At or near the time of Down's account are possible references to the syndrome by Duncan (1866), by Séguin (1846, 1866) and perhaps also by Esquirol (1838). A few hundred years prior to that, some suggestive visual evidence of the syndrome is provided in paintings by Jacob Jordaens in the seventeenth century (Zellweger 1968) and by Andrea Mantegna in the fifteenth (Ruhräh 1935). Additional possible indications of the syndrome's presence in former eras are Brothwell's (1960) report of a skull excavated from a Saxon burial site in Leicestershire, England, and, according to Stratford (1989), the appearances of particular Olmec figurine artefacts discovered in Mexico. Whatever uncertainty there may be concerning the diagnostic validity of these descriptions and depictions, there is no reason to doubt that DS would have occurred since the earliest periods of human history, albeit, as Richards (1968) pointed out, much less frequently than at present.

Reproduction and life expectancy

Reproduction

Apart from maternal age considerations, the overwhelming majority of prospective parents of a child with DS do not have identifiable characteristics that would indicate that they are at increased risk of having an affected child. The greater likelihood of birth of such a child as maternal age advances has long been known with, in most instances, a chance of less than 1/1000 and more than 1/100 of a live-born offspring with the syndrome at maternal ages of under 30 years and over 40 years respectively (Thompson *et al.* 1991, p. 215). Findings as to whether or not advanced paternal age *per se* also is a risk factor for DS have been inconsistent (e.g. Stene *et al.* 1977; Erickson 1979) but, if such a risk does exist, it seems clear that it is substantially lower than that associated with maternal age. Once a child with DS has been born to parents with apparently normal karyotypes, estimates of the chance of recurrence of DS in a subsequent offspring of those parents have varied, but have generally been found empirically to be of the order of one to two per cent, with the lower figure applying at maternal ages under 40 years and the higher one at later ages. Though surveys have not yielded entirely identical results (Tamaren *et al.* 1983; Abuelo *et al.* 1986), normal sibs and more distant relatives of a single individual with regular trisomy 21 do not appear to be at significantly increased risk of having an affected child.

In a minority of cases, a phenotypically normal potential parent is a carrier of an identifiable chromosome anomaly or variant which increases the chances of that individual having a child with DS irrespective of parental age. A well-established circumstance of that kind is the presence in the parent of a balanced Robertsonian translocation involving a 21 chromosome. Thus, for instance, women with such a translocation of the long arms of chromosome 14 and 21 [t(14q21q)] have been found empirically to have a chance of 10–15 per cent of producing a child with the syndrome, with that chance also being increased, though substantially less so, when the father is the carrier of the same translocation. An exceptionally bleak, though fortunately rare, outlook applies when either potential parent carries a 21/21 translocation or isochromosome 21—the outcome of a pregnancy is then always prenatal demise or a live-birth usually with DS and sometimes with monosomy 21.

Besides parental karyotype findings exemplified above, which definitely increase the risk of having a child with DS, some other chromosomal anomalies or variants in phenotypically normal parents have been postulated, but not conclusively established, as risk factors. For instance, conflicting observations in that regard have been reported concerning parental double nucleolus organizer regions (dNORs) on acrocentric chromosomes (Jackson-Cook *et al.* 1985; Serra and Bova 1990) and it has been suggested (Serra *et al.* 1990) that a pericentric inversion of the heterochromatic region of chromosome 9 [inv(9)(qh)] in a parent can increase the likelihood of offspring with DS.

A further important consideration involving a high risk for producing offspring with DS applies when a potential parent has chromosome 21 trisomy in some or all cells examined. The former circumstance, referred to as mosaicism (most often in the form of 46,XX or XY/47,XX or XY+21 mosaicism) may or may not be associated with parental clinical features suggestive or indicative of DS. Jagiello (1981) has summarized some data from publications on 36 such parents, 24 female and 12 male, among whom 38 of 50 progeny had DS. This statistical outcome is probably not representative since parents are more likely to be karyotyped if they have affected in contrast to normal children. Furthermore, mosaicism may be cryptic and a crucial factor is whether the germ line is or is not involved. The prevalence of mosaicism among couples who have had children with regular trisomy 21 DS is not entirely clear. Priest *et al.* (1973), on the basis of dermatoglyphic analyses, felt that as high as 20 per cent of cases of trisomy 21 DS could be due to parental mosaicism. A probably more reliable estimate of about three per cent, derived from cytogenetic studies, was provided by Harris *et al.* (1982). Mosaicism in one or other parent is most likely to be present if the couple concerned have had more than one child with trisomy 21.

Besides parents who are mosaic with an extra chromosome 21 cell line, the question of biologically fully affected persons with DS possibly also becoming parents has taken on increased practical importance in view of their much improved likelihood for survival well into child-bearing ages (see Life expectancy section below) and their more frequent integration within the general community than was the case in the past. In such individuals, particularly males, infertility or sterility appears usually to be present, although psychosocial influences will additionally affect reproductive prospects. However, pregnancies have been reported in at least 20 females with karyotypically proven DS, each of whom had regular (standard) trisomy 21 (Bovicelli *et al.* 1982; Shobha Rani 1990). They produced seven seemingly normal live-born children and eight with DS as well as several malformed and/or mentally retarded offspring with apparently normal karyotypes; the last-mentioned occurrences were presumably not directly due to the maternal DS. It is of interest, in regard to these pregnancies, that nearly all the known fathers (about half of them, none with DS, had been identified) were mentally handicapped and/or close blood relatives of the mothers.

By contrast to females with DS becoming mothers, there appears to be only one clearly documented case to-date of an apparently non-mosaic male with DS fathering a child. This 29-year-old man with regular trisomy 21 had a relationship with a karyotypically normal, retarded female that resulted in a pregnancy ending in spontaneous abortion at about 17 weeks of gestation, some nine weeks after transcervical chorionic villus sampling; the fetus had a normal 46,XY chromosome constitution and no apparent anomalies (Sheridan *et al.* 1989). About two years later, the same couple had a second pregnancy, with birth at term of a karyotypically and phenotypically normal boy (Bobrow *et al.* 1992).

Life expectancy

A very high proportion of embryos and fetuses with chromosome aberrations end in spontaneous abortions. DS is no exception. Kajii *et al.* (1973) estimated that 80 per cent of conceptuses with trisomy 21 do not survive to term and a subsequent estimate by Creasy and Crolla (1974) was not much lower than that (i.e. 65 per cent). Among live-births in general, one in 600-1,000 usually have been found to have DS (Smith and Berg 1976, pp. 234-5); the precise prevalence at birth in different populations is dependent on a number of variables, most notably trends with respect to maternal age when parents have their children and availability/use of prenatal diagnostic resources with their elective pregnancy termination implications—even within relatively circumscribed regions of the world, such as western Europe, marked differences exist between and within countries concerning both maternal age at child-bearing and prenatal diagnostic undertakings, as evidenced by recent data provided by Dolk *et al.* (1990).

Investigators	Study region	Years of birth	Percentage alive at:		
			1 year	5 years	10 years
Carter (1958)	London and neighbouring counties, UK	1944–55[a]	47	40	37
Collmann and Stoller (1963)	Victoria, Australia	1948–57	69	49	46
Fabia and Drolette (1970)	Massachusetts, USA	1950–66	76	69	65
Gallagher and Lowry (1975)	British Columbia, Canada	1952–71	89	85	82
Masaki *et al.* (1981)	Japan	1966–75	94	87	86
Fryers (1984)	Salford, UK	1961–80	81	77	71[b]

[a]Years of first attendance at Hospital for Sick Children, London
[b]Ten-year percentage based on 1961–70 births

Table 2.1. Percentages of live births with Down syndrome surviving through childhood (adapted from Fryers 1986)

Investigators		Malone (1988)	Baird and Sadovnick (1988,1989)
Study region		Western Australia	British Columbia, Canada
Years of Birth		1916–75	1908–81
Percentage alive at:	10–19 years	83	78–75
	20–29 years	76	75–72
	30–39 years	59	72–70
	40–49 years	42	70–63
	50–59 years	25	61–47
	60–69 years	7	44–13[a]

[a]Comparable percentages for general population = 86–78

Table 2.2. Percentages of persons with Down syndrome surviving beyond childhood

Life expectancy for infants born with DS has increased markedly in the last several decades, especially among those surviving the first weeks and initial year when life-threatening congenital malformations of various kinds (particularly cardiac ones) and severe infections (particularly respiratory ones) had previously taken a heavy toll and still do to some extent. Life expectancy for persons with DS in England was estimated to be eight to nine years in 1929 and about 12 years by 1949 (Penrose 1949). Dramatic improvements since then, at least in relatively affluent societies, are illustrated by data provided in Tables 2.1 and 2.2.

The improvements shown in the tables are to-date very unlikely to have taken place to the same extent in many underdeveloped countries or even in different communities within some individual developed countries. In the latter regard, for example, Mastroiacovo *et al.* (1990) reported recent five-year survival for children with DS of 59 per cent in the southern, compared with 80 per cent in the northern-central, regions of Italy, a state of affairs that the writers felt probably indicated poorer medical care and restricted availability of specialized medical services in the former region. An additional factor that could well have an impact on survival prospects in different jurisdictions is prevailing attitudes to handicapped persons, particularly mentally handicapped ones, in terms of recognizing their needs for suitable living conditions, including social, educational and vocational opportunities, and the will and means to provide these for all citizens. Improvements in life expectancy, as an end in itself without attention to making the quality of life better than dismal, is perhaps something of a Pyrrhic victory no less for people with DS than anyone else.

Clinical manifestations

Physical features

The high prevalence of DS and the extensive documentation of the disorder have resulted in its phenotypic manifestations becoming widely known. It needs to be emphasized that none of these manifestations is in itself pathognomonic of the syndrome, that all are not always present or present to the same degree, and that some (see for instance Farkas *et al.* 1991) are age-related. A clinical diagnosis is thus often made on the basis of the overall gestalt or impression rather than on a necessary occurrence of selected individual features. Nevertheless, depending on the comparative frequency in the syndrome and in comparable (matched) segments of the general population or groups with mental retardation excluding DS, some features will have greater diagnostic value than others. With this in mind, various observers have designated what they consider to be the most characteristic physical signs of DS. For instance, the 10 selected a few decades ago by Hall (1964) for infants and by Øster (1953) for children or adults, respectively, are listed in Table 2.3. Other investigators, for example Jackson *et al.* (1976), have given a weighting to particular physical signs based on their findings concerning the discriminative value of those signs. Difficulties arise with signs that are not readily measurable, such as several in Table 2.3, so that a conclusion as to their presence or absence may be subjective. This consideration has led to a number of anthropometric approaches to quantifying surface anatomy characteristics in the syndrome (Farkas *et al.* 1991). Besides that, the fact that dermatoglyphic hand and foot patterns lend themselves readily to precise analysis and the recognition of characteristic (albeit not unique) dermatoglyphic features in DS, led to the development of indices based on a combination of pattern distributions and measurements. These, as exemplified by the indices devised by Ford Walker (1958) and by Reed *et al.* (1970), yield a composite score that provides *per se* a high probability that the individuals examined do or do not have DS. In parenthesis, mention is made here that there have been reports of dermatoglyphic patterns in AD

Sign	Observer
Flat occiput	Ø
Abundant neck skin	H
Flat facial appearance	H
Oblique palpebral fissures	H Ø
Epicanthic folds	Ø
Dysplastic ears	H
Small teeth	Ø
Furrowed tongue	Ø
High arched palate	Ø
Short, broad hands	Ø
Curved 5th finger	Ø
Dysplastic middle phalanx of 5th finger	H
Four-finger palmar crease	H Ø
Lack of Moro reflex	H
Muscle hypotonia	H
Hyperextensibility or hyperflexibility	H Ø
Dysplastic pelvis	H

H, signs selected by Hall (1964) in newly born infants with Down syndrome; Ø, signs selected by Øster (1953) in children or adults with Down syndrome.

Table 2.3. Ten most characteristic physical signs of Down syndrome selected by each of two observers (After Smith and Berg 1976, pp. 152-3)

approximating to those found in DS (e.g. Weinreb 1986); however, others (e.g. Luxenberg *et al.* 1988) have failed to confirm a relationship.

There are multiple physical findings, in addition to those mentioned in Table 2.3, that are relatively common in DS. Some (e.g. speckling of the iris often referred to as Brushfield's spots, a wide space between the first and second toes and characteristic dermatoglyphic patterns on the hands and feet mentioned above) are essentially innocuous. Others (e.g. congenital cardiac and gastro-intestinal anomalies, cervical spine instability and a tendency to develop acute leukaemia) are important to recognize, apart from their possible value in helping to identify the syndrome, because they are incapacitating or life-threatening and generally amenable to beneficial medical or surgical treatment. Readers interested in details of these and additional clinical signs (over 300 have been described, according to Coleman 1978) will find them well documented in comprehensive textbooks on DS published in the last couple of decades (e.g. Smith and Berg 1976; Pueschel and Rynders 1982; Lane and Stratford 1985; Lott and McCoy 1992; Pueschel and Pueschel 1992) and in specific papers extensively referenced in these volumes; the books mentioned are also a useful source of further data on other facets of the syndrome considered in this chapter.

From the perspective of the present book, reference needs to be made to physical findings in DS that are prone to be especially common in adulthood, though not limited to that age group. Pueschel (1990) draws attention in that category to acquired cataracts, hearing loss, hypothyroidism and mitral valve prolapse. Other physical problems common in adults with DS, as well as in younger persons with the syndrome, include susceptibility to infections, sleep apnoea and disturbances of the spine, hip and knee joints (Pary 1992; see also Chapter 4). Prompt detection of these conditions is, of course, important for the purposes of beneficial therapeutic interventions. Besides that, in the context of evaluation for the presence of AD symptomatology in older persons with DS, their effects need to be taken into account as they can result in apparent decline of intellectual function and/or behavioural changes that may erroneously be attributed to AD. A case report by Thase (1982), by no means atypical, illustrates these points and the need for clinicians to be alert to them. Thase described a moderately retarded 38-year-old woman with DS who developed dementia attributed by a physician to 'ageing'. Investigation elsewhere showed her to have hypothyroidism, which, on laevothyroxine medication, resulted in reversal of the dementia and restoration of the woman's previously satisfactory functional state.

In addition, particular physical abnormalities that appear to be more common in older adults than in younger ones or children with DS may have a direct link with AD. Clinically recognizable seizures involving loss of consciousness with a clonic phase is an example. Veall (1974) reported an increased prevalence of such seizures of about 12 per cent in adults with DS aged over 55 years, compared with about six per cent in most age groups from 20–55 years (apart from a peak, for unknown reasons, of 11.7 per cent at ages 30–34 years) and about two per cent in those under 20 years. The precise relationship of these findings to AD is uncertain. However, it is notable that Lai and Williams (1989) found that 41 of 49 (84 per cent) adults with DS who had dementia developed seizures, most often of the generalized tonic–clonic kind; some of those with seizures also developed myoclonus, a finding of particular interest in view of localization on the long arm of chromosome 21 (q22.3 region) of a gene for a progressive form of myoclonus epilepsy (Unverricht-Lundborg type) (Lehesjoki *et al.* 1991). It seems relevant to add that seizures and/or myoclonus are not unusual in persons with AD in the general population (Hauser *et al.* 1986).

Mental characteristics

Mental deficit is a cardinal manifestation of DS and, apart from some instances of trisomy 21/normal chromosome complement mosaicism (Rosecrans 1968; Fishler and Koch 1991), is virtually always present although it may not be readily apparent in early infancy. Developmental 'milestones' such as sitting, standing and walking, bowel and bladder control and speech generally show some delay, and assessment with formal psychometric tests of various kinds throughout life reveal reduction in IQ and related scores (Smith and Berg 1976, pp. 65–72). It is important to remember that particular mental impairments and capabilities in DS, whether due to environmental, genetic or both types of factors, are not globally uniform (i.e. present to the same degree) in individual cases. For example, as Wisniewski *et al.* (1988) noted, there is considerable evidence, including their own findings in adults as well as children, that persons with DS frequently show relative weakness in language as compared to non-verbal functions.

There was a tendency in the past to stereotype persons with DS as being very similar, and even virtually identical, to each other in these respects, but it has become increasingly apparent that developmental and intellectual characteristics vary quite widely from case to case. Furthermore, although early structural alterations and irregularities in the brain are directly related to mental deficit in the syndrome (Becker *et al.* 1991), the functional levels attained are by no means entirely organically determined; it has become clear that they can be substantially influenced by the environmental milieu and the availability or otherwise of stimulating social, educational and, in later life, vocational opportunities. Abilities displayed and attained by individuals with DS are often impressive with steady progress during childhood and young adulthood. Berry *et al.* (1984), for instance, using measures of cognitive, verbal and functional performance, provided evidence that mental development of adults with DS continued well into the third and fourth decades of life. Later on, mental decline, compared with previous functional levels, is likely to become apparent at an earlier age than in the general population. Also, as exemplified in a study by Thase *et al.* (1984) of residents in an Ohio institution, there is substantial evidence that as age advances adults with DS show greater deficits on various neuropsychological measures than do closely matched mentally retarded adults who do not have DS. Among the important determining factors in these regards is the prospect for persons with DS of developing AD relatively early (clinically in some cases and neuropathologically in practically all over the age of 40 years), an issue not dwelt on here as it is comprehensively addressed from various perspectives in chapters throughout this book.

It has also been frequently stated that persons with DS display fairly distinctive temperaments and personality traits. Down (1866) in his initial description referred to them as 'humorous' with 'a lively sense of the ridiculous' and many adjectives, usually positive (e.g. affectionate, good-natured, cheerful, lively) but sometimes unfavourable (e.g. hostile, aggressive, stubborn, over-sensitive), have subsequently been applied. Claims have also been made, and disputed, that persons with DS have a special appreciation of music, an enhanced sense of rhythm and a particular talent for mimicry (for a review see Gibson 1978, pp. 194-201). Observations in these regards have ranged from anecdotal accounts to comparative studies of both institutionalized and non-institutionalized groups with DS and matched mentally retarded controls (e.g. Domino 1965; Gibbs and Thorpe 1983). One large study of over 500 individuals with DS and an equal number of retarded individuals without the syndrome (residing in Arizona and Washington State institutions) focused particularly on maladaptive behaviours and found these to be relatively infrequent in the DS sample (Moore *et al.* 1968). Nevertheless, behavioural and a wide range of other psychiatric disorders in persons of all ages with DS are not unusual (Menolascino 1965; Gath and Gumley 1986; Lund 1988; Myers and Pueschel 1991). Of particular note in the context of the theme of this book is the occurrence of major depression in adults with DS, which may be mistakenly diagnosed, in view of apparent dementia, as AD (Warren *et al.* 1989). The distinction is of practical importance from a therapeutic perspective, as illustrated in case reports by Szymanski and Biederman (1984) of individuals with DS, aged in their early thirties, whose severe depressive disorders were effectively treated with medication and psychological interventions.

Cytogenetic and molecular genetic considerations

Three years after Tjio and Levan (1956) established that the normal human diploid chromosome number is 46, Lejeune *et al.* (1959a, b) reported the presence of an extra small chromosome in DS, thus confirming notions expressed a couple of decades earlier by Waardenburg (1932), Bleyer (1934), Fanconi (1939) and Penrose (1939) that a chromosomal aberration of some kind might be the biological basis for DS. As a historical aside, it seems of interest to quote from the translation by Opitz and Gilbert-Barness (1990) of the relevant passage concerning DS in Waardenburg's (1932) book: "I should like to suggest ...that this is a human example of a specific chromosome aberration...why, if not lethal, should it not produce an extensive 'constitutional anomaly'?... one should determine whether mongolism is due to a 'chromosomal deficiency' through 'non-disjunction' or, conversely, a 'chromosome duplication'." It may be of interest also to mention that when Dr Waardenburg was asked in the mid-1960s what had initially prompted him to suggest a connection between non-disjunction and DS, he responded that only a chromosomal defect could adequately explain the many separate clinical features of the syndrome (Waardenburg 1967).

Following the accounts by Lejeune and his colleagues mentioned above and subsequent improvements in techniques for visualizing the morphological appearance of chromosomes, particularly banding procedures, it soon became amply apparent that the supernumerary chromosome, designated as a No. 21, was uniformly present (i.e. in virtually all cells examined) as a separate 47th chromosome in over 90 per cent of cases of DS. The phenomenon, often referred to as regular or standard trisomy 21, occurs as a result of meiotic non-disjunction of the chromosome 21 pair, usually during maternal meiosis at the meiotic I stage (first meiotic division). Initial evidence, based on cytogenetic studies of chromosome heteromorphisms, suggested that the non-disjunction occurred during paternal meiosis in about 20 per cent of cases, but introduction of more precise analyses of DNA polymorphisms, in multi-centre collaborative studies, indicated a lower percentage in the range of about 5–10 per cent (Takaesu *et al.* 1990; Antonarakis 1991). There has been much speculation concerning the aetiology of non-disjunction (for a review see Jagiello *et al.* 1987). Both genetic and environmental causal influences have been postulated, but the basic reasons for the occurrence of the phenomenon are, at present, uncertain.

The two most common trisomic chromosome 21 variants, accounting for most instances of DS other than regular (standard) trisomy 21 are mosaicism and translocations. Mosaicism, usually with 46,XX or XY and 47,XX or XY,+21 chromosome complements in variable proportions, is frequently reported in about two to three per cent of cases, with this percentage differing in accordance with such factors as diagnostic criteria for suspecting DS (i.e. the range and severity of DS manifestations) and the extent of the search for mosaicism in one or more tissues. A rather higher percentage is often found for translocations (i.e. in the neighbourhood of five per cent); the overwhelming majority of these are Robertsonian in type, most frequently Dq/21q or Gq/21q and either occurring sporadically or inherited from a carrier parent. Hamerton (1982) has provided a more detailed summary of the frequency of mosaicism, translocations and other rarer variants of trisomy 21. Among important developments was the accumulating evidence in the 1970s that triplication of a particular portion of the long arm of chromosome 21 (the distal q22 segment) is crucial for the occurrence of the characteristic DS phenotype (for a review see Summitt 1982). Since then, highly significant additional data have emerged concerning further delineation, especially at the molecular level, of the 'critical' (also called 'obligate') DS region, as well as other regions of chromosome 21. These considerations have been addressed in detail from various perspectives, with particular reference to relationships or possible relationships with AD, in a number of chapters in this book. Here, brief general observations are made concerning the molecular analysis of chromosome 21, an analysis that is proceeding apace.

Identification of genes and DNA sequences on chromosome 21, using a variety of innovative techniques, have been extensively documented during the past decade (for details see Patterson and Epstein 1990), with continuing publication of a veritable avalanche of new observations. A recent committee report on the genetic constitution of chromosome 21 (Cox and Shimizu 1991) comprehensively tabulates findings to-date and provides an extensive bibliography. The report indicates that a total of 24 genes and/or pseudogenes and 176 DNA segments had thus far been regionally assigned to chromosome 21. Gardiner (1990) notes that the long arm of that chromosome has sufficient DNA to contain 500–1,000 genes, so that much remains to be elucidated. However, the advances already made have been of great significance in the process of unravelling the mysteries of the molecular structure and functions of chromosome 21 that constitute the biological bases of the multiple phenotypic facets of DS, including links with AD.

Among important developments are investigations of relatively rare instances of partial trisomy 21 aimed at defining regions of the chromosome and eventually the particular genes within these regions that determine specific phenotypic characteristics of DS. A phenotypic map of chromosome 21 is thus emerging (Epstein *et al.* 1991; Korenberg 1991; Korenberg *et al.* 1992) and contributing to further understanding of the various (and to a considerable extent variable) individual and combined manifestations of the syndrome.

Conclusion

A vast body of data on DS, both in quantity and range, has accumulated since John Langdon Haydon Down (see Frontispiece) put the syndrome on the map, as it were, with his initial seminal paper of 1866. Much has been learned about the condition since then, even though Bleyer's insightful comment half a century ago quoted at the beginning of this chapter, including his suggestion that 'much more remains hidden from view', still seems valid today. Nevertheless, study of the syndrome in phenotypic and genotypic terms has already shed some light on human disorders besides DS. In that regard and with reference to the theme of the present volume, a greater survival of people with DS into adulthood (including middle age and beyond), more precise recognition of clinical and pathological features at these and earlier ages and advances through the application of relatively new cytogenetic and molecular biological techniques (aspects of which are briefly outlined in the chapter) have provided bases for significant contributions to elucidating enigmas of AD, in addition to clarifying further the nature and characteristics of DS. More progress in these respects in the immediately foreseeable future can reasonably be anticipated with improved therapeutic, and even primary preventive, prospects for both conditions.

References

Abuelo, D., Barsel-Bowers, G., Busch, W., Pueschel, S., and Pezzullo, J. (1986). Risk for trisomy 21 in offspring of individuals who have relatives with trisomy 21. *American Journal of Medical Genetics,* 25, 365-7.

Antonarakis, S. E. (1991). Parental origin of the extra chromosome in trisomy 21 as indicated by analysis of DNA polymorphisms. *New England Journal of Medicine,* 324, 872-6.

Baird, P. A., and Sadovnick, A. D. (1988). Life expectancy in Down syndrome adults. *Lancet,* 2, 1354-6.

Baird, P. A., and Sadovnick, A. D. (1989). Life tables for Down syndrome. *Human Genetics,* 82, 291-2.

Becker, L., Mito, T., Takashima, S., and Onodera, K. (1991). Growth and development of the brain in Down syndrome. In *The morphogenesis of Down syndrome,* (ed. C. J. Epstein), pp. 133-52. Wiley-Liss, New York.

Berry, P., Groeneweg, G., Gibson, D., and Brown, R. I. (1984). Mental development of adults with Down syndrome. *American Journal of Mental Deficiency,* 89, 252-6.

Bleyer, A. (1934). Indications that mongoloid imbecility is a gametic mutation of degressive type. *American Journal of Diseases of Children,* 47, 342-8.

Bobrow, M., Barby, T., Hajianpour, A., Maxwell, D., and Yau, S. C. (1992). Fertility in a male with trisomy 21. *Journal of Medical Genetics,* 29, 141.

Bovicelli, L., Orsini, L. F., Rizzo, N., Montacuti, V., and Bacchetta, M. (1982). Reproduction in Down syndrome. *Obstetrics and Gynecology,* 59, No. 6 (Suppl.), 13S-17S.

Brothwell, D. R. (1960). A possible case of mongolism in a Saxon population. *Annals of Human Genetics,* 24, 141-50.

Carter, C. O. (1958). A life-table for mongols with the causes of death. *Journal of Mental Deficiency Research,* 2, 64-74.

Coleman, M. (1978). Down's syndrome. *Pediatric Annals,* 7, 90-103.

Collmann, R. D., and Stoller, A. (1963). A life-table for mongols in Victoria, Australia. *Journal of Mental Deficiency Research,* 7, 53-9.

Cox, D. R., and Shimizu, N. (1991). Report of the committee on the genetic constitution of chromosome 21. *Cytogenetics and Cell Genetics*, **58**, 800-26.

Creasy, M. R., and Crolla, J. A. (1974). Prenatal mortality of trisomy 21 (Down's syndrome). *Lancet*, **1**, 473-4.

Dolk, H., De Wals, P., Gillerot, Y., Lechat, M. F., Aymé, S., Beckers, R., *et al.* (1990). The prevalence at birth of Down syndrome in 19 regions of Europe, 1980-86. In *Key issues in mental retardation research*, (ed. W.I. Fraser), pp. 3-11. Routledge, London.

Domino, G. (1965). Personality traits in institutionalized mongoloids. *American Journal of Mental Deficiency*, **69**, 568-70.

Down, J. L. H. (1866). Observations on an ethnic classification of idiots. *London Hospital Reports*, **3**, 259-62.

Duncan, P. M. (1866). *A manual for the classification, training and education of the feeble-minded, imbecile and idiotic*. Longmans, Green & Co., London.

Epstein, C. J., Korenberg, J. R., Annerén, G., Antonarakis, S. E., Aymé, S., Courchesne, E., *et al.* (1991). Protocols to establish genotype-phenotype correlations in Down syndrome. *American Journal of Human Genetics*, **49**, 207-35.

Erickson, J. D. (1979). Paternal age and Down syndrome. *American Journal of Human Genetics*, **31**, 489-97.

Esquirol, J. E. D. (1838). *Des maladies mentales considérés sous les rapports médical, hygiénique et médico-légal* (2 Vols). Baillière, Paris.

Fabia, J., and Drolette, M. (1970). Life-tables up to age 10 for mongols with and without congenital heart defect. *Journal of Mental Deficiency Research*, **14**, 235-42.

Fanconi, G. (1939). Die Mutationstheorie des Mongolismus. *Schweizerische Medizinische Wochenschrift*, **69**, 81-3.

Farkas, L. G., Posnick, J. C., and Hreczko, T. (1991). Anthropometry of the head and face in ninety-five Down syndrome patients. In *The morphogenesis of Down syndrome*, (ed. C.J. Epstein), pp. 53-9. Wiley-Liss, New York.

Fishler, K., and Koch, R. (1991). Mental development in Down syndrome mosaicism. *American Journal on Mental Retardation*, **96**, 345-51.

Ford Walker, N. (1958). The use of dermal configurations in the diagnosis of mongolism. *Pediatric Clinics of North America*, **5**, 531-43.

Fryers, T. (1984). *The epidemiology of severe intellectual impairment: the dynamics of prevalence*, pp. 100-22. Academic Press, London.

Fryers, T. (1986). Survival in Down's syndrome. *Journal of Mental Deficiency Research*, **30**, 101-10.

Gallagher, R. P., and Lowry, R. B. (1975). Longevity in Down's syndrome in British Columbia. *Journal of Mental Deficiency Research*, **19**, 157-63.

Gardiner, K. (1990). Physical mapping of the long arm of chromosome 21. In *Molecular genetics of chromosome 21 and Down syndrome*, (ed. D. Patterson and C. J. Epstein), pp. 1-14. Wiley-Liss, New York.

Gath, A., and Gumley, D. (1986). Behaviour problems in retarded children with special reference to Down's syndrome. *British Journal of Psychiatry*, **149**, 156-61.

Gibbs, M. V., and Thorpe, J. G. (1983). Personality stereotype of noninstitutionalized Down syndrome children. *American Journal of Mental Deficiency*, **87**, 601-5.

Gibson, D. (1978). *Down's syndrome: the psychology of mongolism*. Cambridge University Press, Cambridge.

Hall, B. (1964). Mongolism in newborns: a clinical and cytogenetic study. *Acta Paediatrica*, Suppl. **154**, 1-95.

Hamerton, J. L. (1982). Frequency of mosaicism, translocation, and other variants of trisomy 21. In *Trisomy 21 (Down syndrome): research perspectives*, (ed. F. F. de la Cruz and P. S. Gerald), pp. 99-107. University Park Press, Baltimore.

Harris, D. J., Begleiter, M. L., Chamberlin, J., Hankins, L., and Magenis, R. E. (1982). Parental trisomy 21 mosaicism. *American Journal of Human Genetics*, **34**, 125-33.

Hauser, W. A., Morris, M. L., Heston, L. L., and Anderson, V. E. (1986). Seizures and myoclonus in patients with Alzheimer's disease. *Neurology*, **36**, 1226-30.

Jackson, J. F., North, E. R. III, and Thomas, J. G. (1976). Clinical diagnosis of Down's syndrome. *Clinical Genetics*, **9**, 483-7.

Jackson-Cook, C.K., Flannery, D.B., Corey, L.A., Nance, W.E., and Brown, J.A. (1985). Nucleolar organizer region variants as a risk factor for Down syndrome. *American Journal of Human Genetics*, **37**, 1049-61.

Jagiello, G. (1981). Reproduction in Down syndrome. In *Trisomy 21 (Down syndrome): research perspectives*, (ed. F. F. de la Cruz and P. S. Gerald), pp. 151-62. University Park Press, Baltimore.

Jagiello, G. M., Fang, J. -S., Ducayen, M. B., and Sung, W. K. (1987). Etiology of human trisomy 21. In *New perspectives on Down syndrome*, (ed. S. M. Pueschel, C. Tingey, J. E. Rynders, A. C. Crocker and D. M. Crutcher), pp. 23-38. Paul H. Brookes, Baltimore.

Kajii, T., Ohama, K., Niikawa, N., Ferrier, A., and Avirachan, S. (1973). Banding analysis of abnormal karyotypes in spontaneous abortion. *American Journal of Human Genetics*, **25**, 539-47.

Kanner, L. (1964). *A history of the care and study of the mentally retarded*, pp. 87-109. Charles C. Thomas, Springfield, Illinois.

Koch, G. (1973). *Down-syndrom: mongolismus*. Bibliographica Genetica Medica, Vol. 1. J. Hogl, Erlangen.

Korenberg, J. R. (1991). Down syndrome phenotypic mapping. In *The morphogenesis of Down syndrome*, (ed. C.J. Epstein), pp. 43-52. Wiley-Liss, New York.

Korenberg, J. R., Bradley, C., and Disteche, C. M. (1992). Down syndrome: molecular mapping of the congenital heart disease and duodenal stenosis. *American Journal of Human Genetics*, **50**, 294-302.

Lai, F., and Williams, R. S. (1989). A prospective study of Alzheimer disease in Down syndrome. *Archives of Neurology*, **46**, 849-53.

Lane, D., and Stratford, B. (ed.) (1985). *Current approaches to Down's syndrome*. Holt, Rinehart and Winston, London.

Lehesjoki, A. -E., Koskiniemi, M., Sistonen, P., Miao, J., Hästbacka, J., Norio, R., and de la Chapelle, A. (1991). Localization of a gene for progressive myoclonus epilepsy to chromosome 21q22. *Proceedings of the National Academy of Sciences of the United States of America*, **88**, 3696-9.

Lejeune, J., Gautier, M., and Turpin, R. (1959a). Les chromosomes humains en culture de tissus. *Comptes Rendus Hebdomadaires des Séances de l'Académie des Sciences*, **248**, 602-3.

Lejeune, J., Gautier, M., and Turpin, R. (1959b). Étude des chromosomes somatiques de neuf enfants mongoliens. *Comptes Rendus Hebdomadaires des Séances de l'Académie des Sciences*, **248**, 1721-2.

Lott, I. T., and McCoy, E. E., (ed.) (1992). *Down syndrome: Advances in medical care*. Wiley-Liss, New York.

Lund, J. (1988). Psychiatric aspects of Down's syndrome. *Acta Psychiatrica Scandinavica*, **78**, 369-74.

Luxenberg, J. S., Plato, C. C., Fox, K. M., Friedland, R. P., and Rapoport, S. I. (1988). Digital and palmar dermatoglyphics in dementia of the Alzheimer type. *American Journal of Medical Genetics*, **30**, 733-40.

Malone, Q. (1988). Mortality and survival of the Down's syndrome population in Western Australia. *Journal of Mental Deficiency Research*, **32**, 59-65.

Masaki, M., Higurashi, M., Iijima, K., Ishikawa, N., Tanaka, F., Fujii, T., *et al.* (1981). Mortality and survival for Down's syndrome in Japan. *American Journal of Human Genetics*, **33**, 629-39.

Mastroiacovo, P., Bertollini, R., and Corchia, C. (1990). Survival trends in Down syndrome. *Lancet*, **335**, 1278-9.

Menolascino, F. J. (1965). Psychiatric aspects of mongolism. *American Journal of Mental Deficiency*, **69**, 653-60.

Moore, B. C., Thuline, H. C., and Capes, L. (1968). Mongoloid and non-mongoloid retardates: a behavioral comparison. *American Journal of Mental Deficiency*, **73**, 433-6.

Myers, B. A., and Pueschel, S. M. (1991). Psychiatric disorders in persons with Down syndrome. *Journal of Nervous and Mental Disease*, **179**, 609-13.

Opitz, J. M., and Gilbert-Barness, E. F. (1990). Reflections on the pathogenesis of Down syndrome. *American Journal of Medical Genetics*, Suppl. 7, 38-51.

Øster, J. (1953). *Mongolism: a clinicogenealogical investigation comprising 526 mongols living in Seeland and neighbouring islands in Denmark*. Danish Science Press, Copenhagen.

Pary, R. (1992). Differential diagnosis of functional decline in Down's syndrome. *The Habilitative Mental Healthcare Newsletter*, 11, 37-41.

Patterson, D., and Epstein, C. J. (ed.)(1990). *Molecular genetics of chromosome 21 and Down syndrome*. Wiley-Liss, New York.

Penrose, L. S. (1939). Maternal age, order of birth and developmental abnormalities. *Journal of Mental Science*, 85, 1141-50.

Penrose, L. S. (1949). The incidence of mongolism in the general population. *Journal of Mental Science*, 95, 685-8.

Priest, J. H., Verhulst, C., and Sirkin, S. (1973). Parental dermatoglyphics in Down's syndrome: a ten-year study. *Journal of Medical Genetics*, 10, 328-32.

Pueschel, S. M. (1990). Clinical aspects of Down syndrome from infancy to adulthood. *American Journal of Medical Genetics*, Suppl. 7, 52-6.

Pueschel, S. M., and Pueschel, J. K. (ed.) (1992). *Biomedical concerns in persons with Down syndrome*. Paul H. Brookes, Baltimore.

Pueschel, S. M., and Rynders, J. E. (ed.) (1982). *Down syndrome: advances in biomedicine and the behavioral sciences*. The Ware Press, Cambridge, MA.

Reed, T. E., Borgaonkar, D. S., Conneally, P. M., Yu, P., Nance, W. E., and Christian, J. C. (1970). Dermatoglyphic nomogram for the diagnosis of Down's syndrome. *Journal of Pediatrics*, 77, 1024-32.

Richards, B. W. (1968). Is Down's syndrome a modern disease? *Lancet*, 2, 353-4.

Rosecrans, C. J. (1968). The relationship of normal/21-trisomy mosaicism and intellectual development. *American Journal of Mental Deficiency*, 72, 562-6.

Ruhräh, J. (1935). Cretin or mongol or both together. *American Journal of Diseases of Children*, 49, 477-8.

Séguin, E. (1846). *Le traitement moral, l'hygiène et l'éducation des idiots*. J. B. Baillière, Paris.

Séguin E. (1866). *Idiocy and its treatment by the physiological method*. William Wood, New York.

Serra, A., and Bova, R. (1990). Acrocentric chromosome double NOR is not a risk factor for Down syndrome. *American Journal of Medical Genetics*, Suppl. 7, 169-74.

Serra, A., Brahe, C., Millington-Ward, A., Neri, G., Tedeschi, B., Tassone, F., and Bova, R. (1990). Pericentric inversion of chromosome 9: prevalence in 300 Down syndrome families and molecular studies of nondisjunction. *American Journal of Medical Genetics*, Suppl. 7, 162-8.

Sheridan, R., Llerena, J., Matkins, S., Debenham, P., Cawood, A., and Bobrow, M. (1989). Fertility in a male with trisomy 21. *Journal of Medical Genetics*, 26, 294-8.

Shobha Rani, A., Jyothi, A., Reddy, P. P., and Reddy, O. S. (1990). Reproduction in Down's syndrome. *International Journal of Gynecology and Obstetrics*, 31, 81-6.

Smith, G. F., and Berg, J. M. (1976). *Down's anomaly*, (2nd edn). Churchill Livingstone, Edinburgh.

Stene, J., Fischer, G., Stene, E., Mikkelsen, M., and Petersen, E. (1977). Paternal age effect in Down's syndrome. *Annals of Human Genetics*, 40, 299-306.

Stratford, B. (1989). *Down's syndrome: past, present and future*. Penguin Books, London.

Summitt, R. L. (1982). Chromosome 21: specific segments that cause the phenotype of Down syndrome. In *Trisomy 21 (Down syndrome): research perspectives*, (ed. F. F. de la Cruz and P. S. Gerald), pp. 225-35. University Park Press, Baltimore.

Szymanski, L. S., and Biederman, J. (1984). Depression and anorexia nervosa of persons with Down syndrome. *American Journal of Mental Deficiency*, 89, 246-51.

Takaesu, N., Jacobs, P.A., Cockwell, A., Blackston, R.D., Freeman, S., Nuccio, J., *et al.* (1990). Nondisjunction of chromosome 21. *American Journal of Medical Genetics*, Suppl. 7, 175-81.

Tamaren, J., Spuhler, K., and Sujansky, E. (1983). Risk of Down syndrome among second- and third-degree relatives of a proband with trisomy 21. *American Journal of Medical Genetics*, 15, 393-403.

Thase, M. E. (1982). Reversible dementia in Down's syndrome. *Journal of Mental Deficiency Research*, **26**, 111-3.

Thase, M. E., Tigner, R., Smeltzer, D. J., and Liss, L. (1984). Age-related neuropsychological deficits in Down's syndrome. *Biological Psychiatry*, **19**, 571-85.

Thompson, M. W., McInnes, R. R., and Willard, H. F. (1991). *Genetics in medicine.* (5th edn). W. B. Saunders, Philadelphia.

Tjio, J.H., and Levan, A. (1956). The chromosome number of man. *Hereditas(Lund)*, **42**, 1-6.

Veall, R. M. (1974). The prevalence of epilepsy among mongols related to age. *Journal of Mental Deficiency Research*, **18**, 99-106.

Waardenburg, P. J. (1932). *Das menschliche Auge und seine Erbanlagen*, pp. 47-8. Martinus Nijhoff, The Hague,.

Waardenburg, P. J. (1967). General discussion. In *Mongolism*, (ed. G. E. W. Wolstenholme and R. Porter), pp. 91-2. Churchill, London.

Warren, A. C., Holroyd, S., and Folstein, M. F. (1989). Major depression in Down's syndrome. *British Journal of Psychiatry*, **155**, 202-5.

Weinreb, H. J. (1986). Dermatoglyphic patterns in Alzheimer's disease. *Journal of Neurogenetics*, **3**, 233-46.

Wisniewski, K. E., Miezejeski, C. M., and Hill, A. L. (1988). Neurological and psychological status of individuals with Down syndrome. In *The psychobiology of Down syndrome*, (ed. L. Nadel), pp. 315-43. MIT Press, Cambridge, MA.

Zellweger, H. (1968). Is Down's syndrome a modern disease? *Lancet*, **2**, 458.

II

Alzheimer disease and Down syndrome: evidence of an association

Alzheimer disease and Down syndrome: scientific symbiosis—a historical commentary

T. G. Beach

Summary

Persons with Down syndrome surviving beyond the third decade invariably develop the characteristic histopathological brain lesions of Alzheimer disease. An adequate conceptual framework of dementing diseases and the necessary histological technology to diagnose these became available in the first decade of this century, but it was not until 1948 that George A. Jervis first recognized that the occurrence of senile plaques and neurofibrillary tangles in Down syndrome was more than just a coincidence. Jervis immediately realized that comparison of these two conditions could contribute to an understanding of both. In subsequent years this comparison has been extended from the morphological through to the neurochemical and genetic levels, culminating in a plausible hypothesis of molecular pathogenesis encompassing both disorders. The Alzheimer-type lesions that occur in Down syndrome have been shown to be identical to those in Alzheimer disease and are accompanied by the major neurochemical changes, including the cholinergic deficit. Clinically, deterioration in behavioural parameters has been found to be common in elderly individuals with Down syndrome, but not as common as would be expected from autopsy studies. The gene for amyloid precursor protein, which contains the coding region for the amyloid β protein of senile plaque amyloid, has been localized to chromosome 21, and mutations of the gene have been found to be causative in some families with early-onset Alzheimer disease. This has led to the hypothesis that various alterations of amyloid precursor protein metabolism may lead, by a final common pathway, to plaques, tangles and dementia.

Introduction

We are truly fortunate to have handed to us, through the diligent work of our predecessors, the opportunity to learn from the natural experiment which is the subject of this book. The importance of this opportunity can only be appreciated by realizing that the study of human pathology is often handicapped by the inability to experiment. Without the chance to alter parameters, thereby evaluating their influence, it is often necessary to use processes more closely allied to a historical discipline rather than a scientific one (Gould 1980a). The sequence of pathogenetic change must often be inferred from the end results. While this method can bring some success, it lacks the certainty that the experimental method provides. Down syndrome (DS), a disorder clearly resulting from a chromosomal change, with a range of pathological effects that includes Alzheimer changes as only a subset, has become a natural experiment with which to evaluate hypotheses about Alzheimer disease (AD)—and vice versa. The knowledge gained is sure to be of benefit to individuals with both conditions.

The history of the association between DS and AD can be broken down into three broad phases. The initial phase lasted from 1866, when DS was first clearly described (Down 1866), until 1948 (Jervis 1948), when recognition of its relationship with AD was first attained. During this phase, knowledge of the association was lacking, but the facts that would eventually allow recognition to occur were collecting. The second phase lasted from 1948 until the late 1960s. During this period of dormancy, knowledge of the association was limited to a few specialists, with its usefulness largely unexploited. The third phase takes us to the present time; the relatively sudden appreciation of the importance of the association has sparked an explosive inquiry that still continues. In this chapter, some of the more interesting and notable trends and revelations to arise out of this research are discussed.

Recognition of the association

The pivotal point in the history of the relationship between DS and AD is undoubtedly its recognition by Jervis (1948). He was the first to recognize that a genuine association did exist and that the study of DS might offer 'some clue' to the causes of senile dementia. Prior to that time there is a tantalizing foreshadowing of this achievement, but an ultimate failure to reach it. As early as 1876, Fraser and Mitchell are reported to have discovered an association between DS and dementia (Mann 1988a), but AD had not yet been described and dementia was, at the time, a relatively non-differentiated term. Investigators of the nineteenth century were greatly hindered in their understanding of dementia by the lack of a workable clinical approach and the virtual absence of the crucial histopathological methods. The crystallization of clinical and pathological classifications of dementia, which occurred at the turn of the century, provided effective tools for a meaningful exploration of what had previously been an inpenetrable thicket (Beach 1987). In particular, advances in histological staining had revealed the existence of those peculiar lesions of the ageing brain, the senile plaque and the neurofibrillary tangle. Thus Struwe (1929) appears to have been the first investigator who, being in possession of both knowledge and technique, might have discovered the association when he reported the presence of senile plaques in a 37-year-old individual with DS. However, it is apparent that he did not: rather than giving rise to a conceptual linkage between DS and AD, this case was interpreted as establishing an association between tuberculosis and AD (Jervis and Soltz 1936). Generalized tuberculosis was a common cause of death amongst persons with DS in that era and Davidoff (1928) gave explicit warning not to confuse the lesions of tuberculosis with other pathology in the DS brain. The next chance did not occur for almost 20 years, when Bertrand and Koffas (1946) found senile plaques in a 34-year-old individual with DS. These authors were apparently unaware of Struwe's report and did not offer a comment on the overall significance of their finding. Although Struwe, Bertrand and Koffas had not realized the generality of the relationship they had reported on, their work may have alerted Jervis, for he cited both papers in his 1948 article.

Jervis was additionally favoured with the advantage of a prepared mind, as he had already in his career contributed seminally to the literature on AD (Jervis and Soltz 1936). As mentioned previously, when he learned of Struwe's 1929 paper, he regarded the presence of senile plaques in that DS case as being due to tuberculosis. But when Bertrand and Koffas added their case in 1946, this may have been enough to make him reject his earlier interpretation and conduct an investigation of his own. His paper of 1948 presented three individuals with DS, aged 37, 42 and 47 years, each of whom had shown a profound emotional and intellectual deterioration in the last few years of life. At autopsy, all were found to have senile plaques and two also displayed neurofibrillary tangles. This was the first demonstration of tangles in DS and the first full clinicopathological correlation supporting an Alzheimer-like syndrome in DS. Jervis made no mistake as to the significance of his finding. 'Although this condition appears of interest to the study of both mongoloidism and senility, it has escaped attention thus far.'; 'Since mongoloid patients show a marked tendency to develop this type of reaction, it is suggested that the study of it offers some information which may contribute to a better understanding of the causes of senile dementia.'

Before proceeding to the post-Jervis years, it is of interest to note, in DS research of this century, the vestiges of what Gould (1980b) has called 'scientific racism', and Gibson (1978) termed a 'scientific nightmare'. This stemmed directly from Down's landmark paper of 1866. Under the influence of the over-extended concept that 'ontogeny recapitulates phylogeny', Down proposed that mental retardation could arise from arrests of development; furthermore, afflicted individuals would exemplify the evolutionary stage they were passing through at the time of arrest. On the conceptual evolutionary ladder of the time, European Caucasians occupied the uppermost rung, with the 'lower' races strung out on the rungs immediately below, followed by the apes and the rest of the animal kingdom. Thus, in his original 1866 paper, he described Caucasian 'idiots' that reminded him of African, Malay, American Indian and Oriental peoples.

Although Down's classification system was never widely adopted, the idea that individuals with DS might show some physical trace of an earlier evolutionary existence persisted well into this century. Down's own son rejected the idea that DS was an arrest at the racial level, but persisted in arguing for evolutionary reversion: '...if this is a case of reversion it must be reversion to a type even further back than Mongol stock...' (Gould 1980b). Davidoff (1928), reviewing the literature on brain anatomy in DS, noted the earlier work of Gans, who considered certain cerebellar heteropias (tuber flocculi), often seen in persons with DS, to resemble similar formations that he believed to be normally seen in the chimpanzee and orangutan. Gans also ventured that defects in the calcarine cortex and corpora quadrigemina represented a regression to the state of mammals with laterally placed eyes. Davidoff appears to have challenged these observations; a footnote states that he examined serial sections of the cerebellum from both chimpanzee and orangutan without finding any evidence of the structure described by Gans, and he failed to find, in his own 10 cases, the described defects of the visual pathway. The corrective efforts were continued by Benda (1940), who wrote, 'It has been one of the main subjects of this study to clarify the conception of earlier investigators who considered the broadening of the convolutions as a biological reversion toward the pattern seen in lower primates.' He proposed that the simplification of gyral pattern sometimes seen in DS was due, not to developmental arrest, but to fusion of fissures after birth. Finally, in 1972, Colon apparently sought to close the chapter on this peculiar story by proclaiming the brain in DS to be '...essentially human, but pathologic.', but not until he had done a detailed comparative morphometric analysis of the cerebral cortex in 'mongolism', 'rabbit', 'chimpanzee' and 'human'.

Confirmation of the association

Returning to the contributions of Jervis, it is curious that despite his clear enunciation of the potential importance of the association between DS and AD, his findings languished in relative obscurity for over 10 years. It was the neuropathologists who revived flagging interest in the association. Their first objective was to determine just how valid this association was. Malamud (1964) started the process with a study of 251 DS cases, 20 of which were aged 37 years or older. The results were convincing: all 20 had plaques and/or tangles. Malamud's findings were quickly replicated by a host of others, notably Solitaire and Lamarche (1966), Neumann (1967), Haberland (1969), Olson and Shaw (1969), and Burger and Vogel (1973). The electron microscope, which had so recently brought excitement to the study of AD (Kidd 1963; Terry 1963), confirmed that the similarities persisted at the ultrastructural level (Burger and Vogel 1973; Ellis *et al.* 1974).

This avalanche of neuropathology caught clinicians by surprise. Those who worked with DS patients were, for the most part, unaware of a late-onset cognitive disorder in DS. Owens *et al.* (1971) wrote, 'A connection...is not evident, and it is a rather startling surprise to the initiate that a relationship has been reported to exist.' These authors go on to report that in their clinical study of 35 adults with DS (19 aged between 36 and 50 years and the rest younger), not a single case was demented. They then asked the question, which remains perhaps the most puzzling aspect of the association, 'Why doesn't the widespread pathology unmistakably manifest itself as a dementia?'

Fifteen years later Wisniewski and Rabe (1986) reviewed the discrepancy between neuropathology and dementia; numerous intervening studies had confirmed its existence but its significance was still uncertain. It was clear that in many areas older subjects performed worse than younger ones, but most studies still failed to demonstrate fully fledged dementia. The authors offered two reasons to account for this discrepancy. First, they pointed out the difficulty of diagnosing deterioration in a group that already functioned at the lower end of many rating methods; more sensitive tests and longitudinal studies were needed. Secondly, they suggested that persons with DS might have a greater tolerance for the lesions of AD. Oliver and Holland (1986) concurred that testing methods were inadequate. Not only were people with DS functioning at the lower end of test sensitivity, but there was great inter-individual variation in baseline functioning, further complicating matters. They noted that the reported age-related changes were predominantly behavioural: loss of sociability and self-care skills were repeatedly noted as early signs. As these changes are usually a later, rather than an earlier, development in AD, it would appear that earlier, more subtle changes were being missed altogether.

Because the baseline functioning of persons with DS varies so widely, many investigators of the 1970s and 1980s recognized that longitudinal studies were essential to uncover intellectual declines in ageing individuals. Detailed and thorough longitudinal studies of ageing in DS have begun to appear only in the last few years; of these perhaps the most important has been the study of Lai and Williams (1989). Ninety-six cases, all over the age of 35, were followed for varying periods up to nine years, with a combination of standard neurological examination and subjective staff assessment of level of functioning. A significant aspect of the authors' approach was that, for the purpose of the study, the definition of dementia was broadened considerably, being defined as a decline in one or more of the assessed skills, rather than as an absolute level of function. The prevalence of dementia in the institutionalized part of the study population was found to be eight per cent in those aged 35–49, 55 per cent in those aged 50–59, and 75 per cent in those over 60 years. In agreement with Oliver and Holland (1986), the authors noted that the most frequent initial clinical signs of dementia in DS corresponded roughly to the later stages of AD; they ventured that the earliest stages, such as simple memory difficulties, were probably masked by poor language skills. Also in agreement with earlier studies was the observation that in most cases the first symptom was a change in personality, such as irritability or emotional lability. The results, in combination with the pathological studies available, confirm the existence of a latent period between the appearance of the histopathological lesions and the onset of clinical symptoms. The average age of onset of clinical deterioration was 54.2 years, which is approximately 20–30 years after the occurrence of the first pathological lesions. While this study must be considered a landmark, the development and application of more formalized neuropsychiatric assessment methodology might improve sensitivity and objectivity in future studies. It is also interesting to note that despite the relaxed definition of dementia, its prevalence still fails to approach that predicted by the neuropathology. It remains to be seen whether this discrepancy is an artefact of insufficiently sensitive clinical testing or a biological difference between AD and DS, as suggested by Wisniewski and Rabe (1986).

The initial pathomorphological and clinical comparisons indicated that the AD-type pathology in DS was strikingly similar to that in AD, with differences being minor or questionable; the comparisons have subsequently been extended to the neurochemical and molecular levels. One of the most important objectives was to determine whether or not the well-known cholinergic deficit, the most consistent neurochemical change in AD, occurred in ageing persons with DS. Early reports (Yates *et al.* 1980, 1983) of a matching cortical cholinergic deficit have been replicated by others (Godridge *et al.* 1987), although the data still suffer from having a relatively small control population. Additional evidence for cholinergic pathology has come from findings of neuronal loss in the nucleus basalis of Meynert, the source of cortical cholinergic innervation (Price *et al.* 1982; Mann *et al.* 1984a, 1985). That the observed decrements in cholinergic indices are a result of degenerative, rather than congenital, processes is supported by recent work demonstrating that infants with DS have normal levels (Kish *et al.* 1989). The neurochemical match has been further filled out to include the cerebral noradrenergic and serotonergic deficits (Yates *et al.* 1981, 1983, 1986; Nyberg *et al.* 1982; Reynolds and Godridge 1985; Godridge *et al.* 1987), with neuronal depletion in the appropriate brainstem nuclei (Mann *et al.* 1982, 1984a, 1985). Although a detailed neurochemical comparison is not yet complete (lacking, for example, an analysis of neuropeptides), the major changes of AD appear to be present in DS.

Explanation of the association

Recent genetic hypotheses

The last remaining level to which the comparison has been pushed has been that of the DNA itself. Here the natural experiment was particularly useful. Since 1959, it had been known that DS was the consequence of triplication of chromosome 21 (Lejeune *et al.* 1959). It was logical, therefore, to look for chromosomal anomalies in AD. Searches for these anomalies brought initially alluring results (Nielsen 1970; Jarvik *et al.* 1971) and epidemiological investigations increased enthusiasm with reports of an increased incidence of trisomy 21 in AD families (Heston *et al.* 1966; Heston 1977; Heyman *et al.* 1983). These developments spawned unitary pathogenetic theories involving meiotic mistakes such as chromosomal non-disjunction (Heston and Mastri 1977), but reinforcing evidence failed to appear, leading to the conclusion that gross chromosomal changes were not likely to be the cause of AD (Schweber 1985; Kay 1986).

Therefore the search focused down on smaller fragments of the chromosome. Schweber (1985) suggested that duplication of a single gene on chromosome 21, which would be inherited in familial AD (FAD) and produced *de novo* in sporadic AD, could be the cause of AD. A favoured early candidate was the gene coding for superoxide dismutase-1 (SOD-1), localized to chromosome 21 (Tan *et al.* 1973). This enzyme potentially plays a key role in the regulation of cellular ageing by performing the first reaction in a sequence that removes damaging oxygen metabolites (Sinet 1982). It was quickly established that the enzymatic activity of SOD-1 was elevated in DS, as expected by the presence of an extra copy of the gene (Sinet 1982). Theoretically, increased activity of SOD-1 could lead to harmful effects generated by the reaction product, hydrogen peroxide. Peroxidation of cellular lipids and proteins, which could conceivably include self-inactivation of SOD-1 itself, were possibilities considered. A key point was whether or not the next enzyme in the sequence, glutathione peroxidase (GSHP), which converts hydrogen peroxide to water, would be able to increase its activity to keep up with that of SOD-1. If not, hydrogen peroxide would accumulate and the hypothesized damage might ensue. The activity of GSHP in DS cells has been assessed by several groups; most agree that it is upregulated (Sinet *et al.* 1975; Anneren *et al.* 1985; Anneren and Epstein 1987; Crosti *et al.* 1989). There is also no evidence of significant peroxidation effects in adult DS tissue, but fetal cells have been reported to show both increased lipid peroxidation and a lack of GSHP upregulation (Brooksbank and Balazs 1984). The levels of SOD-1 and GSHP in AD have variously been reported as increased, decreased or normal (Marklund *et al.* 1985; Sulkava *et al.* 1986; Perrin *et al.* 1990). Possible reasons for these conflicting studies include regional variation in the basal and reactive levels of these enzymes (Delacourte *et al.* 1988; Zemlan *et al.* 1989), as well as considerable inter-individual variation (Beutler *et al.* 1991). Evidence for peroxidation effects in AD is also inconclusive (Hajimohammadreza and Brammer 1990; Subbarao *et al.* 1990). The gene for SOD-1 is apparently not duplicated in AD (Delabar *et al.* 1986), but this does not eliminate the possibility of its increased expression due to changes in *cis*-acting regulatory elements. The issue is therefore still open (Volicer and Crino 1990) and the recent production of transgenic mice carrying an extra copy of SOD-1 (Avraham *et al.* 1988; Ceballos-Picot *et al.* 1991) should prove invaluable for delineating more precisely the effects of increased dosage of this gene.

Other chromosome 21 genes (Cooper and Hall 1988) of interest include those coding for interferon alpha and beta receptors, heat shock protein 70, cystathionine-β-synthetase, and proto-oncogene *ets2*. Of these, the last-mentioned has shown the greatest potential for a dual pathogenetic role, since it is contained within the obligatorily replicated segment of chromosome 21 in DS, and was reportedly duplicated in AD (Delabar *et al.* 1986). Confirmation of this duplication has not yet appeared, however (Glenner and Murphy 1989). The extra copy of cystathionine-β-synthetase has been suggested to account for the reportedly low incidence of atherosclerosis in DS (Brattstrom *et al.* 1987); the same authors noted that some evidence exists of a decreased occurrence of atherosclerosis in AD, forging another possible link between the two disorders.

One chromosome 21 gene, however, has recently risen above the others as the clear frontrunner for a crucial role in pathogenesis. This is the gene for amyloid precursor protein (APP). Within this protein is the amyloidogenic fragment, named amyloid β-protein by Glenner and Wong (1984) after they isolated it from vascular amyloid. The metabolism of this protein has come to be regarded as central to the understanding of AD. As well as being present in vessel wall amyloid, amyloid β-protein is the principal constituent of senile plaques in both AD and DS (Glenner and Murphy 1989). Therefore there was tremendous excitement when several groups announced that the APP gene was located on chromosome 21 (Goldgaber *et al.* 1987; Kang *et al.* 1987; Tanzi *et al.* 1987a). This followed the report, only months earlier, of the linkage of AD to an anonymous chromosome 21 fragment (St George-Hyslop *et al.* 1987). Since the APP gene and the anonymous fragment were in the same general area of the chromosome, it briefly appeared that they might be one and the same. It rapidly became apparent, however, that the two markers were physically separate (Tanzi *et al.* 1987b; Van Broeckhoven *et al.* 1987) and the situation was further muddied by reports, in other lineages, of absence of linkage to chromosome 21 probes as well as possible linkages to other chromosomes (Pericak-Vance *et al.* 1988; Schellenberg *et al.* 1988). This, together with the failure to find duplications or mutations of the APP gene (Selkoe 1989), momentarily dampened the fire.

A second wave of examination, however, has recently detected mutations in codon 717 of the APP gene in several families with early-onset FAD (Chartier-Harlin *et al.* 1991; Goate *et al.* 1991; Hardy *et al.* 1991; Murrell *et al.* 1991). Because the mutations appear thus far only in affected family members and are absent in hundreds of control subjects, it is very likely that they are causative. Furthermore, a different mutation of the APP gene has been found in hereditary cerebral haemorrhage with angiopathy, Dutch type (Levy *et al.* 1990) and, in parallel developments, point mutations in the prion protein gene have been linked to amyloidogenesis and dementia in Gerstmann-Sträussler-Scheinker syndrome (Hsiao and Prusiner 1991), while mutations in the cystatin C gene have been found to cause amyloidogenesis and cerebral haemorrhage in hereditary cystatin C amyloid angiopathy (Palsdottir *et al.* 1989).

Other types of evidence have also pointed directly at amyloid formation as a primary event. Life history autopsy studies of DS have identified amyloid deposition as the earliest microscopic lesion, preceding neurofibrillary tangle formation by about 10 years (Burger and Vogel 1973; Giaccone *et al.* 1989; Mann and Esiri 1989). Transgenic mice that overexpress amyloid β-protein sequences have been reported to develop extracellular amyloid deposits (Quon *et al.* 1991). While most sporadic and familial AD cases do not have abnormalities of the APP gene, the new evidence is so compelling that most theorists are now favouring a unitary hypothesis in which alteration of APP metabolism is the crucial event (Hardy and Allsop 1991; Kosik 1991; Selkoe 1991; Gandy and Greengard 1992). To account for the apparently varying localization of the DNA lesion, it has been postulated that the defect may occur not only in the structural gene for APP, but also in genes coding for its regulation or catabolism. These would yield a 'final common pathway' to plaques, tangles and dementia.

The premature ageing hypothesis

One concept that has swept up students of both AD and DS is that both conditions could be due to some form of premature ageing. This idea had been advanced for AD almost since Alzheimer's first description of the disease (Beach 1987). The possibility that individuals with DS undergo precocious ageing was also raised early (Fraser and Mitchell 1876), but only became popular after Jervis first linked AD-type pathology to DS in his paper of 1948. The issue had, as with AD, another side, which was the possibility that the Alzheimer lesions are the expression of a specific disease process, rather than ageing *per se*. Jervis raised this question first in 1948 and again in a testimonial to his tenacity in 1970: 'The question may be raised as to whether the pathological lesions here described are simply the manifestations of early physiological senescence of the brain in individuals of short life span or are the result of morbid processes of the type underlying senile dementia.' (Jervis 1948). Besides the plaques and tangles, other physical signs of premature ageing in DS have been offered, including calcification of the basal ganglia (Jervis 1948), hair loss, greying of hair, deterioration of skin, cataract formation, skeletal changes, premature menopause and testicular atrophy (Lott 1982; Oliver and Holland 1986).

The case for generalized ageing in DS is in fact weak upon closer examination. Lipofuscinosis, a marker of ageing processes in the brain, has been found to be no more severe in AD and DS than in age-matched controls (Mann *et al.* 1984b). Calcification of the basal ganglia does appear to be more common in DS than in the general population (Wisniewski *et al.* 1982), but is not actually associated with dementia (Mann 1988b) and may be most severe in individuals with congenital heart disease dying in their first decade (Wisniewski *et al.* 1982). In addition, basal ganglia calcification may have a higher than normal prevalence among the mentally retarded population in general (Malamud 1964). Therefore, it may not be related to ageing or Alzheimer changes, but more likely represents the presence of another inciting factor, such as congenital heart or vascular disease. The putative skin deterioration has also been called into question (Murdoch and Evans 1978) and the increased prevalence of cataracts in DS may be caused by the extra copy of the crystallin gene (Cooper and Hall 1988). Atherosclerosis, an archtypical ageing phenomenon, is, as mentioned earlier, less common in DS.

The most direct evidence for premature ageing in DS has been the low average life expectancy; in 1929, for example, the average age at death was about nine years (see Chapter 2). This, however, seems to have been due largely to the high prevalence of congenital heart disease, with its sequelae of early cardiorespiratory failure and attendant pneumonias, and in the high early mortality rates associated with institutionalization in general (Thase 1982). The introduction of antibiotics and corrective heart surgery greatly improved the chances of survival to middle age, but average lifespan still falls some 20–30 years short of the general population. There remains a steep mortality curve after the age of 40, which may be due largely to the effects of the Alzheimer neuropathology (Thase 1982).

The premature ageing hypothesis has long been under fire in its application to AD (Beach 1987) and the emerging consensus that AD-type neuropathology is caused by disorders of APP metabolism appears to confirm the views of many that AD is a disease rather than a direct and invariable manifestation of ageing. If the AD-type lesions in DS are also attributable to perturbations of APP metabolism, this removes the major piece of evidence for accelerated ageing in that condition.

The undeniable age-associated increased incidence of AD-type dementia in both AD and DS remains to be explained, however. It may be that the extra copy of the APP gene in DS and the codon 717 mutation in some FAD families sets in motion a process so slow that it takes decades to become evident. Alternatively, it may be that pathological expression requires ageing changes as cofactors. For FAD, where the disease is inherited in an autosomal dominant fashion, another possibility is that it is an example of a disease of ageing caused by what Medawar (1981) has termed a temporally pleiotropic gene. This concept envisions a gene that has deleterious effects on the organism in post-reproductive years, but whose frequency increases in the population nevertheless, due to beneficial effects taking place in earlier life. Since the defects occurring in an individual after the reproductive years have no impact on reproductive 'fitness', the beneficial effects allow it to increase by natural selection. Alternatively, there may be no early beneficial effects, but the gene may still persist or even increase in frequency if it is essentially 'neutral' in terms of natural selection. It is at present unclear what fraction of AD is inherited as a single gene defect; estimates range from less than five per cent to 75 per cent (Fitch *et al.* 1988). In sporadic cases with multifactorial inheritance, it is doubtful whether the temporally pleiotropic gene concept would apply.

Other aspects of the association

There are many additional interesting aspects of the association between DS and AD, some of which may prove to be fruitful avenues for future investigation. These include the high incidence of epilepsy, Parkinson disease and hypothyroidism (Lai and Williams 1989) in both DS and AD. Numerous anatomical anomalies in the DS brain have been described (Scott *et al.* 1983), including abnormal cortical gyration and layering, as well as early cessation of dendritic spine proliferation. A careful search for these and other anatomical alterations in DS and AD brains might conceivably yield an anatomical marker that precedes the development of plaques and tangles. The prospects for more revealing animal models of both conditions are good; the effects of extra copies of specific genes can now be isolated using transgenic animals and trisomy 16 in mice, which is a homologue of trisomy 21 in humans (Epstein *et al.* 1985), and are being exploited with some success as well (see Chapters 13 and 14).

Conclusion

The discovery by George A. Jervis in 1948 that the brains of individuals with DS and AD hold in common the presence of a distinctive histopathology has afforded a natural experiment that has been of great value to those attempting to understand the two disorders. The work of many subsequent investigators has confirmed Jervis' hope and prophecy that his finding '...may offer some clue...' to the nature of senile dementia, and that it might be responsible for '...suggesting lines of future research.' Advances in technology unimagined in 1948 have enabled a full exploitation of this natural experiment and the results have been key elements of an emerging pathogenetic hypothesis. The 1959 discovery that DS was caused by triplication of chromosome 21 led directly to a search for abnormalities of this chromosome in AD. Localization of the APP gene and genetic linkage of FAD to chromosome 21 increased the probability that abnormalities of a locus or loci within chromosome 21 are the cause of AD-type pathology. Finally, causative abnormalities have been found in some pedigrees of FAD, in the form of mutations of the APP gene. This has led to a model of AD pathogenesis that postulates that alterations of APP metabolism are the primary abnormality in AD. Coming full circle, this model would predict that it is the extra copy of APP that is responsible for the AD-type pathology seen in DS. The validity of the model remains to be proven, but it is eminently testable and should serve as a great stimulant to further research.

References

Anneren, K. G., and Epstein, C. J. (1987). Lipid peroxidation and superoxide dismutase-1 and glutathione peroxidase activities in trisomy 16 fetal mice and human trisomy 21 fibroblasts. *Pediatric Research*, 21, 88-92.

Anneren, G., Edqvist, L. E., Gebre-Medhin, M., and Gustavson, K. H. (1985). Glutathione peroxidase activity in erythrocytes in Down's syndrome. Abnormal variation in relation to age and sex through childhood and adolescence. *Trisomy* 21, 1, 9-17.

Avraham, K. B., Schickler, M., Sapoznikov, D., Yarom, R., and Groner, Y. (1988). Down's syndrome: abnormal neuromuscular junction in tongue of transgenic mice with elevated levels of human Cu-Zn superoxide dismutase. *Cell*, 54, 823-9.

Beach, T. G. (1987). The history of Alzheimer's disease: three debates. *Journal of the History of Medicine and Allied Sciences*, 42, 327-49.

Benda, C. E. (1940). The central nervous system in mongolism. *American Journal of Mental Deficiency*, 45, 42-7.

Bertrand, I., and Koffas, D. (1946). Cas d'idiotie mongolienne adulte avec nombreuses plaques seniles et concretions calcaires pallidales. *Revue Neurologique (Paris)*, 78, 338-45.

Beutler, E., Curnutte, J. T., and Forman, L. (1991). Glutathione peroxidase deficiency and childhood seizures. *Lancet*, 338, 700.

Brattstrom, L., Englund, E., and Brun, A. (1987). Does Down syndrome support homocysteine theory of arteriosclerosis? *Lancet*, 1, 391-2.

Brooksbank, B. W. L., and Balazs, R. (1984). Superoxide dismutase, glutathione peroxidase and lipoperoxidation in Down's syndrome fetal brain. *Developmental Brain Research*, 16, 37-44.

Burger, P. C., and Vogel, F. S. (1973). The development of the pathologic changes of Alzheimer's disease and senile dementia in patients with Down's syndrome. *American Journal of Pathology*, 73, 457-76.

Ceballos-Picot, I., Nicole, A., Briand, P., Grimber, G., Delacourte, A., Defossez, A., *et al.* (1991). Neuronal-specific expression of human copper-zinc superoxide dismutase gene in transgenic mice: animal model of gene dosage effects in Down's syndrome. *Brain Research*, 552, 198-214.

Chartier-Harlin, M. -C., Crawford, F., Houlden, H., Warren, A., Hughes, D., Fidani, L., *et al.* (1991). Early-onset Alzheimer's disease caused by mutations at codon 717 of the β-amyloid precursor protein gene. *Nature*, 353, 844-6.

Colon, E. J. (1972). The structure of the cerebral cortex in Down's syndrome. *Neuropediatrie*, 3, 362-76.

Cooper, D. N., and Hall, C. (1988). Down's syndrome and the molecular biology of chromosome 21. *Progress in Neurobiology*, 30, 507-30.

Crosti, N., Bajer, J., Gentile, M., Resta, G., and Serra, A. (1989). Catalase and glutathione peroxidase activity in cells with trisomy 21. *Clinical Genetics*, 36, 107-16.

Davidoff, L. M. (1928). The brain in mongolian idiocy. *Archives of Neurology and Psychiatry (Chicago)*, 20, 1229-57.

Delabar, J. M., Lamour, Y., Gegonne, A., Davous, P., Roudier, M., Nicole, A., *et al.* (1986). Rearrangement of chromosome 21 in Alzheimer's disease. *Annales de Génétique (Paris)*, 29, 226-8.

Delacourte, A., Defossez, A., Ceballos, I., Nicole, A., and Sinet, P. M. (1988). Preferential localization of copper zinc superoxide dismutase in the vulnerable cortical neurons in Alzheimer's disease. *Neuroscience Letters*, 92, 247-53.

Down, J. L. H. (1866). Observations on an ethnic classification of idiots. *London Hospital Reports*, 3, 259-62.

Ellis, W. G., McCulloch, J. R., and Corley, C. L. (1974). Presenile dementia in Down's syndrome: ultrastructural identity with Alzheimer's disease. *Neurology*, 24, 101-6.

Epstein, C. J., Cox, D. R., and Epstein, L. B. (1985). Mouse trisomy 16: an animal model of human trisomy 21 (Down syndrome). *Annals of the New York Academy of Sciences*, 450, 157-68.

Fitch, N., Becker, R., and Heller, A. (1988). The inheritance of Alzheimer's disease: a new interpretation. *Annals of Neurology*, 23, 14-9.

Fraser, J., and Mitchell, A. (1876). Kalmuc idiocy: report of a case with autopsy with notes on 62 cases. *Journal of Mental Science*, 22, 161-9.

Gandy, S., and Greengard, P. (1992). Amyloidogenesis in Alzheimer's disease: some possible therapeutic opportunities. *Trends in Pharmacological Sciences*, 13, 108-13.

Giaccone, G., Tagliavini, F., Linoli, G., Bouras, C., Frigerio, L., Frangione, B., and Bugiani, O. (1989). Down patients: extracellular preamyloid deposits precede neuritic degeneration and senile plaques. *Neuroscience Letters*, 97, 232-8.

Gibson, D. (1978). *Down's syndrome: the psychology of mongolism*. Cambridge University Press, Cambridge.

Glenner, G. G., and Murphy, M. A. (1989). Amyloidosis of the nervous system. *Journal of the Neurological Sciences*, 94, 1 -28.

Glenner, G. G., and Wong, C. W. (1984). Alzheimer's disease and Down's syndrome: sharing of a unique cerebrovascular amyloid fibril protein. *Biochemical and Biophysical Research Communications*, 122, 1131-5.

Goate, A., Chartier-Harlin, M. -C., Mullan, M., Brown, J., Crawford, F., Fidani, L., *et al.* (1991). Segregation of a missense mutation in the amyloid precursor protein gene with familial Alzheimer's disease. *Nature*, 349, 704-6.

Godridge, H., Reynolds, G. P., Czudek, C., Calcutt, N. A., and Benton, M. (1987). Alzheimer-like neurotransmitter deficits in adult Down's syndrome brain tissue. *Journal of Neurology, Neurosurgery and Psychiatry*, 50, 775-8.

Goldgaber, D., Lerman, M. I., McBride, O. W., Saffiotti, U., and Gajdusek, D. C. (1987). Characterization and chromosomal localization of a cDNA encoding brain amyloid of Alzheimer's disease. *Science*, 235, 877-80.

Gould, S. J. (1980a). Senseless signs of history. In *The panda's thumb*, pp. 27-34. Norton, New York.

Gould, S. J. (1980b). Dr Down's syndrome. In *The panda's thumb*, pp. 160-8. Norton, New York.

Haberland, C. (1969). Alzheimer's disease in Down syndrome: clinical-neuropathological observations. *Acta Neurologica Belgica,* 89, 369-80.

Hajimohammadreza, I., and Brammer, M. (1990). Brain membrane fluidity and lipid peroxidation in Alzheimer's disease. *Neuroscience Letters,* 112, 333-7.

Hardy, J., and Allsop, D. (1991). Amyloid deposition as the central event in the etiology of Alzheimer's disease. *Trends in Pharmacological Sciences,* 12, 383-8.

Hardy, J., Mullan, M., Chartier-Harlin, M. -C., Brown, J., Goate, A., Rossor, M., *et al.* (1991). Molecular classification of Alzheimer's disease. *Lancet,* 337, 1342-3.

Heston, L. L. (1977). Alzheimer's disease, trisomy 21, and myeloproliferative disorders: associations suggesting a genetic diathesis. *Science,* 196, 322-3.

Heston, L. L., and Mastri, A. R. (1977). Genetics of Alzheimer's disease: associations with hematological malignancies and Down's syndrome. *Archives of General Psychiatry,* 34, 976-81.

Heston, L. L., Lowther, D. L., and Leventhal, C. M. (1966). Alzheimer's disease: a family study. *Archives of Neurology,* 15, 225-33.

Heyman, A., Wilkinson, W. E., Hurwitz, B. J., Schmechel, D., Sigmon, A. H., Weinberg, T., *et al.* (1983). Alzheimer's disease: genetic aspects and associated clinical disorders. *Annals of Neurology,* 14, 507-16.

Hsiao, K., and Prusiner, S. B. (1991). Molecular genetics and transgenic model of Gerstmann-Straussler-Scheinker disease. *Alzheimer's Disease and Associated Disorders,* 5, 155-62.

Jarvik, L. F., Altschuler, K. Z., Kato, K., and Blumner, B. (1971). Organic brain syndrome and chromosome loss in aged twins. *Diseases of the Nervous System,* 32, 159-70.

Jervis, G. A. (1948). Early senile dementia in mongoloid idiocy. *American Journal of Psychiatry,* 105, 102-6.

Jervis, G. A. (1970). Premature senility in Down's syndrome. *Annals of the New York Academy of Sciences,* 171, 559-61.

Jervis, G. A., and Soltz, S. E. (1936). Alzheimer's disease—the so-called juvenile type. *American Journal of Psychiatry,* 93, 39-56.

Kang, J., Lemaire, H. -G., Unterbeck, A., Salbaum, J. M., Masters, C. L., Grzeschik, K. -H., *et al.* (1987). The precursor of Alzheimer's amyloid A4 protein resembles a cell-surface receptor. *Nature,* 325, 733-6.

Kay, D. W. K. (1986). The genetics of Alzheimer's disease. *British Medical Bulletin,* 42, 19-23.

Kidd, M. (1963). Paired helical filaments in electron microscopy of Alzheimer's disease. *Nature,* 197, 192-3.

Kish, S., Karlinsky, H., Becker, L., Gilbert, J., Rebbetoy, M., Chang, L. -J., *et al.* (1989). Down's syndrome individuals begin life with normal levels of brain cholinergic markers. *Journal of Neurochemistry,* 52, 1183-7.

Kosik, K. S. (1991). Alzheimer plaques and tangles: advances on both fronts. *Trends in Neuroscience,* 14, 218-9.

Lai, F., and Williams, R. S. (1989). A prospective study of Alzheimer disease in Down syndrome. *Archives of Neurology,* 46, 849-53.

Lejeune, J., Turpin, R., and Gautier, M. (1959). Le mongolisme, premier exemple d'aberration autosomique humaine. *Annales de Génétique (Paris),* 2, 41-9.

Levy, E., Carman, M. D., Fernandez-Madrid, I. J., Power, M. D., Lieberburg, I., Van Duinen, S. G., *et al.* (1990). Mutation of the Alzheimer's disease amyloid gene in hereditary cerebral hemorrhage, Dutch type. *Science,* 248, 1124-6.

Lott, I. T. (1982). Down's syndrome, aging, and Alzheimer's disease: a clinical review. *Annals of the New York Academy of Sciences,* 396, 15-27.

Malamud, N. (1964). Neuropathology. In *Mental retardation: a review of research,* (ed. H.A. Stevens and R. Heber), pp. 429-52. University of Chicago Press, Chicago.

Mann, D. M. A. (1988a). Alzheimer's disease and Down's syndrome. *Histopathology,* 13, 125-37.

Mann, D. M. A. (1988b). Calcification of the basal ganglia in Down's syndrome and Alzheimer's disease. *Acta Neuropathologica,* 76, 595-8.

Mann, D. M. A., and Esiri, M.M. (1989). The pattern of acquisition of plaques and tangles in the brains of patients under 50 years of age with Down's syndrome. *Journal of the Neurological Sciences*, **89**, 169-79.

Mann, D. M. A., Yates, P. O., and Hawkes, J. (1982). The noradrenergic system in Alzheimer's and multi-infarct dementias. *Journal of Neurology, Neurosurgery and Psychiatry,* **45**, 113-9.

Mann, D. M. A., Yates, P. O., and Marcyniuk, B. (1984a). Alzheimer's presenile dementia, senile dementia of Alzheimer type and Down's syndrome of middle age form an age related continuum of pathological changes. *Neuropathology and Applied Neurobiology*, **10**, 185-207.

Mann, D. M. A., Yates, P. O., and Marcyniuk, B. (1984b). Relationship between pigment accumulation and age in Alzheimer's disease and Down's syndrome. *Acta Neuropathologica,* **63**, 72-7.

Mann, D. M. A., Yates, P. O., and Marcyniuk, B. (1985). Pathological evidence for neurotransmitter deficits in Down's syndrome of middle age. *Journal of Mental Deficiency Research*, **29**, 125-35.

Marklund, S. L., Adolfsson, R., Gottfries, C. G., and Winblad, B. (1985). Superoxide-dismutase isoenzyme in normal brains and in brains from patients with dementia of the Alzheimer type. *Journal of the Neurological Sciences*, **67**, 319-25.

Medawar, P. (1981). *The uniqueness of the individual.* Methuen, London.

Murdoch, J. C., and Evans, J. H. (1978). An objective in vitro study of ageing in the skin of patients with Down's syndrome. *Journal of Mental Deficiency Research*, **22**, 131-5.

Murrell, J., Farlow, M., Ghetti, B., and Benson, M. D. (1991). A mutation in the amyloid precursor protein associated with hereditary Alzheimer's disease. *Science*, **254**, 97-9.

Neumann, M. A. (1967). Langdon Down syndrome and Alzheimer's disease. *Journal of Neuropathology and Experimental Neurology*, **26**, 149.

Nielsen, J. (1970). Chromosomes in senile, presenile, and arteriosclerotic dementia. *Journal of Gerontology*, **25**, 312-5.

Nyberg, P., Carlssen, A., and Winblad, B. (1982). Brain monoamines in cases of Down's syndrome with and without dementia. *Journal of Neural Transmission*, **55**, 289-99.

Oliver, C., and Holland, A. J. (1986). Down's syndrome and Alzheimer's disease: a review. *Psychological Medicine*, **16**, 307-22.

Olson, M. I., and Shaw, C. -M. (1969). Presenile dementia and Alzheimer's disease in mongolism. *Brain*, **92**, 147-56.

Owens, D., Dawson, J. C., and Losin, S. (1971). Alzheimer's disease in Down's syndrome. *American Journal of Mental Deficiency*, **75**, 606-12.

Palsdottir, A., Abrahamson, M., Thorsteinsson, L., Arnason, A., Olafsson, I., Grubb, A., and Jensson, O. (1989). Mutation in the cystatin C gene causes hereditary brain hemorrhage. In *Alzheimer's disease and related disorders*, (ed. K. Iqbal, H. M. Wisniewski, and B. Winblad), pp. 241-6. Liss, New York.

Pericak-Vance, M. A., Yamaoka, L. H., Haynes, C. S., Speer, M. C., Haines, J. L., Gaskell, P. C., *et al.* (1988). Genetic linkage studies in Alzheimer's disease families. *Experimental Neurology*, **102**, 271-9.

Perrin, R., Brancon, S., Jeandel, C., Artur, Y., Minn, A., Penin, F., and Siest, G. (1990). Blood activity of Cu/Zn superoxide dismutase, glutathione peroxidase and catalase in Alzheimer's disease: a case-control study. *Gerontology*, **36**, 306-13.

Price, D. L., Whithouse, P. J., Struble, R. G., Coyle, J. T., Clark, A. W., Delong, M. R., *et al.* (1982). Alzheimer's disease and Down's syndrome. *Annals of the New York Academy of Sciences*, **396**, 145-64.

Quon, D., Wang, Y., Catalano, R., Scardina, J. M., Murakami, K., and Cordell, B. (1991). Formation of β-amyloid protein deposits in brains of transgenic mice. *Nature*, **352**, 239-41.

Reynolds, G. P., and Godridge, H. (1985). Alzheimer-like monoamine deficits in adults with Down's syndrome. *Lancet*, **2**, 1368-9.

Schellenberg, G. D., Bird, T. D., Wijsman, E. M., Moore, D. K., Boehnke, M., Bryant, M., *et al.* (1988). Absence of linkage of chromosome 21q21 markers to familial Alzheimer's disease. *Science*, **241**, 1507-10.

Schweber, M. (1985). A possible unitary genetic hypothesis for Alzheimer's disease and Down syndrome. *Annals of the New York Academy of Sciences*, 450, 223-38.

Scott, B. S., Becker, L. E., and Petit, T. L. (1983). Neurobiology of Down's syndrome. *Progress in Neurobiology*, 21, 199-237.

Selkoe, D. J. (1989). The deposition of amyloid proteins in the aging mammalian brain: implications for Alzheimer's disease. *Annals of Medicine*, 21, 73-6.

Selkoe, D. J. (1991). The molecular pathology of Alzheimer's disease. *Neuron*, 6, 487-96.

Sinet, P. M. (1982). Metabolism of oxygen derivatives in Down syndrome. *Annals of the New York Academy of Sciences*, 396, 83-94.

Sinet, P. M., Michelson, A. M., Bazin, A., Lejeune, J., and Jerome, H. (1975). Increase in glutathione peroxidase activity in erythrocytes from trisomy 21 subjects. *Biochemical and Biophysical Research Communications*, 67, 910-5.

Solitaire, G. B., and Lamarche, J. B. (1966). Alzheimer's disease and senile dementia as seen in mongoloids: neuropathological observations. *American Journal of Mental Deficiency*, 70, 840-8.

St George-Hyslop, P. H., Tanzi, R., Polinsky, R. J., Haines, J. L., Nee, L., Watkins, P. C., *et al.* (1987). The genetic defect causing familial Alzheimer's disease maps on chromosome 21. *Science*, 235, 885-9.

Struwe, F. (1929). Histopathologische Untersuchungen über Entstehung und Wesen der senilen Plaques. *Zentralblatt für die gesamte Neurologie und Psychiatrie*, 122, 291-307.

Subbarao, K. V., Richardson, J. S., and Ang, L. C. (1990). Autopsy samples of Alzheimer's cortex show increased peroxidation in vitro. *Journal of Neurochemistry*, 55, 342-5.

Sulkava, R., Nordberg, U. R., Erkinjuntti, T., and Westermarck, T. (1986). Erythrocyte glutathione peroxidase and superoxide dismutase in Alzheimer's disease and other dementias. *Acta Neurologica Scandinavica*, 73, 487-9.

Tan, Y. H., Tischfield, J., and Ruddle, F. H. (1973). The linkage of genes for the human interferon-induced anti-viral protein and indophenol oxidase-B traits to human chromosome G-21. *Journal of Experimental Medicine*, 137, 317-30.

Tanzi, R. E., Gusella, J. F., Watkins, P. C., Bruns, G. A. P., St George-Hyslop, P., Van Keuren, M. L., *et al.* (1987a). Amyloid β protein gene: cDNA, mRNA distribution, and genetic linkage near the Alzheimer locus. *Science*, 235, 880-4.

Tanzi, R. E., St George-Hyslop, P. H., Haines, J., Polinsky, R. J., Nee, L., Foncin, J. F., *et al.* (1987b). The genetic defect in Alzheimer's disease is not tightly linked to the amyloid β protein gene. *Nature*, 329, 156-7.

Terry, R. D. (1963). The fine structure of neurofibrillary tangles in Alzheimer's disease. *Journal of Neuropathology and Experimental Neurology*, 22, 629-42.

Thase, M. E. (1982). Longevity and mortality in Down's syndrome. *Journal of Mental Deficiency Research*, 26, 177-92.

Van Broeckhoven, C., Genthe, A. M., Vandenberghe, A., Horsthemke, B., Backhovens, H., Raeymaekers, P., *et al.* (1987). Failure of familial Alzheimer's disease to segregate with the A4-amyloid gene in several European families. *Nature*, 329, 153-5.

Volicer, L., and Crino, P. B. (1990). Involvement of free radicals in dementia of the Alzheimer type: a hypothesis. *Neurobiology of Aging*, 11, 567-71.

Wisniewski, H. M., and Rabe, A. (1986). Discrepancy between Alzheimer-type neuropathology and dementia in persons with Down's syndrome. *Annals of the New York Academy of Sciences*, 477, 247-59.

Wisniewski, K. E., French, J. H., Rosen, J. F., Kozlowski, P. B., Tenner, M., and Wisniewski, H. M. (1982). Basal ganglia calcification in Down's syndrome—another manifestation of premature aging. *Annals of the New York Academy of Sciences*, 396, 179-89.

Yates, C. M., Simpson, J., Maloney, A. F. J., Gordon, A., and Reid, A. H. (1980). Alzheimer-like cholinergic deficiency in Down's syndrome. *Lancet*, 2, 979.

Yates, C. M., Titchie, I. M., Simpson, J., Maloney, A. F. J., and Gordon, A. (1981). Noradrenaline in Alzheimer-type dementia and Down's syndrome. *Lancet*, 2, 39-40.

Yates, C. M., Simpson, J., Gordon, A., Maloney, A. F., Allison, Y., Ritchie, I. M., and Urquhart, A. (1983). Catecholamines and cholinergic enzymes in presenile and senile Alzheimer-type dementia and Down's syndrome. *Brain Research*, **280**, 119-26.

Yates, C., Simpson, J., and Gordon, A. (1986). Regional brain 5-hydroxytryptamine levels are reduced in senile Down's syndrome as in Alzheimer's disease. *Neuroscience Letters*, **65**, 189-92.

Zemlan, F. P., Thienhaus, O. J., and Bosmann, H. B. (1989). Superoxide dismutase activity in Alzheimer's disease: possible mechanism for paired helical filament formation. *Brain Research*, **476**, 160-2.

4

Association between Alzheimer disease and Down syndrome: clinical observations

A. J. Dalton, G. B. Seltzer, M. S. Adlin, and H. M. Wisniewski

Summary

Some adults with Down syndrome undergo clinical deterioration consistent with the dementia of Alzheimer disease. However, there are significant methodological problems in assigning an accurate clinical diagnosis of Alzheimer disease in this population. A particular difficulty concerns the lack of appropriate psychological tests able to detect longitudinal changes in performance in the presence of pre-existing mental retardation. Furthermore, a wide range of health problems associated with ageing in Down syndrome may mimic or mask the presence of Alzheimer disease. Despite these diagnostic difficulties, both retrospective and prospective studies have demonstrated the presence of clinical dementia in individuals with Down syndrome, with prevalence increasing as age advances. Clinical characteristics of the dementia, especially when elicited with prospective studies, are consistent with the signs of Alzheimer disease observed in persons who do not have Down syndrome. In particular, changes in various functional abilities over time appear to be a key diagnostic consideration. Although individuals with Down syndrome are at increased risk to develop the dementia of Alzheimer disease, the clinical expression of the disease in Down syndrome is substantially less frequent than would be expected from the results of post-mortem neuropathological observations. The reasons for this intriguing discrepancy between the clinical and neuropathological findings of Alzheimer disease in Down syndrome are not yet fully understood.

Introduction

The growing interest in the clinical aspects of Alzheimer disease (AD) in individuals with Down syndrome (DS) is reflected in the large number of reviews of this association which have appeared over the past decade (e.g. Lott, 1982; Sinex and Merril 1982; Dalton and Crapper-McLachlan 1986; Dalton and Wisniewski 1990; Rabe *et al.* 1990; Schupf *et al.* 1990; Dalton 1992; Lai 1992; Zigman *et al.* 1993). Unfortunately, fundamental understanding of the nature

and aetiology of AD remains limited. As a result, inconsistencies and contradic-
tions in the published clinical data are difficult to interpret, with at least four
hypotheses possible: (i) the neuropathological features of AD and the associated clinical dementia could be a single clinicopathological disease with a common aetiology; (ii) AD could be a clinicopathological syndrome with several
aetiologies, each leading to similar, though not identical, clinical and pathological findings; (iii) AD could have several aetiologies, each producing a syndrome
that is not distinguishable on clinical, pathological or biochemical grounds; (iv)
AD could be due to a particular response of the host to specific aetiological factors that produce the same clinical disease with different pathological findings
in individuals (for detailed discussion see Mortimer and Hutton 1985).

Methodological problems in making a clinical diagnosis of AD are noteworthy. For example, there is no test or procedure that can anticipate the development of clinical features of AD before the onset of manifestations.
Furthermore, there is no clinical instrument that adequately monitors the total
range of deficits induced by AD and no single test is useful throughout the
entire course of the disease (Crapper-McLachlan *et al.* 1984). A clinical diagnosis of AD in individuals with DS will be particularly limited by test 'floor
effects'. Persons with DS suspected of having AD are frequently evaluated by
examining changes in performances on an IQ test. Scores on IQ tests by such
persons will often be unhelpful because many of them, like other mentally
retarded individuals, will consistently score very poorly or even zero on many
items, thereby making comparisons uninformative (Dalton and Wisniewski
1990; Dalton 1992; Lai 1992). The Wechsler Memory Scale (Wechsler and
Stone 1945), widely used for decades in the assessment of memory impairments, has similar limitations. In addition, this Scale does not validly measure
memory functions (see Kear-Colwell 1973 for a review of factor analytic studies). Other frequently used instruments, such as the Stanford-Binet test and the
Mental Scale of the Bayley Scales of Infant Development (Bayley 1969), are
limited in usefulness because they have been standardized for use with children
and not adults.

An additional contributing factor to the difficulties of identifying and char-
acterizing AD in individuals with DS is that changes are superimposed on a
background of highly limited verbal and communication skills (see Chapter 9).
Moreover, impaired physical and physiological development associated with
the DS genotype, a life-time of limited social, educational and other experiences, and life-long retardation of intellectual capacities combine to produce a
clinical picture in older persons with DS that may differ significantly from what
is observed in affected individuals with AD from the general population.
Furthermore, age-related illnesses over-represented in DS may confound or
mimic a clinical diagnosis of AD.

A clinical diagnosis of AD in persons with DS therefore necessitates an
awareness of not only the manifestations of dementia consistent with an AD
process, but also the health changes associated with ageing in DS. Each of these
factors is considered below.

Manifestations of dementia in Down syndrome consistent with Alzheimer disease

Behavioural manifestations

Retrospective studies

The interpretation of clinical studies of AD in persons with DS is difficult in the absence of a neuropathological diagnosis of the former condition. Similarly, neuropathological studies without data on clinical manifestations of AD have limited usefulness. In an attempt to circumvent these problems, Dalton and Crapper-McLachlan (1986) reviewed all relevant studies conducted between 1946 and 1985 where both post-mortem brain tissues and clinical data were available. Sixteen eligible studies described 35 cases of DS ranging in age from 23–60 years at the time of death. One or more clinical manifestations of AD were reported in 25 (75 per cent) of the 33 persons (aged 35–60 years at the time of death) who showed either senile plaques only or senile plaques and neurofibrillary tangles. The clinical features of AD in decreasing frequencies were: seizures (58 per cent), personality changes (46 per cent), focal neurological signs (46 per cent), apathy/inactivity (36 per cent), loss of conversational skills (36 per cent), incontinence (36 per cent), electroencephalographic abnormalities (33 per cent), loss of self-help skills (30 per cent), tremors/myoclonus (24 per cent), visual/auditory deficits (24 per cent), walking/mobility problems (21 per cent), stubborn/uncooperative behaviour (21 per cent), depression (18 per cent), memory loss (18 per cent), posture of flexion (15 per cent), increased muscle tone (12 per cent), disorientation (12 per cent) and hallucinations/delusions (3 per cent).

Burt *et al.* (1992) provided a similar list of 17 clinical features of AD in persons with DS that had been most frequently reported across studies. They did not specify the criteria employed in selecting studies for review, nor did they give actual frequencies of the reported clinical features. However, they noted that the most frequently reported features were seizures, personality changes, apathy/inactivity and loss of self-help skills. Burt *et al.* also provided a list of manifestations common to AD and depression in adults with DS. The list includes apathy/inactivity, loss of self-help skills, depression, urinary incontinence, irritability, slowing, uncooperative/unmanageable behaviour, loss of housekeeping skills, greater dependency, loss of interest in surroundings, weight loss, emotional deterioration, destructiveness, hallucinations/delusions and sleep difficulties. The relationships between these manifestations and the clinical diagnosis of depression and AD have not been elucidated. These uncertainties are further evident when considering other signs of AD, described in persons from the general population but not in DS, such as dysarthria, aphasia, agnosia, perseveration, dyslexia, acalculia, echolalia, logoclonia, verbigeration, anasognosia, groping, grasping and cognitive abulia.

Overall, the retrospective data suggest that there may be both subtle and substantial differences in the clinical manifestations of AD in persons with DS when compared with AD-affected individuals without DS. These differences may represent differences in behavioural repertoires rather than 'pathological' differences or simply differences in the ease of assessment.

Prospective studies

These studies provide a significantly different clinical picture of AD in persons with DS than that based on retrospective descriptions. Prospective studies, involving follow-up for several years, reveal a clinical picture in individuals with DS consistent with signs of AD observed in persons from the general population (e.g. Lai and Williams 1989; Evenhuis 1990; Dalton 1992; Lai 1992). Memory loss, temporal disorientation and reduced verbal skills were the earliest signs of AD found in ageing persons with DS studied longitudinally by Lai and Williams (1989). For those functioning at particularly low intellectual levels, apathy, inattention and decreased social interaction were among the early signs of dementia. Loss of self-help skills, motor impairment and seizures were also noted. Evenhuis (1990) found declines in daily functional skills in 15 of 17 ageing persons with DS who appeared to be showing signs of AD. Similar neuropathology to that of AD in the general population was noted in the eight cases available for autopsy. However, clinical expression of AD was not demonstrable in two of these eight cases, aged 49 and 50 years at the time of death. Dalton (1992) found signs of AD at some time in 10 of 12 ageing persons with DS who were followed prospectively for periods from 4–18 years using the same repeated neurobehavioural tests of learning and memory (match-to-sample, delayed match-to-sample) and test environments. The two persons still alive who are, as yet, clinically asymptomatic for AD are now aged 57 and 58 years. AD was confirmed in the six individuals whose brains became available for autopsy; all six cases had shown clinical evidence of AD. In a recent follow-up report, Lai (1992) described 68 of 116 individuals with DS who showed clinical signs of AD, most of whom were evaluated neurologically and followed over a decade. Standard neurological examinations, serial assessments for orientation, memory, verbal and motor skills, and self-care abilities were performed regularly. The clinical dementia of AD was considered to be present when there was convincing functional decline in one or more of the test items and prevalence rates demonstrated the expected increase with advancing age.

Memory functions

The presence of visual recognition memory impairment as one of the earliest signs of AD has been consistently reported by Dalton and his colleagues in longitudinal studies (Dalton *et al.* 1974; Dalton and Crapper 1977; Dalton and Crapper-McLachlan 1984, 1986; Dalton and Wisniewski 1990). This impairment was identified using tests of delayed match-to-sample (DMTS) performances, which appear to be particularly sensitive for detecting the early stage of AD. The recognition of previously shown stimuli in the DMTS paradigm is a much easier test of memory for mentally retarded persons than the use of recall and the level of test difficulty can be controlled experimentally with relative ease. Simple, coloured shapes of familiar objects are employed using positive reinforcements for correct responses with memory intervals ranging from 0″ to 30″. These tests have been effectively used to compare cognitive deterioration with the distribution and number of senile plaques and neurofibrillary tangles in the brains of persons with DS (Wisniewski *et al.* 1985), abnormalities in superoxide dismutase (Percy *et al.* 1990a) and auto-immune thyroiditis (Percy *et al.* 1990b). Modified versions of DMTS tests have also been used in studies of depression and memory impairment in ageing adults with DS (Burt *et al.* 1992).

Dyspraxia

Dyspraxia is a characteristic feature of the middle, late and terminal stages of AD in the general population, but has rarely been reported in persons with DS. The infrequent reports of dyspraxia attributable to AD in persons with DS may reflect the lack of appropriate tests and may also be related to the extensive training and experience with affected individuals required for the effective observation and analysis of human movement (Ulrich *et al.* 1989).

The development and preliminary evaluation of the usefulness of a three-part, video-recorded, structured performance test of dyspraxia has been recently described (Dalton 1992). The test was successfully used previously as an outcome instrument for the evaluation of desferrioxamine in the treatment of patients with AD (Crapper-McLachlan *et al.* 1991). In another context, Dalton (1992) reported that the average performances of 32 adults with DS (17 men and 15 women) with a mean age of 43 years (range = 32–58 years) were significantly poorer (Student's $t = 3.07$, $P < 0.01$) than the average for a group of 11 mentally retarded persons without DS whose mean age was 72 years (range = 49–86 years). There was no statistically reliable relationship between age and dyspraxia scores for these two groups and neither group could be distinguished from the performances of 50 patients with a clinical diagnosis of AD but who were neither mentally retarded nor affected with DS. The data suggest that this dyspraxia test initially developed for patients with AD may also be appropriate for persons with DS. It remains to be determined whether the test will identify deterioration during longitudinal follow-up.

Psychiatric aspects

There are relatively few studies of the psychiatric aspects of ageing mentally retarded persons with DS with or without a diagnosis of AD (e.g. Reid and Aungle 1974; Reid *et al.* 1978; Hogg *et al.* 1988; Moss *et al.* 1991). Nevertheless, diagnosis of the dementia of AD in persons with DS is particularly complicated by the presence of psychiatric disorders in this population. In general, major affective disorders appear to be relatively common (Senatore *et al.* 1985; Harper and Wadsworth 1990; Zigman *et al.* 1991), with estimates of between 10 and 40 per cent prevalence rates (Reiss 1982). Clinical evidence suggests that older persons with DS and depression exhibit a complex of symptoms including memory loss, incontinence and loss of self-help skills that may have a devastating effect on their cognitive and adaptive functioning (Sovner and Hurley 1983; Warren *et al.* 1989). Depressive episodes have been sufficiently severe to be life threatening (Szymanski and Biederman 1984). A diagnostic challenge is faced by both researchers and clinicians in differentiating dementia from depression in DS. A case study report of five depressed adults with DS indicated that when disturbances in cognition were severe and vegetative symptoms were present, a diagnosis of AD was erroneously assigned (Warren *et al.* 1989).

The difficulties of performing psychiatric evaluations in individuals with DS may be partly overcome by the use of informant-based questionnaires. The Multi-dimensional Observation Scale for Elderly Subjects (MOSES) (Hersch *et al.* 1978a,b; Short 1982; Helmes *et al.* 1985, 1987) may provide such an instrument and is currently under evaluation (A. J. Dalton, B. Fedor and P. J. Patti, unpublished data) for use with ageing mentally retarded persons with and without DS.

The ageing process in Down syndrome

AD is clearly an age-related disease since 90 per cent of all AD cases are 65 years of age or older (Mortimer 1983). It has been argued that AD is premature ageing, probably at a genetically determined rate (Wright and Whalley 1984); others have proposed that it is a disease, especially early-onset AD (Roth 1986).

Wisniewski and Merz (1984) suggested that ageing, as a gradual loss of reserve, provides an environment within the body in which certain pathological changes can readily develop, implying that there is an intricate relationship between ageing of the brain and the disease (Rabe *et al.* 1990). Haveman *et al.* (1989), in comparisons of ageing between institutionalized severely retarded persons with and without DS, found that those with DS showed a greater increase in the following conditions as age advanced: auditory and visual handicaps, arthrosis, osteoporosis, dementia and epilepsy. However, individuals with DS were less prone, with advancing age, to develop myocardial infarction, heart decompensation, emphysema, chronic lung disease, bone fractures, spinal curvature, prostate enlargement and diabetes mellitus. Since both groups had spent most of their lives in the same environments, these age-related differences between them could not have been due to differences in environmental and/or life-style events. Instead, such differences may reflect characteristic uniformities in the expression of the phenotypic changes associated with normal ageing in persons with DS.

The physical changes which individuals with DS experience as they age can have a 'cascading' effect on other functions. The early co-occurrence and/or co-morbidity of these changes in DS, frequently observed later in the general ageing population, have often been denoted as premature ageing in adults with DS. These handicaps become more extensive and severe with increasing age, producing significant effects on the skills of daily living. The physical, behavioural and mental health changes associated with ageing among individuals with DS have been recently reviewed (Zigman *et al.* 1991). Some of these issues particularly relevant to the accurate clinical diagnosis of AD in persons with DS are as follows:

Hearing loss and visual problems

Many individuals with DS acquire a conductive hearing loss in childhood due to a propensity for middle-ear infections (Davies 1988). Sensorineural hearing loss (presbycusis) often begins to develop in persons with the syndrome in their early twenties (Buchanan 1990). This type of hearing loss, also experienced by the general elderly population, results in a loss of ability to hear higher pitched sounds. In addition, occlusion of the ear canal by wax is common among older individuals and can further impair hearing (Rees and Duckert 1990).

Diagnosis of hearing loss in persons with DS is difficult. Affected individuals generally do not complain about this problem and service providers, as well as family members, are usually unaware of it. Keiser *et al.* (1981) estimated that 40–77 per cent of adults with DS suffer from a hearing loss. Hearing losses of 20 dB and over at the 3 kHz level, detected by otoscopy, impedance audiometry and brain-stem response audiometry, were found by Evenhuis *et al.* (1992) in 33 of 35 (94 per cent) moderately to profoundly mentally retarded middle-aged persons with DS ranging in age from 35–62 years, with only two showing normal hearing sensitivity using bilateral brain-stem response audiometry. Cochlear losses were noted in 60 per cent of these cases. However, Evenhuis *et al.* were unable to confirm the expected relationship between severity of hearing loss with age, possibly because they did not include persons younger than 35 years of age.

Visual problems are also common in persons with DS. Up to 46 per cent of adults with DS develop cataracts (Hiles *et al.* 1990), and it has been suggested that this condition should be ruled out as a cause of deterioration before a diagnosis of AD is made (Odell 1988). Keratoconus has been reported (Hestnes *et al.* 1991) among individuals with DS. Aitchison *et al.* (1990) noted the prevalence of strabismus, cataracts and refractive errors among institutionalized persons with DS over 40 years of age. Visual problems, together with higher rates of hearing impairments in DS, contribute extensively to observed functional or behavioural deterioration that may be incorrectly interpreted as AD.

Cardiac abnormalities

Congenital heart defects are present in 40 per cent of infants with DS (Rowe and Uchida 1961). Most corrective cardiac operations are usually performed within the first year of life. Since cardiac surgery was generally not available before the 1960s, persons who have had such repairs have not yet reached middle age. Therefore, it is uncertain whether they will have special needs as they grow older. About 30–50 per cent of adults with DS have mitral valve prolapse and 10–15 per cent have aortic regurgitation (Goldhaber *et al.* 1986), conditions that may be asymptomatic and require expert evaluation. In addition, some individuals with uncorrected congenital heart defects may experience cardiac decompensation later in life. This physical deterioration may be misinterpreted as secondary to an AD process.

Thyroid function

Hypothyroidism occurs in 20-30 per cent of persons with DS (Baxter *et al.* 1975). Among 20 adults with DS referred for evaluation to the Aging and Developmental Disabilities Clinic of the University of Wisconsin-Madison Waisman Center, seven were found to have this condition (G. B. Seltzer and M. S. Adlin, unpublished observations). Percy *et al.* (1990b) found that auto-immune thyroiditis associated with subclinical hypothyroidism was characteristic of adults with DS. In a statistical analysis, uncorrected for sex and age effects, persons with DS who had clinical manifestations of AD, compared with those without such manifestations, had lower serum T_3 and a higher titre of antithyroid auto-antibody. Lai (1992) reported hypothyroidism in 63 per cent of persons with DS showing signs of AD compared with 49 per cent in the non-demented group of persons with DS (data not provided). Manifestations of hypothyroidism include lethargy, functional decline, confusion, constipation, dry skin and hair, fatigue and depression. Untreated hypothyroidism can lead to hallucinations and coma. These findings may mimic or mask AD symptomatology.

Immune system

Persons with DS seem to experience 'premature ageing' of the immune system (Levin *et al.* 1975; Whittingham *et al.* 1977) and demonstrate abnormal humoral and cell-mediated responses. Changes in the immune system seen in older individuals in the general population have been identified in persons with DS at a much earlier age (Rabinowe *et al.* 1989). In a recent study (P. D. Mehta, M. Percy, S. P. Mehta, and A. J. Dalton, unpublished observations), ageing persons with DS (n = 24, mean age = 55 years, range = 31–68 years) showed significantly higher levels of IgG_1 and IgG_3 ($P < 0.0001$) and lower levels of IgG_2 and IgG_4 ($P < 0.0001$) in their sera compared with age- and sex-matched normal controls. The lower levels of IgG_2 and IgG_4 subclasses are consistent with levels found in individuals susceptible to frequent infections. In a related study (Mehta *et al.* 1993), a group of 41 ageing persons with DS (median age = 56 years, range = 31–68 years) was found to have significantly higher levels of the cytokine interleukin-6 (IL-6) than 41 age- and sex-matched normal controls. In addition, there were significantly higher levels of IL-6 and beta-2-microglobulin in those persons with DS who showed clinical signs of AD (n = l0, mean age = 56 years) than in persons with DS without signs of AD (n = 18, mean age = 48 years). Elevated levels of IL-6 may have resulted from an AD-associated inflammatory reaction in the older persons with DS. Should these findings be replicated in additional studies they may provide insight into the link between the pathophysiology of AD and DS.

Cancer

Leukaemia is more common in children and adults with DS than in the general population (Odell 1988). It can present as tiredness, weight loss and an overall general decline in functioning (Pary 1992), findings that may be mistaken as due to AD. Adult cancers such as colon cancer seem to occur at the same rates as in the general population. Breast and cervical cancer may occur less commonly (Oster and Van Den Temple 1975).

Sleep apnoea

Obstructive sleep apnoea has been reported both in children and adults with DS (Telakivi *et al.* 1987; Silverman 1988; Hultchrantz and Svanholm 1991; Marcus *et al.* 1991). Predisposing factors among those with DS include an abnormally small upper airway, increased secretions, obesity, generalized hypotonia causing a collapse of the airway during inspiration, tongue hypotonia, and adenoid and tonsillar enlargement caused by frequent infections. Consequences of sleep apnoea include excessive daytime inactivity, behavioural disturbances, failure to thrive, declining functional skills and disrupted sleep patterns. Diagnosis of sleep apnoea may be difficult to confirm because many individuals with DS will not tolerate sleeping connected to a machine all night in a laboratory (Pary 1992).

'Early signs' of AD in moderately to profoundly mentally retarded persons with DS (e.g. apathy, inattention, decreased social interaction, daytime sleepiness, gait deterioration, myoclonus and seizures) reported by several investigators (e.g. Evenhuis 1990) could be due to sleep apnoea rather than AD. Untreated, sleep apnoea can lead to lung and heart disease including congestive heart failure. In the general population, the prevalence of sleep apnoea has been noted to increase with age. It was diagnosed in 60 per cent of elderly persons complaining of sleeping difficulties (Roehrs *et al.* 1985) and in 24 per cent of a random sample of elderly persons (Ancoli-Israel *et al.* 1987). Since persons with DS have many predisposing factors, an increased prevalence of obstructive sleep apnoea among persons with the syndrome can be expected as they grow older. Clearly, it is important to evaluate sleep apnoea in individuals with DS experiencing behavioural or functional changes.

Musculoskeletal changes

Many orthopaedic conditions of the lower extremities have been reported in persons with DS, including severe flat feet, metatarsus varus, complex bunions and dislocation of the patella and hip. These conditions frequently result in pain and impaired mobility (Diamond *et al.* 1981), which lead to reduced activity levels (Pary 1992). Pain, as a cause of gait disturbance, may be unrecognized due to limited self-reporting. Cope and Olson (1987) reported cervical spine abnormalities in 14 of 35 individuals with DS aged two to 40 years. Atlanto-axial subluxation was present in 18 per cent and abnormalities of the cervical vertebrae were found in the remainder. Cervical degenerative disc disease was noted on X-rays in 14.7 per cent of participants in a retrospective survey of 107 individuals with DS (Diamond *et al.* 1981); none of these persons was reported to have symptoms (the ages are not provided). Others have found that adults with DS are susceptible to joint problems in the neck, knee or hips (Collacott *et al.* 1989; Tangerud *et al.* 1990). Abnormal cervical vertebrae probably predispose to degenerative disc disease and cervical stenosis, which can lead to pain and neurological impairments later in life, the latter conceivably misinterpreted as secondary to AD.

Polypharmacy

With age, the body becomes more sensitive to the effects of medication because of alterations in the metabolism of drugs (Everitt and Avorn 1986). Although pharmacokinetics appears not to have been studied in older adults with DS, some of the same alterations in renal and hepatic function seen in the ageing general population can be anticipated in DS as a result of the apparent 'premature ageing' phenomenon. Individuals with DS in their fifties may be more susceptible to adverse effects from medications that could be mistaken as clinical manifestations of AD. In addition, many older persons with DS who are, or had been, institutionalized may have received neuroleptic drugs, with long-term side effects such as tardive dyskinesia and tardive akathesia (Gualtieri 1991). Continued use of neuroleptics can adversely affect function, particularly mobility.

Relationship between neuropathology and dementia in Down syndrome

Many investigators have noted that the clinical expression of AD in persons with DS is substantially less frequent than would be expected from the results of neuropathological investigations. The characteristic neuropathological features of AD are virtually always present in DS if death occurs after the age of 40 years (see Chapter 5), whereas clinical signs may be apparent in only 15–30 per cent of these cases. This discrepancy has been reviewed elsewhere (Wisniewski and Rabe 1986; Dalton and Wisniewski 1990; Rabe *et al.* 1990; Zigman *et al.* 1993).

AD was reported in 12 of 49 adults with DS who were followed longitudinally for eight years using the same test procedures (Dalton and Crapper-McLachlan 1984), while Lai and Williams (1989) noted that 40 per cent of their cases showed signs of dementia. With the exception of the findings of Blessed *et al.* (1968), relatively low correlations have generally been obtained between counts of senile plaques and/or neurofibrillary tangles with severity of dementia (e.g. Wilcock and Esiri 1982). Wisniewski and Rabe (1986) concluded that factors other than plaques and tangles must be involved in the dementia, that these neuropathological changes differ in their significance in the development of dementia and that the brain regions in which they occur may also be an important factor. Wisniewski and Rabe (1986) noted that plaques and tangles develop in the brains of normal non-demented elderly persons, but that the brains of age-matched persons with AD have many more of these changes. Thus, there appears to be a quantitative rather than a qualitative difference between the brains of AD patients in comparison to normal aged individuals. Dalton *et al.* (1974) suggested that the density and distribution of plaques and tangles in individuals with DS may be too low to produce measurable changes on psychological tests of memory. Lott and Lai (1982) speculated that the discrepancy between the clinical and neuropathological findings of AD in individuals with DS may be due to differences in the topographic distribution of the lesions. However, Wisniewski and Rabe (1986) argued that there is no convincing evidence to invoke such topographic differences. Instead, they proposed that the discrepancy may be due to a higher threshold for dementia in persons with DS. Similarly, Mann (1988) speculated that there may be a 'pathological threshold' to dementia that only some ageing persons with DS cross despite the universal prevalence of AD pathology in the brain. Lai and Williams (1989) have suggested that there may be an asymptomatic 'incubation period' of 10–20 years before clinical signs appear.

Dalton and Crapper-McLachlan (1984) proposed that all persons with DS would eventually develop dementia if they lived long enough and that the discrepancy between neuropathological and clinical findings could be explained as a survival effect in which persons without dementia do not come to autopsy, whereas those with AD neuropathology do. This 'healthy survivor effect' proposition also has been advanced more recently by Zigman *et al.* (1993), who suggest that this effect could lead to an underestimation of both the extent of regression and the rate of AD in surviving elderly persons. Rabe *et al.* (1990) disagreed with this interpretation, suggesting that Dalton and Crapper-McLachlan may have overestimated the prevalence of AD because some of the observed decline in function may have been due to factors related to early mortality. This issue was specifically addressed by Schupf *et al.* (1989). They examined data from 99 deceased individuals with DS and 99 controls matched for age, sex, level of mental retardation and residential placement, in a historical cohort study of adaptive behaviour for the three years preceding death. Causes of death did not differentiate the DS and control cases, but the two oldest groups with DS (50–59 and 60–69 years) showed more regression in behavioural skills when compared with both younger persons with DS and with their age-matched controls. These authors argued that the proportion of people with DS who displayed regression in excess of that observed for the controls can serve as an estimate of regression specifically attributable to an increased prevalence of AD. The age-specific prevalence estimates were: 0.12 (40-49 years), 0.36 (50–59 years) and 0.46 (60–69 years). These results must be interpreted with caution. The instrument employed for the assessment of daily living skills (an abbreviated version of the Minnesota Developmental Programming System

Behavioral Scales of Joiner and Krantz 1979) was not specifically designed for the measurement of AD and may have failed to recognize signs of AD despite being appropriate for identifying more global changes associated with the events preceding death. Furthermore, Fenner *et al.* (1987) found no statistically reliable changes in adaptive daily living skills as a function of age in 39 persons with DS measured with the Adaptive Behavior Scale on a two-year longitudinal basis.

The above results also need to be reconciled with much higher clinical prevalence rates of AD reported by Lai and Williams (1989), Dalton and Wisniewski (1990) and Evenhuis (1990) in their longitudinal studies. Such a reconciliation may not be possible currently because of significant differences in methodologies and sampling practices between studies. The discrepancy between prevalence of AD neuropathology and clinical dementia in ageing persons with DS suggests that additional factors influencing risk for the clinical manifestations of AD need to be identified (Zigman *et al.* 1993).

Conclusions

Particular signs or symptoms that could be suggestive of the dementia of AD in individuals with DS must be interpreted with caution. The phenotypic expression of the DS genotype in late life and normal as well as pathological changes associated with ageing present an array of possibilities requiring careful scrutiny. Limited sensory, motor, language, communication and intellectual capabilities in these individuals provide a complex background upon which is superimposed the development of subtle, insidious alterations in function. These decrements in function may be caused by AD or other unrelated processes.

The problems of accurate diagnosis are even more apparent when the results of retrospective and prospective studies are compared. It has been frequently proposed that criteria for the diagnosis of AD in DS should be based on change in functions over a specified interval of time (e.g. Wisniewski and Rabe 1986; Zigman *et al.* 1987; Holden *et al.* 1991). For example, an analytic method for defining criteria for the diagnosis of AD, based upon a longitudinal analysis of the differential patterns of regression in adaptive functions between those with DS and those with other aetiological diagnoses, has been advanced (Zigman *et al.* 1991). Deterioration in one or more functions, on the basis of a careful definition of a clinically significant change, and using suitable tests and cut-off scores, could be defined as 'indicative' of AD. An initial clinical evaluation showing significant impairment in one or more functions susceptible to AD could be defined as 'suggestive' of AD. Ultimately the NINCDS-ADRDA criteria for clinical diagnosis of AD (McKhann *et al.* 1984) should be modified to take the presence of mental retardation into account.

Acknowledgements

The authors gratefully acknowledge the contributions of families, staff and friends involved with agencies in the Province of Ontario and in New York State (particularly in Columbia, Oneida and Suffolk counties) and with the University of Wisconsin-Madison Aging and Developmental Disabilities Clinic. Funding for work reported here was provided by grants from the National Institute on Aging (2RO1-AGO8849), the Velleman Foundation and the Administration on Developmental Disabilities (HHS/AFC grant #07DD027317).

References

Aitchison, C., Easty, D. L., and Jancar, J. (1990). Eye abnormalities in the mentally handicapped. *Journal of Mental Deficiency Research*, 34, 41-8.

Ancoli-Israel, S., Kripke, D. F., and Mason, W. (1987). Characteristics of obstructive and central sleep apnea in the elderly: *An interim report. Biological Psychiatry*, 22, 741-50.

Bayley, N. (1969). Manual for the Bayley scales of infant development. Psychological Corporation, New York.

Baxter, R. G., Larkins, R. G., Martin, F. I., Heyman, P., Myles, L., and Ryan, L. (1975). Down syndrome and thyroid dysfunction in adults. *Lancet*, 2, 794-6.

Blessed, G., Tomlinson, B. E., and Roth, M. (1968). The association between quantitative measures of dementia and of senile changes in the cerebral grey matter of elderly subjects. *British Journal of Psychiatry*, 114, 797-817.

Buchanan, L. H. (1990). Early onset of presbycusis in Down syndrome. *Scandinavian Audiology*, 19, 103-10.

Burt, D. B., Loveland, K. A., and Lewis, K. R. (1992). Depression and the onset of dementia in adults with mental retardation. *American Journal on Mental Retardation*, 96, 502-11.

Collacott, R. A., Ellison, D., Harper, W., Newland, C., and Ray-Chaudhurt, K. (1989). Atlanto-occipital instability in Down's syndrome. *Journal of Mental Deficiency Research*, 33, 499-505.

Cope, R., and Olson, S. (1987). Abnormalities of the cervical spine in Down's syndrome: diagnosis, risks, and review of the literature, with particular reference to the Special Olympics. *Southern Medical Journal*, 80, 33-6.

Crapper-McLachlan, D. R., Dalton, A. J., Galin, H., Schlotterer, G.R., and Daicar, E. (1984). Alzheimer's disease: clinical course and cognitive disturbances. *Acta Neurologica Scandinavica*, 69, 83-90.

Crapper-McLachlan, D. R., Dalton, A. J., Kruck, T. P. A., Bell, M. Y., Smith, W. L., Kalow, W., *et al.* (1991). Intramuscular desferrioxamine in patients with Alzheimer's disease. *Lancet*, 337, 1304-8.

Dalton, A. J. (1992). Dementia in Down syndrome: methods of evaluation. In *Down syndrome and Alzheimer disease* (ed. L. Nadel and C.J. Epstein), pp. 51–76. Wiley-Liss, New York.

Dalton, A. J., and Crapper, D. R. (1977). Down's syndrome and aging of the brain. In *Research to practice in mental retardation: biomedical aspects*, Vol. 3, (ed. P. Mittler), pp. 391-400. University Park Press, Baltimore.

Dalton, A. J., and Crapper-McLachlan, D. R. (1984). Incidence of memory deterioration in aging persons with Down's syndrome. In *Perspectives and progress in mental retardation: biomedical aspects*, Vol. 2, (ed. J. M. Berg), pp. 55-62. University Park Press, Baltimore.

Dalton, A. J., and Crapper-McLachlan, D. R. (1986). Clinical expression of Alzheimer's disease in Down's syndrome. *Psychiatric Clinics of North America*, 9, 659-70.

Dalton, A. J., and Wisniewski, H. M. (1990). Down syndrome and the dementia of Alzheimer disease. *International Review of Psychiatry*, 2, 43-52.

Dalton, A. J., Crapper, D. R., and Schlotterer, G. R. (1974). Alzheimer's disease in Down's syndrome: visual retention deficits. *Cortex*, 10, 366-77.

Davies, B. (1988). Auditory disorders in Down's syndrome. *Scandinavian Audiology Supplement,* 30, 65-8.

Diamond, L. S., Lynne, D., and Sigman, B. (1981). Orthopedic disorders in patients with Down syndrome. *Orthopedic Clinics of North America,* 12, 57-71.

Evenhuis, H. M. (1990). The natural history of dementia in Down's syndrome. *Archives of Neurology,* 47, 263-7.

Evenhuis, H. M., van Zanten, G. A., Brocaar, M. P., and Roerdinkholder, W. H. M. (1992). Hearing loss in middle-age persons with Down syndrome. *American Journal on Mental Retardation,* 97, 47-56.

Everitt, D. E., and Avorn, J. (1986). Drug prescribing for the elderly. *Archives of Internal Medicine,* 146, 2393-6.

Fenner, M. E., Hewitt, K. E., and Torpy, D. M. (1987). Down's syndrome: intellectual and behavioural functioning during adulthood. *Journal of Mental Deficiency Research,* 31, 241-9.

Goldhaber, S. Z., Rubin, I. L., Brown, W., Robertson, N., Stubblefield, F., and Sloss, L. J. (1986). Valvular heart disease (aortic regurgitation and mitral valve prolapse) among institutionalized adults with Down's syndrome. *American Journal of Cardiology,* 57, 278-81.

Gualtieri, C. T. (1991). TMS: a system for prevention and control. In *Mental retardation: developing pharmacotherapies,* (ed. J.J. Ratey), pp. 35-49. American Psychiatric Press, Washington, DC.

Harper, D. C., and Wadsworth, J. S. (1990). Depression and dementia in elders with mental retardation: a pilot study. *Research in Developmental Disabilities,* 11, 177-98.

Haveman, N., Masskant, M. A., and Sturmans, F. (1989). Older Dutch residents of institutions with and without Down syndrome: comparisons of mortality and morbidity trends and motor/social functioning. *Australian and New Zealand Journal of Developmental Disabilities,* 15, 241-55.

Helmes, E., Csapo, K. G., and Short, J. -A. (1985). History, development and validation of a new rating scale for the institutionalized elderly. The multidimensional observation scale for elderly subjects (MOSES). Bulletin, no. 8501. Department of Psychiatry, University of Western Ontario, London, Canada.

Helmes, E., Csapo, K. G., and Short, J. -A. (1987). Standardization and validation of the multidimensional observation scale for elderly subjects. *Journal of Gerontology,* 42, 395-403.

Hersch, E. L., Csapo, K. G., and Palmer, R. B. (1978a). Development of the London Psychogeriatric Rating Scale. *London Psychiatric Hospital Research Bulletin,* 1, 3-21.

Hersch, E. L., Kral, V. A., and Palmer, R. B. (1978b). Clinical value of the London Psychogeriatric Rating Scale. *Journal of the American Geriatrics Society,* 26, 348-54.

Hestnes, A., Sand, T., and Fostad, K. (1991). Ocular findings in Down's syndrome. *Journal of Mental Deficiency Research,* 35, 194-203.

Hiles, D. A., Pettapiece, M. C., Biglan, A. W., and Cheng, K.P. (1990). *Down Syndrome Group Western Pennsylvanian Newsletter,* 8, 27-30.

Hogg, J., Moss, S., and Cooke, D. (1988). *Ageing and mental handicap.* Chapman and Hall, London.

Holden, J. J. A., Chalifoux, M., Clairman, C., Dalziel, F., Ditullio, K., McGeer, D., et al. (1991). Down's syndrome and Alzheimer dementia: clinical evaluation and genetic association. In *Alzheimer's disease: basic mechanisms, diagnosis and therapeutic strategies,* (ed. K. Iqbal, D. R. C. McLachlan, B. Winblad and H.M. Wisniewski), pp. 435-41. Wiley, Chichester.

Hultchrantz, E., and Svanholm, H. (1991). Down syndrome and sleep apnea: a therapeutic challenge. *International Journal of Developmental Disabilities,* 21, 263-8.

Joiner, L. M., and Krantz, G. C. (1979). The assessment of behavioral competence of developmentally disabled individuals: The MDPS. University of Minnesota Press, Minneapolis.

Kear-Colwell, J. J. (1973). The structure of the Wechsler Memory Scale and its relationship to `brain damage.' *British Journal of Social and Clinical Psychology,* 12, 384-92.

Keiser, H., Montague, J., Wold, D., Maune, S., and Pattison, D. (1981). Hearing loss of Down syndrome adults. *American Journal of Mental Deficiency,* 85, 467-72.

Lai, F. (1992). Clinicopathologic features of Alzheimer disease in Down syndrome. In *Down syndrome and Alzheimer disease,* (ed. L. Nadel and C.J. Epstein), pp. 15–34. Wiley-Liss, New York.

Lai, F., and Williams, R. S. (1989). A prospective study of Alzheimer disease in Down syndrome. *Archives of Neurology*, **46**, 849-53.

Levin, S., Nir, E., and Mogilner, B. M. (1975). T-system immune-deficiency in Down's syndrome. *Pediatrics*, **56**, 123-6.

Lott, I. T. (1982). Down syndrome, aging and Alzheimer's disease: a clinical review. *Annals of the New York Academy of Sciences*, **396**, 15-27.

Lott, I. T., and Lai, F. (1982). Dementia in Down's syndrome: Observations from a neurology clinic. *Applied Research in Mental Retardation*, **3**, 233-9.

McKhann, G., Drachman, D., Folstein, M., Katzman, R., Price, D., and Stadlan, E. M. (1984). Clinical diagnosis of Alzheimer's disease: Report of the NINCDS-ADRDA Work Group under the auspices of the Department of Health and Human Services Task Force on Alzheimer's disease. *Neurology*, **34**, 939-44.

Mann, D. M. A. (1988). Alzheimer's disease and Down's syndrome. *Histopathology*, **13**, 125-37.

Marcus, C. L., Keens, T. G., Bautista, D. B., von Pechmann, W. S., and Ward, S. L. (1991). Obstructive sleep apnea in children with Down syndrome. *Pediatrics*, **88**, 132-9.

Mehta, P. D., Dalton, A. J., Percy, M. E., and Wisniewski, H. M. (1993). Increased Beta-2 microglobulin (β2-M) and interleukin-6 (IL–6) in sera from older persons with Down syndrome. In *Alzheimer's disease and related disorders,* (ed. M. Nicolini, G. F. Zatta and B. Corain). Pergamon Press, London. (In press).

Mortimer, J. A. (1983). Alzheimer's disease and senile dementia: prevalence and incidence. In Alzheimer's disease, (ed. B. Reisberg), pp. 141-8. Free Press, New York.

Mortimer, J. A., and Hutton, J. T. (1985). Epidemiology and etiology of Alzheimer's disease. In *Senile dementia of the Alzheimer type*, (ed. J. T. Hutton and A. D. Kenny), pp. 177-96. Alan R. Liss, New York.

Moss, S., Goldberg, D., and Patel, P. (1991). *Psychiatric and physical morbidity in older people with severe mental handicap*. HARC., University of Manchester, Manchester.

Odell, J. D. (1988). Medical considerations. In *Down syndrome: a resource handbook,* (ed. C. Tingey), pp. 33-45. College-Hill, Boston.

Oster, J., and Van Den Temple, A. (1975). Mortality and life-table in Down's syndrome. *Acta Pediatrica Scandinavica*, **64**, 322.

Pary, R. (1992). Differential diagnosis of functional decline in Down's syndrome. *The Habilitative Mental Healthcare Newsletter,* **11**, (6), 37-41.

Percy, M. E., Dalton, A. J., Markovic, V. D., Crapper-McLachlan, D.R., Hummel, J.T., Rusk, A. C. M., et al. (1990a). Red cell superoxide dismutase, glutathione peroxidase and catalase in Down syndrome patients with and without manifestations of Alzheimer disease. *American Journal of Medical Genetics*, **35**, 459-67.

Percy, M. E., Dalton, A. J., Markovic, V. D., Crapper-McLachlan, D. R., Gera, E., Hummel, J. T., et al. (1990b). Autoimmune thyroiditis associated with mild "subclinical" hypothyroidism in adults with Down syndrome: A comparison with and without manifestations of Alzheimer disease. *American Journal of Medical Genetics*, **36**, 148-54.

Rabe, A., Wisniewski, K. E., Schupf, N., and Wisniewski, H. M. (1990). Relationship of Down's syndrome to Alzheimer's disease. In *Application of basic neuroscience to child psychiatry,* (ed. S.I. Deutsch, A. Weizman and R. Weizman), pp. 325-40. Plenum, New York.

Rabinowe, S. L., Rubin, I. L., George, K. L., Adri, M. N., and Eisenbarth, G. S. (1989). Trisomy 21 (Down syndrome): autoimmunity, aging and monoclonal antibody-defined T-cell abnormalities. *Journal of Autoimmunity*, **2**, 25-30.

Rees, T. S., and Duckert, L. G. (1990). Auditory and vestibular dysfunction in aging. In *Principles of geriatric medicine and gerontology,* (2nd edn), (ed. W. R. Hazzard, R. Andres, E. L. Bierman and J. P. Blass), pp. 432-44. McGraw-Hill, New York.

Reid, A. H., and Aungle, P. G. (1974). Dementia in ageing mental defectives: A clinical psychiatric study. *Journal of Mental Deficiency Research,* **18**, 15-23.

Reid, A. H., Maloney, A. F. J., and Aungle, P. G. (1978). Dementia in ageing mental defectives: A clinical and neuropathological study. *Journal of Mental Deficiency Research*, **22**, 233-41.

Reiss, S. (1982). Psychopathology and mental retardation: survey of a developmental disabilities mental health program. *Mental Retardation*, **20**, 128-32.

Roehrs, T., Zorick, F., Wittig, R., and Roth, T. (1985). Efficacy of a reduced triazolam dose in elderly insomniacs. *Neurobiology of Aging*, 6, 293-6.

Roth, M. (1986). The association of clinical and neurobiological findings and its bearing on the classification and aetiology of Alzheimer's disease. *British Medical Bulletin*, 42, 42-50.

Rowe, R. D., and Uchida, I. A. (1961). Cardiac malformation in mongolism: a prospective study of 184 mongoloid children. *American Journal of Medicine*, 31, 726-35.

Schupf, N., Silverman, W. P., Sterling, R. C., and Zigman, W. B. (1989). Down syndrome, terminal illness and risk for dementia of the Alzheimer type. *Brain Dysfunction*, 2, 181-8.

Schupf, N., Zigman, W. B., Silverman, W. P., Rabe, A., and Wisniewski, H. M. (1990). Genetic epidemiology of Alzheimer's disease. In *Aging brain and dementia: new trends in diagnosis and therapy,* (ed. L. Battistin), pp. 57-78. Alan R. Liss, New York.

Senatore, V., Matson, J. L., and Kazdin, A. E. (1985). An inventory to assess psychopathology of mentally retarded adults. *American Journal of Mental Deficiency*, 89, 459-66.

Short, J. A. C. (1982). The rationale, construction and development of the M.O.S.E.S. - a multidimensional observation scale for elderly subjects. In *Proceedings of the Ninth Annual Meeting of the Ontario Psychogeriatric Association,* (ed. R. W. Hopkins), pp. 286-90. Ontario Psychogeriatric Association, Kingston.

Silverman, M. (1988). Airway obstruction and sleep disruption in Down's syndrome. *British Medical Journal*, 296, 1618-9.

Sinex, F. M., and Merril, C. R. (eds.)(1982). Alzheimer's disease, Down's syndrome, and aging. *Annals of the New York Academy of Sciences*, 396, 1-199.

Sovner, A., and Hurley A. D. (1983). Do the mentally retarded suffer from affective illness? *Archives of General Psychiatry*, 40, 61-7.

Szymanski, L. S., and Biederman, J. (1984). Depression and anorexia nervosa of persons with Down syndrome. *American Journal of Mental Deficiency,* 89, 246-51.

Tangerud, A., Hestnes, A. S., and Sunndalsfoll, S. (1990). Degenerative changes in the cervical spine in Down's syndrome. *Journal of Mental Deficiency Research,* 34, 179-85.

Telakivi, T., Partinen, M., Salmi, T., Leinonen, L., and Harkonen, T. (1987). Nocturnal periodic breathing in adults with Down's syndrome. *Journal of Mental Deficiency Research,* 31, 31-9.

Ulrich, D. A., Riggen, K. J., Ozmun, J. C., Screws, D.P., and Cleland, F.E. (1989). Assessing movement control in children with mental retardation: a generalizability analysis of observers. *American Journal of Mental Retardation,* 94, 161-8.

Warren, A. C., Holroyd, S., and Folstein, M. F. (1989). Major depression in Down's syndrome. *British Journal of Psychiatry,* 155, 202-5.

Wechsler, D., and Stone, C. P. (1945). *The Wechsler Memory Scale*. Psychological Corporation, New York.

Whittingham, S., Pitt, D. B., Shaema, D. L., and Mackay, I.R. (1977). Stress deficiency of the T-lymphocyte system exemplified by Down syndrome. *Lancet*, 1, 163-6.

Wilcock, G. K., and Esiri, M. M. (1982). Plaques, tangles, and dementia: a quantitative study. *Journal of the Neurological Sciences*, 56, 343-56.

Wisniewski, H. M., and Merz, G. S. (1984). Aging, Alzheimer's disease, and developmental disabilities. In *Aging and developmental disabilities,* (ed. M. P. Janicki and H. M. Wisniewski), pp.177-84. Brookes, Baltimore.

Wisniewski, H. M., and Rabe, A. (1986). Discrepancy between Alzheimer-type neuropathology and dementia in persons with Down syndrome. *Annals of the New York Academy of Sciences,* 477, 247-60.

Wisniewski, K. E., Dalton, A. J., Crapper-Mclachlan, D. R., Wen, G. Y., and Wisniewski, H. M. (1985). Alzheimer's disease in Down's syndrome: clinicopathologic studies. *Neurology*, 35, 957-61.

Wright, A. F., and Whalley, L. F. (1984). Genetics, ageing, and dementia. *British Journal of Psychiatry,* 145, 20-38.

Zigman, W. B., Schupf, N., Lubin, R. A., and Silverman, W. P. (1987). Premature regression of adults with Down syndrome. *American Journal of Mental Deficiency*, 92, 161-8.

Zigman, W. B., Seltzer, G. B., Adlin, M., and Silverman, W. P. (1991). Physical, behavioral, and mental health changes associated with aging. In *Aging and developmental disabilities:*

challenges for the 1990s, (ed. M. P. Janicki and M. M. Seltzer), pp. 52-75. American Association on Mental Retardation, Washington, DC.

Zigman, W. B., Schupf, N., Zigman, A., Silverman, W. P., and Wisniewksi, H. M. (1993). Aging and Alzheimer disease in people with mental retardation. *International Review of Research in Mental Retardation* **19**, 41–70.

5

Association between Alzheimer disease and Down syndrome: neuropathological observations

D. M. A. Mann

Summary

Persons with Down syndrome who live beyond 40 years of age almost invariably show the same kinds of pathological changes in their brains as do others in the general population suffering from Alzheimer disease. Hence, it is possible to use the predictability of this association to study the brains of younger individuals with Down syndrome in order to determine the sites and nature of the early pathological lesions of Alzheimer disease and to investigate how these destructive processes progress with time. We have examined 43 persons with Down syndrome aged between 13 and 71 years, using various molecular markers, for the presence of senile plaques and neurofibrillary tangles, the extent of nerve cell loss and for any changes in cognition or behaviour indicative of the onset of dementia. The earliest changes detectable by these methods occurred in the second and third decades and were characterized by a deposition of amyloid β/A4 protein. Neurofibrillary alterations within plaques (neurites) and within nerve cell bodies (tangles) did not appear until the fourth or fifth decades and nerve cell loss and dementia did not occur until after 50 years of age. The breakdown of amyloid precursor protein and the deposition of β/A4 protein may be seminal events in the initiation and progression of the pathological process of Alzheimer disease in persons with Down syndrome. Triplication of chromosome 21 in Down syndrome leading to an overexpression of amyloid precursor protein may trigger this pathological cascade.

Introduction

An association between Down syndrome (DS) and dementia was noted over a century ago by Fraser and Mitchell (1876) who wrote that 'in not a few instances, however, death was attributed to nothing more than a general decay—a sort of precipitated senility'. Much later, linkage between this 'senile decay' and the presence within the brain of pathological lesions characteristic of Alzheimer disease (AD), namely senile plaques (SP) and neurofibrillary tangles (NFT), was noted (Struwe 1929; Jervis 1948). However, only over the past two decades has it become firmly established that nearly all persons with DS older than 40 years develop SP and NFT within their brains that appear to essentially mimic, in terms of form, distribution and number, those found in AD in the general population.

The prevalence and distribution of Alzheimer type changes in the brain in Down syndrome

The finding of SP, NFT or both, within one or more regions of the brain in persons with DS of all ages, has been the subject of numerous case reports and extensive surveys, particularly in recent years (Struwe 1929; Bertrand and Koffas 1946; Jervis 1948; Solitaire and Lamarche 1966; Neumann 1967; Haberland 1969; Olson and Shaw 1969; Malamud 1972; O'Hara 1972; Burger and Vogel 1973; Schochet *et al.* 1973; Ellis *et al.* 1974; Reid and Maloney 1974; Crapper *et al.* 1975; Murdoch and Adams 1977; Rees 1977; Wisniewski *et al.* 1979, 1985a,b; Ball and Nuttall 1980; Ropper and Williams 1980; Blumbergs *et al.* 1981; Pogacar and Rubio 1982; Whalley 1982; Sylvester 1983; Yates *et al.* 1983; Mann *et al.* 1984, 1985a, 1986, 1987a, 1990 a,b; Ross *et al.* 1984; Belza and Urich 1986; Giaccone *et al.* 1989; Mann and Esiri 1989; Motte and Williams 1989; De La Monte and Hedley-White 1990; Ferrer and Gullotta 1990). These 39 studies included 434 subjects, ranging in age from under 10 years to over 70 years, 260 (59.9 per cent) of whom showed SP and/or NFT when classical silver staining (e.g. Bodian, Palmgren, Bielschowsky) methods were employed (Table 5.1). When analysed by decade of life, typical (classical) SP with or without NFT first appeared, infrequently, during the second decade, rose rapidly in prevalence through the third and fourth decades, reached nearly 100 per cent prevalence in subjects aged 40–60 years and was always present in those over 60 years. In summary, 208 of the 211 subjects (98.6 per cent) over 40 years of age showed SP and NFT, whereas only 52 of 223 subjects (23.3 per cent) under 40 years showed either or both changes. It seems therefore that in DS there is a transitional period, usually between ages 20–40 years, during which the absence of SP and NFT in any part of the brain changes into a widespread presence in virtually all cases.

Only three individuals with DS over 40 years of age have so far been claimed not to show any SP or NFT within their brains (Murdoch and Adams 1977; Whalley 1982). Murdoch and Adams (1977) briefly referred to a 56-year-old man whose brain was said not to contain either SP or NFT; karyotype details were not presented and the extent of neuropathological investigations is unclear. Whalley (1982) reported two cases: a 55-year-old woman with mosaicism and a 49-year-old woman with full trisomy 21 whose brain weight of 1520 g is unusual and in whom histological examination was insufficiently extensive to definitely exclude the presence of SP and NFT anywhere in the brain. From these limited data it is uncertain to what extent, if at all, exceptions from the usual association between AD pathology and DS might exist.

Age range (years)	Total number of persons	Number showing plaques and tangles	Percentage of persons affected
0–9	38	0	0
10–19	81	6	7.4
20–29	59	10	16.9
30–39	45	36	80.0
All <39	223	52	23.3
40–49	62	61	98.4
50–59	98	96	98.0
60–69	48	48	100.0
70–79	3	3	100.0
All >40	211	208	98.6
All 0–79	434	260	59.9

Table 5.1. Prevalence of senile plaques and neurofibrillary tangles in the brains of persons with Down syndrome (reviewed data analysed by age)

Another consideration that emerges from the above studies is that while in most (if not all) people with DS over 50 years of age SP and NFT always occur together in high numbers, in those under this age a much more variable picture is seen. For example, in the 16 studies (Struwe 1929; Jervis 1948; Olson and Shaw 1969; Burger and Vogel 1973; Schochet et al. 1973; Wisniewski et al. 1979, 1985a; Ball and Nuttall 1980; Ropper and Williams 1980; Sylvester 1983; Yates et al. 1983; Ross et al. 1984; Mann et al. 1986; Giaccone et al. 1989; Mann and Esiri 1989; Motte and Williams 1989) of subjects aged between six and 40 years in whom the occurrence of SP and NFT in various brain regions was investigated separately (a total of 94 brains), 43 subjects showed neither SP nor NFT in any area examined, 41 showed SP and NFT together in all areas examined or in the hippocampus alone, and at least 10 subjects in five of the studies (Struwe 1929; Jervis 1948; Burger and Vogel 1973; Giaccone et al. 1989; Motte and Williams 1989) were considered to show SP alone in one or more regions. However, because in most instances the investigations were limited predominantly to frontal cortex or hippocampus or both, it is possible that some NFT occurred in areas of brain not investigated. In no subject did NFT occur in the absence of SP.

From all reported studies to date (Solitaire and Lamarche 1966; Olson and Shaw 1969; Burger and Vogel 1973; Reid and Maloney 1974; Crapper et al. 1975; Rees 1977; Wisniewski et al. 1985b; Motte and Williams 1989) and also from our own experience (Mann et al. 1984, 1985a, 1986, 1987a, 1990b), it appears that the pattern of involvement of brain structures by SP and NFT, in persons with DS who live beyond 50 years of age, closely follows that typically seen in AD (for reviews see Mann 1985, 1988a). For example, the amygdala, hippocampus and association areas of frontal, temporal and parietal cortex are all strongly favoured by SP formation, whereas visual, motor and somatosensory cortex are less affected (Mann et al. 1986; Motte and Williams 1989).

	Age (years)	Temporal cortex		Hippocampus	
		Senile[a] plaques	Neurofibrillary[b] tangles	Senile[b] plaques	Neurofibrillary[b] tangles
Down syndrome (n =15)	60.3 ± 1.3	19.6* ± 1.8	26.6 ± 3.1	130.8* ± 20.4	298.3** ± 34.2
Alzheimer disease (n =10)	62.0 ± 2.1	27.0 ± 2.7	27.0 ± 2.9	71.6 ± 23.1	148.2 ± 25.4

*,** denotes significantly different from Alzheimer disease mean value, P <0.05 and <0.01, respectively
[a]Number/mm^2
[b]Number per section profile

Table 5.2. Mean (± SEM) numbers of senile plaques and neurofibrillary tangles in temporal cortex and hippocampus of 15 persons over 50 years of age with Down syndrome and 10 persons of similar age range with Alzheimer disease

Nerve cells in the olfactory nuclei and tracts are also often affected by NFT (Mann *et al.* 1986, 1988a) and sometimes by SP as well. Typical SP are not observed in the cerebellar cortex (see later), nor are NFT usually present in Purkinje cells though occasional nerve cells in the dentate nucleus may contain NFT (Mann *et al.* 1990a).

Whether the density of SP and NFT within affected brain regions is similar in middle-aged persons with DS to that seen in people with AD in the general population is not clear. Ball and Nuttall (1980) estimated NFT density in the hippocampus of two individuals with DS to fall within the range encountered in eight individuals with AD; SP were not quantified in this study. Ropper and Williams (1980) estimated SP and NFT densities in the hippocampus of eight persons with DS over 50 years old and, although no comparative data were presented, the authors stated them to be comparable to levels seen in 'demented old people'. In our earlier studies (Mann *et al.* 1984, 1985a) SP and NFT densities in the temporal cortex and hippocampus of six middle-aged persons with DS seemed to match closely those values seen in these regions in persons of similar age with AD. However, our later work, which included further subjects (see Mann 1988a), revealed that the mean SP density in 15 persons with DS over 50 years of age was less in the temporal cortex than in 10 persons of that age with AD, whereas NFT densities were equivalent (Table 5.2). In the hippocampus, however, both SP and NFT density in the DS cases far exceeded that usually seen in AD. Comparisons of this kind may reveal other subtle differences in distribution or severity of neuropathological change in DS.

The morphological and microchemical appearance of SP and NFT in Down syndrome

Plaques

Most reports have stressed the close structural similarities between the appearances of SP and NFT in conventional silver-stained sections of tissues from persons with DS compared with those typically seen in AD (for light and electron microscope appearances of SP and NFT in AD see Wisniewski and Terry 1973), both at light (Struwe 1929; Jervis 1948; Solitaire and Lamarche 1966; Olson and Shaw 1969; Burger and Vogel 1973) and electron (O'Hara 1972; Schochet et al. 1973; Ellis et al. 1974) microscopic levels. However, this apparent identity has more recently been challenged. Allsop et al. (1986), using brain samples from two individuals with DS aged 59 and 64 years, noted a high proportion of amorphous plaque cores that were dissimilar to those typical of AD. These amorphous plaque cores in the DS cases were larger and the amyloid fibrils less compact than those of the classical plaque core of AD and, under Congo red birefringence, lacked the well-defined typical polarization crosses. Masters et al. (1985) also noted plaque cores in DS to be largely of an amorphous type, whereas in AD the predominant form was of a dense spherical core with a well-defined polarization cross.

While these studies suggest that SP structure in DS at middle-age may not exactly mimic that found in AD, it should be noted that the latter observations were made on biochemically isolated plaque cores rather than on SP *in situ*. In our series of DS cases over 50 years of age, such large amorphous-type SP were indeed more commonly seen than in AD cases, especially in the amygdala and entorhinal cortex (Mann et al. 1986). Assuming that persons with DS begin to form and accumulate SP (and NFT) at 30–40 years of age, and perhaps earlier in some individuals, the pathological picture seen at death will often have been evolving for at least 20 years. In AD, the duration of illness from onset of clinical signs to death averages between five and 10 years, and only occasionally exceeds 15 years. Differences in SP morphology between DS and AD may only represent variations in 'stages' in its natural history consequent upon a longer pathological course in DS.

Extensive effort has been made over the past few years to characterize the molecular and cellular elements that constitute SP, both in AD and DS. The proteinaceous, β-pleated, congophilic material that constitutes much of the 'core' of typical SP in AD and DS, formerly known as 'amyloid', comprises a polypeptide of 39–42 amino acids (Masters et al. 1985) that is now termed β/A4 peptide (according to its molecular weight, 4.2 kDa, and its ability to polymerize spontaneously into a β-pleated configuration). β/A4 is formed through an alternative cleavage of a larger precursor molecule, amyloid precursor protein (APP) (Esch et al. 1990; Sisodia et al. 1990). This precursor spans the nerve cell membrane with a small intracellular domain and a much larger extracellular domain (Kang et al. 1987); it may function as a cell surface receptor or have secretory or cell adhesion properties.

This same, or at least a very similar, β/A4 molecule is also present within the walls of arteries in AD that display 'congophilic angiopathy' (Glenner and Wong 1984a), although in other investigations slight sequence variations have been detected (Joachim *et al.* 1988). Glenner and Wong (1984b) have shown that the same β/A4 molecule is present in arterial walls in DS, and Masters *et al.* (1985) reported that the β/A4 of plaque cores in DS is identical to that of plaque cores in AD. Microchemical analysis of the non-proteinaceous residue of plaque cores in both AD and DS (Masters *et al.* 1985; Edwardson *et al.* 1986) reveals much aluminium and silicon, possibly co-localized in the form of an aluminosilicate.

Surrounding the amyloid core are various cellular and non-cellular elements, including more β/A4. Unusual accumulations of glycoproteins, probably as oligosaccharides, can be detected by lectin histochemistry (Szumanska *et al.* 1987; Mann *et al.* 1988b, 1989a). These are recognized particularly strongly by lectins that bind to α-D-mannose residues, in varying states of molecular complexity [e.g. Concanavilin A (Con A), wheat germ agglutinin (WGA), Pisum sativum (PSA) and Phaseolus vulgaris erythroagglutinin (ePHA)]. However, the precise molecular nature of the glycoproteins, as well as the cellular elements within which they are contained, remain to be characterized. Snow *et al.* (1990) showed accumulations of heparan sulphate proteoglycan within SP, both in AD and DS. Various glial cells are usually associated with SP, in both AD and DS; microglia are often intimately associated with the amyloid cores (Wegiel and Wisniewski 1990; Mann *et al.* 1992c), whereas astrocytes and their processes are more abundant around the periphery of the SP, though frequently extending throughout the plaque region as well (Murphy *et al.* 1990; Mann *et al.* 1992c).

Recently, antibodies directed against β/A4 have been used to detect amyloid deposits throughout the brain in AD and DS. Such antibody staining has revealed that in AD, besides the typical cored SP (easily demonstrable by conventional silver methods or Congo red stains), there are within the cortex further and often more numerous 'diffuse' types of deposits (Tagliavini *et al.* 1988; Yamaguchi *et al.* 1988; Ikeda *et al.* 1989a; Ogomori *et al.* 1989; Wisniewski *et al.* 1989; Mann *et al.* 1990a). Similar deposits are also widely seen in the cortex, hippocampus and amygdala in DS (Allsop *et al.* 1989; Giaccone *et al.* 1989; Ikeda *et al.* 1989b; Mann and Esiri 1989; Mann *et al.* 1989a, 1990a; Rumble *et al.* 1989; Murphy *et al.* 1990; Snow *et al.* 1990; Spargo *et al.* 1990). These do not seem to be associated with a neuritic element (as detectable by conventional silver or antibody methods, see later), or much astrocytic reaction (Murphy *et al.* 1990; Mann *et al.* 1992c), though they often do contain microglia (Mann *et al.* 1992c) and display oligosaccharide accumulation (Mann *et al.* 1989a). Importantly, β/A4 immunostaining has shown other non-cortical areas, such as the cerebellum (Ikeda *et al.* 1989a; Joachim *et al.* 1989; Ogomori *et al.* 1989; Wisniewski *et al.* 1989; Yamaguchi *et al.* 1989; Mann *et al.* 1990a) and striatum (Ogomori *et al.* 1989; Suenaga *et al.* 1990), also to contain much similarly diffuse β/A4 deposition in both AD and DS.

Tangles

Immunohistochemical studies indicate that the microtubule-associated protein, tau, is probably the major antigenic determinant of the paired helical filaments (PHF) of the NFT in AD (Delacourte and Defossez 1986; Ihara *et al.* 1986; Kosik *et al.* 1986; Wood *et al.* 1986). The presence of tau within PHF has been confirmed by direct protein analysis (Goedert *et al.* 1988; Kondo *et al.* 1988; Wischik *et al.* 1988); this, in conjunction with other immunohistochemical studies (e.g. Kosik *et al.* 1988), implies that the whole molecule of tau is incorporated into PHF via its carboxyl third (i.e. the microtubule binding region) and that it probably exists in an abnormally phosphorylated state (Wood *et al.* 1986). Immunostaining (Perry *et al.* 1987; Lennox *et al.* 1988; Lowe *et al.* 1988) and direct protein analysis (Mori *et al.* 1987) have shown a further protein, ubiquitin, to form an important part of the NFT, and lectin histochemistry (Szumanska *et al.* 1987; Mann *et al.* 1988b; Sparkman *et al.* 1990) has revealed the NFT to contain, or be associated with, certain saccharide sequences.

Whether these immunochemical and biochemical properties also hold for the NFT in DS is uncertain. Anderton *et al.* (1982), employing an antibody to neurofilament protein, showed cross-reactivities to occur between the NFT of AD and DS. More recent work (Mann *et al.* 1989b; Murphy *et al.* 1990) showed the NFT of DS and of AD to be similarly immunoreactive to tau and ubiquitin. Immunoblotting (Flament *et al.* 1990; Hanger *et al.* 1991) revealed similar mobility profiles for tau proteins in elderly DS brains to those seen in AD. However, NFT in DS do not appear to interact with lectins (D. M. A. Mann, A. M. T. Brown and B. Marcyniuk, unpublished data), as seems to be the case in AD.

Brain atrophy and neuronal fall-out in Down syndrome

While younger persons with DS apparently show normal cerebral values for regional glucose utilization (Schapiro *et al.* 1990) and blood flow (Risberg 1980; Schapiro *et al.* 1988), both parameters are reduced, particularly in temporal and parietal cortex, in subjects beyond 45 years of age as compared with both younger DS individuals (Schapiro *et al.* 1987) or normal age-matched controls (Melamed *et al.* 1987). Similar changes are observed in AD. Serial CT scanning (Schapiro *et al.* 1989) shows that healthy young persons with DS have smaller brains than controls and that older persons with DS, compared with younger ones, have diminished brain size declining further with advancing age and onset of dementia. In comparison with mentally able individuals of similar age, individuals with DS have low brain weight (Solitaire and Lamarche 1966; Whalley 1982; Mann *et al.* 1984, 1986; Wisniewski *et al.* 1985a) and, like younger subjects with AD (De La Monte 1989; Mann 1991), atrophy is globally distributed, though especially in frontal and temporal lobes (De La Monte and Hedley-White 1990).

The cerebral atrophy of AD is brought about by loss of neurones within particular cortical and subcortical structures, and by loss of pathways connecting these areas. Principally affected in AD (for reviews see Mann 1985, 1988a) are the large pyramidal cells of the association cortex and the hippocampus, neurones of the amygdala and entorhinal cortex, as well as the cells of the cholinergic basal forebrain system, the noradrenergic locus caeruleus and the serotonergic raphe. In DS, as in AD, these losses lead to associated reductions in neurochemical markers (i.e. transmitters, enzymes or receptors) for such systems (for a review see Mann and Yates 1986). In elderly persons with DS there is a low (for age) number of pyramidal and non-pyramidal nerve cells in such areas as cerebral cortex (Mann *et al.* 1985a; Kobayashi *et al.* 1990), hippocampus (Ball and Nuttall 1980; Mann *et al.* 1985a), nucleus basalis (Price *et al.* 1982; Mann *et al.* 1984, 1985b; Casanova *et al.* 1985), locus caeruleus (Mann *et al.* 1984, 1985b; Marcyniuk *et al.* 1988; German *et al.* 1992), raphe (Mann *et al.* 1984, 1985b) and ventral tegmentum (Mann *et al.* 1987b; Gibb *et al.* 1989). Biochemical markers for cholinergic (Yates *et al.* 1980, 1985), noradrenergic (Yates *et al.* 1981, 1983; Reynolds and Godridge 1985; Godridge *et al.* 1987), serotonergic (Yates *et al.* 1986; Godridge *et al.* 1987), cortical pyramidal (glutamate) (Reynolds and Warner 1988; Simpson *et al.* 1989) and non-pyramidal (GABA) (Reynolds and Warner 1988) cells are also reduced in elderly persons with DS in line with loss of parent nerve cells and seemingly to an extent comparable with AD at similar age.

While it is appropriate to assess the degree of involvement of brain structures in AD by reference to age-matched mentally able persons drawn from the general population, such comparisons may not strictly hold in DS, since it is clear that the brains of young adults with DS (but without SP and NFT) are from an early age abnormal in both structure and function (Solitaire and Lamarche 1966; Suetsuga and Mehraein 1980; Takashima *et al.* 1981; Whalley 1982; Sylvester 1983; Wisniewski *et al.* 1984, 1985a; Becker *et al.* 1986; Ferrer and Gullotta 1990). Hence, the numbers of nerve cells in the cortex (Ross *et al.* 1984; Mann *et al.* 1985a), hippocampus (Ball and Nuttall 1980) or in subcortical regions (Gandolfi *et al.* 1981; Casanova *et al.* 1985; McGeer *et al.* 1985; Mann *et al.* 1987a) may be less in young persons with DS than in unaffected young adults. Therefore the 'true' level of change in cell number or in biochemical markers in DS at middle-age can be determined with confidence only by reference to baseline values obtained from younger persons with DS and not by comparisons with data drawn from the general population. If the former is done (see Mann *et al.* 1987a, 1990b), changes in many areas do indeed match those seen in AD, though in some regions the degree of cell damage seems to be less in DS than in AD.

Other pathological similarities

As in AD, granulovacuolar degeneration of neurones in the hippocampus (particularly area CA1) is a feature of DS at middle-age (Solitaire and Lamarche 1966; Olson and Shaw 1969; Burger and Vogel 1973; Ellis *et al.* 1974; Ball and Nuttall 1980). Similarly, Hirano bodies are commonly present in this hippocampal region at the same time of life (Burger and Vogel 1973; Ellis *et al.* 1974). Calcification of the walls of the larger arteries of the globus pallidus and deposition of calcified deposits (calcospherites), in the same region, is commonly seen in AD, particularly in late onset cases (Mann 1988c). Elderly persons with DS also often show a particularly excessive (for age) calcification of this part of the basal ganglia (Wisniewski *et al.* 1982; Takashima and Becker 1985; Mann 1988c). Again, as in younger persons with AD, whose illnesses are uncomplicated by single or multiple lacunar or embolic infarcts, atherosclerosis of the major vessels at the base of the brain is conspicuously absent in elderly persons with DS (Olson and Shaw 1969; Burger and Vogel 1973; Murdoch *et al.* 1977; Mann 1988b). However, as indicated earlier, deposition of β/A4 protein within the walls of large meningeal arteries, especially those supplying the posterior hemispheres and cerebellum (Mann *et al.* 1990a), and intraparenchymal arteries, is present in most middle-aged individuals with DS, as well as in many persons with AD.

Overall, it seems that any differences between AD and DS, in terms of SP or NFT structure, neurochemistry or patterns of neuronal damage and loss of transmitters, are slight. Such variations that do seem to occur may reflect differences in life history (e.g. community versus institutional life) or the varying time-course of evolution of pathology and are not necessarily of major aetiological or pathogenetic significance. Furthermore, the excessive occurrence of AD in elderly persons with DS does not reflect changes that result from mental handicap *per se*. Malamud (1972) found SP and NFT in only five of 312 (1.6 per cent) subjects with DS below 40 years of age, but such changes were present in all 35 subjects over this age. In contrast, he noted SP and NFT in none of 588 other mentally handicapped persons under 40 years of age and in only 31 of 225 (14 per cent) above this age; these 31 persons showed only mild changes and then only in the most elderly. The prevalence of such changes in elderly mentally handicapped persons without DS is probably similar to that present in the elderly population in general.

It seems, therefore, that in typical trisomy 21 DS the pathological changes that develop in the brain in late adult life are indeed those of AD and that the development of these changes is intimately related to the chromosomal alterations that characterize and underpin DS.

The relevance of Down syndrome to understanding the aetiology and pathogenesis of Alzheimer disease

Pathogenetic considerations

The structural or biochemical changes noted earlier in young persons with DS, in terms of brain size and appearance, numbers of particular nerve cell types, alterations in nerve cell morphology and connectivity, while of potential importance to the pathophysiology underlying the basic mental handicap of DS, offer no obvious clues as to why such persons (perhaps all of them) ultimately develop the characteristics of AD. However, because there seems to be a transitional period of 20–30 years in DS, during which the absence of Alzheimer-type changes develops into an almost universal and predictable presence, it becomes possible to reconstruct a longitudinal time-course of pathological events from cross-sectional data obtained from examining the brains of persons with DS dying before, during and after this transitional period. This kind of study cannot be carried out on individuals with AD itself, since their tissues are usually available only after death and then mostly from clinical and pathological 'end-stage' cases in whom the early changes of the disease will either no longer be present or will not easily be identifiable. Moreover, it is not possible with certainty to differentiate those non-demented persons showing early pathological stages of AD from others also showing such minimal changes but who may not have developed the full-blown pathological picture of AD, and become demented, had they lived longer.

Therefore, in order to understand how the tissue damage in AD is initiated and how it might progress with time into the pattern associated with typical AD, we have studied the brains of 43 subjects with DS, ranging in age from 13–71 years, and have employed various histological probes designed to detect the presence of many of the molecular and cellular elements described previously which characterize the pathological end-products (i.e. SP and NFT) of AD. In our youngest subject aged 13 years, we were unable to detect any changes using markers for β/A4, tau, ubiquitin, PHF, oligosaccharides or glial cells. However, between this and 50 years of age, we observed (see Mann *et al.* 1989a,b; Mann and Esiri 1989) a sequential progression of changes within the cerebral cortex and hippocampus that were initiated by the deposition of β/A4 protein in the form of diffuse plaques. Soon afterwards, microglial cells became evident within these amyloid deposits and at these same sites granular accumulations of glycoconjugate(s), recognizable by the lectin Con A, and other similar granular material detectable by anti-ubiquitin appeared. Later, some 'cored' amyloid deposits were seen containing many more microglia, much ubiquitinated material and large quantities of oligosaccharide that was now also detectable by other lectins such as WGA, PSA and ePHA (these also detect α-D-mannosyl residues but in more complex and elaborate configurations than those that Con A recognize). The cored deposits were also reactive with anti-tau and anti-GFAP and contained filamentous structures (PHF) that were also immunoreactive with anti-ubiquitin. At this stage, NFT, recognizable with silver stains, anti-tau or anti-ubiquitin, were present only in isolated neurones in the cerebral cortex, but were numerous in certain parts of the hippocampus (e.g. CAI, subiculum and entorhinal cortex) and amygdala. After the age of 50 years a pathological picture indistinguishable from AD was seen in all the DS cases. Loss of neurones from areas rich in NFT occurred after 50 years of age (Mann *et al.* 1990b) and in most cases there was progressive deterioration in behaviour and personality (Mann *et al.* 1990b), signs consistent with the onset and progression of dementia. The neurofibrillary material from dead neurones seems to be 'liberated' into the extracellular space where it 'loses' its tau and ubiquitin immunoreactivity (Mann *et al.* 1989b). This change may be due to the action of extracellular proteases which 'strip off' epitopes from the liberated tangle or it may reflect conformational changes in the PHF which render these epitopes still present but no longer accessible to the antibodies. The extracellular tangle becomes infiltrated by astroglial processes which 'break-up' the

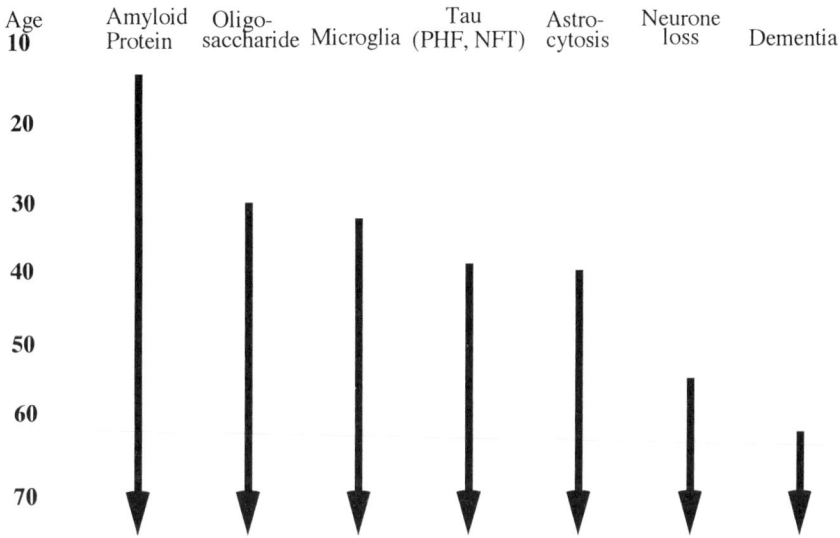

Figure 5.1. Time-course of pathological events in persons with Down syndrome of differing ages

bundles of filaments and make the NFT appear to be GFAP immunoreactive. Although some workers (e.g. Hyman *et al.* 1989; Spillantini *et al.* 1990; Tabaton *et al.* 1991) have claimed that a small proportion of NFT in AD, both extracellular and intracellular, are immunoreactive with antibodies against β/A4, we have not observed this to be so in persons with DS.

It seems, therefore, that using this particular range of molecular markers, the onset and progression of the pathological cascade of AD in persons with DS is triggered by events that result in the deposition of β/A4 protein (Figure 5.1). Other studies on younger persons with DS (Allsop *et al.* 1989; Giaccone *et al.* 1989; Ikeda *et al.* 1989b; Rumble *et al.* 1989; Murphy *et al.* 1990) also point towards this conclusion and the work of Snow *et al.* (1990) accords with our observations of an early accumulation of glycoconjugates.

Aetiological factors

Although studies of persons with DS dying at various ages clearly yield important clues as to the pathogenetic mechanism underlying AD, it is less certain to what extent DS might contribute towards understanding the aetiology of AD. As mentioned earlier, β/A4 is produced from APP via an alternative breakdown mechanism (Esch *et al.* 1990; Sisodia *et al.* 1990). The gene coding for APP is located on the long arm of chromosome 21, and although not within the obligate region for DS, hence will, in cases of full trisomy at least, be present in triplicate. It has been reported (Rumble *et al.* 1989) that APP is indeed overexpressed in DS and possible imbalances in the handling of an overproduction of APP on one hand and a relative insufficiency of normal processing fac-

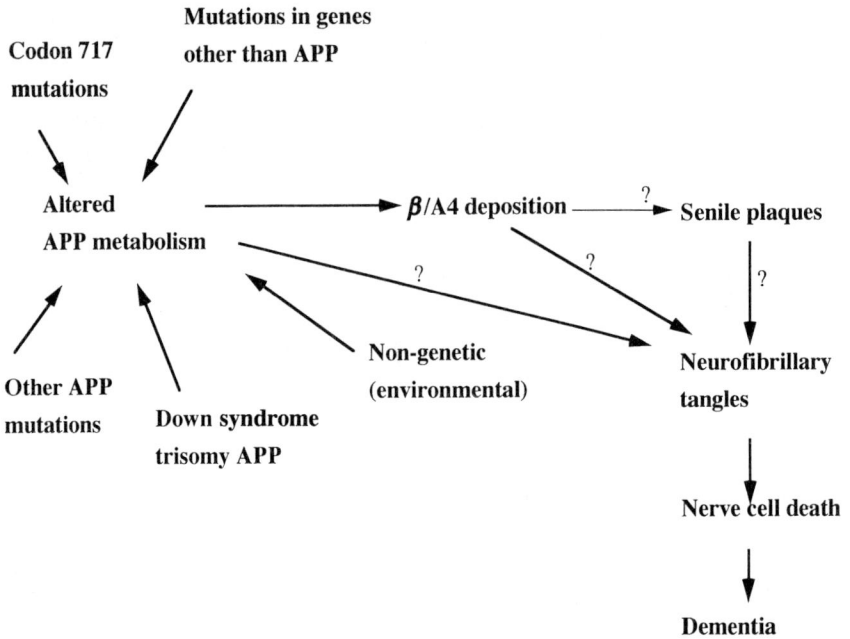

Figure 5.2. Possible pathogenic pathways for the causes and progression of the pathological changes of Alzheimer disease

tors on the other could lead to a progressive β/A4 deposition through alternative catabolic mechanisms (Figure 5.2). If this is so, why β/A4 deposition should not commence before the age of 13 years (see Rumble *et al.* 1989) is not clear.

Recently, it has been demonstrated (Richards *et al.* 1991) that implantation of neural tissue (hippocampus) from the trisomy 16 mouse DS homologue into normal young recipient mice of the same strain leads to the development, within the grafted tissue, of nerve cells that in contrast to those in normal surrounding tissues, express immunoreactivity to β/A4 protein, APP and tau proteins. The APP gene in the mouse is carried on chromosome 16 (Coyle *et al.* 1988) and is presumably overexpressed within the surviving graft tissue within the host animal. These data accord with the chronological progression of the lesions of AD as seen in persons with DS, in whom the full-length APP is known to be overexpressed (Rumble *et al.* 1989), and strengthen the view that an abnormal processing of APP is the cornerstone on which the remainder of the pathological cascade rests (Figure 5.2).

To what extent a triplicated APP gene is mandatory for the development of Alzheimer-type pathology in full trisomy 21 DS is not known, though this could be 'tested' in persons with the DS phenotype due to partial trisomy. Localization of the APP and the FAD genes (see later) on chromosome 21 are proximal, with the FAD gene nearer the centromere (Goldgaber *et al.* 1987;

Kang *et al.* 1987; Robakis *et al.* 1987; Tanzi *et al.* 1987); both loci are distinct from the distal part of chromosome 21 associated with the DS phenotype (Jenkins *et al.* 1983). In some persons with partial trisomy 21 the DS phenotype might have been inherited independently of other genes dictating SP and NFT formation (see previously referred to case of Whalley 1982), whereas in others the regions mandatory for SP and NFT formation might also be inherited [see case 34/68 of Sylvester (1986) which displayed severe 'ageing']. Establishing the precise point of non-disjunction could help to determine which genes may be in excess and thereby pinpoint more accurately which loci are obligatory for development of AD in persons with DS.

In AD, it is now known that the APP gene does not exist in triplicate nor is APP overexpressed in this way. Some cases of AD can be inherited at a relatively early age in an apparently autosomal dominant fashion. While in most such families both the APP gene and its products appear normal, it has recently been discovered in a number of European (Chartier-Harlin *et al.* 1991; Goate *et al.* 1991), American (Murrell *et al.* 1991) and Japanese (Naruse *et al.* 1991) families that there are point mutations at codon 717 of the APP gene that lead to three different single amino acid substitutions in the APP molecule. These are mostly associated with typical pathology of AD (Murrell *et al.* 1991; Naruse *et al.* 1991; Mann *et al.* 1992a), although in another family (Lantos *et al.* 1992) some variations (e.g. presence of Lewy bodies in brain stem neurones) were observed. The implications of these studies is that mutations within codon 717 of the APP molecule are also pathogenic (Figure 5.2) and that, in some way perhaps again involving abnormal processing of APP, these lead to the pathological outcome of AD. Again, as in DS, it is not clear how such a mutation might dictate the time of onset (usually 40–50 years of age) in these cases. In other families not showing changes in the APP gene, an alternate locus on chromosome 21, the so-called familial Alzheimer disease (FAD) gene must presumably be responsible for driving the pathological process perhaps again via an abnormal APP metabolism (Figure 5.2). Finally, there may be cases of AD that relate to mutations in genes on chromosomes other than 21 that in some way modulate the disease, particularly in late onset instances (Pericak-Vance *et al.* 1988; Schellenberg *et al.* 1988). Non-genetic (environmental) factors could also drive the pathological process directly or interplay with genetic alterations to bring about the same end-result. This might be externally triggered (antecedent head injury, viral infections, toxic effects of inorganic or organic compounds have all been suggested), although it has recently been proposed (Johnson *et al.* 1990) that alterations in the ratio between the non-Kunitz (APP_{695}) and the Kunitz (APP_{751}) mRNA isoforms leading to a switch in the molecular balance in favour of the protease inhibitor-containing variant (i.e. APP_{751}) might contribute to β/A4 deposition. Such a change in ratio might be age-related (Konig *et al.* 1989). Recent observations (Quon *et al.* 1991) that experimentally induced imbalances in APP_{751}/APP_{695} expression in transgenic mice yield brain deposits of β/A4, would accord with such a proposal.

However, it is far from clear whether β/A4 protein is itself neurotoxic (and thereby responsible *per se* for the generation of all subsequent pathological events) or whether it simply represents a relatively innocuous marker of a wider ranging process that carries in its wake other changes, some of which might lead to neurofibrillary alterations and cell death. *In vitro* studies using synthetic β/A4 peptides suggest that β/A4 may be neurotoxic (Yankner *et al.* 1989) or neurotrophic (Whitson *et al.* 1989); it is not known whether 'naturally' produced β/A4 holds these same properties '*in vivo*', although the injection of isolated plaque cores into rat brain causing neuronal loss and expression of Alzheimer-related antigens (Frautschy *et al.* 1991) suggests that this could be so.

However, in humans, deposition of β/A4 can frequently occur in the absence of neuritic changes and NFT formation, particularly in areas such as striatum (Suenaga *et al.* 1990) and cerebellum (Joachim *et al.* 1989; Mann *et al.* 1990a) in both AD and DS, and even within the cerebral cortex and hippocampus in conditions other than AD and DS (Mann and Jones 1990; Mann *et al.* 1992b). Hence, PHF formation may proceed in parallel to, but not necessarily as a direct consequence of, β/A4 formation. It is possible that a different fragment of the APP molecule, released during its cleavage, is responsible for or modulates PHF formation. Whatever the factor that triggers PHF formation is, it seems clear that it is an entity specific for AD and, perhaps more importantly, one that is specifically produced by the cerebral cortex in AD; if it is produced in other regions (e.g. cerebellum), it does not affect local neurones in the same way. The fact that the APP molecule is glycosylated (Kang *et al.* 1987) and observations that unusual and excessive accumulations of glycoproteins and proteoglycans occur within the cerebral cortical (Szumanska *et al.* 1987; Mann *et al.* 1988b, 1989a; Snow *et al.* 1990), but not the cerebellar cortical (Mann *et al.* 1990a; Snow *et al.* 1990) amyloid deposits in both AD and DS, suggest that oligosaccharides, possibly derived from APP, might mediate the neuritic changes.

Conclusion

At present, it seems that although DS and AD appear to be largely homogeneous in pathological terms, the mechanisms driving the destructive process(es) in each of these conditions are diverse. The overexpression of APP that occurs in DS is seemingly only one of many potential ways whereby the common pathological end-point that we call 'Alzheimer disease' might be brought about. Like the 'prion diseases' of humans and animals, in which a common end-pathology (spongiform encephalopathy) is driven by diverse genetic and non-genetic aetiologies, DS may simply represent one variant within the wider family of 'Alzheimer diseases' or, as others have suggested, the β-amyloidopathies (Hardy *et al.* 1991).

References

Allsop, D., Kidd, M., Landon, M., and Tomlinson, A. (1986). Isolated senile plaque cores in Alzheimer's disease and Down's syndrome show differences in morphology. *Journal of Neurology, Neurosurgery and Psychiatry*, **49**, 886-92.

Allsop, D., Haga, S. -I., Haga, C., Ikeda, S. -I., Mann, D. M. A., and Ishii, T. (1989). Early senile plaques in Down's syndrome brains show a clear relationship with cell bodies of neurones. *Neuropathology and Applied Neurobiology*, 15, 531-42.

Anderton, B. H., Breinburg, D., Downes, M. J., Green, P. J., Tomlinson, B. E., Ulrich, J., *et al.* (1982). Monoclonal antibodies show that neurofibrillary tangles and neurofilaments share antigenic determinants. *Nature*, 298, 87-96.

Ball, M. J., and Nuttall, K. (1980). Neurofibrillary tangles and granulovacuolar degeneration and neuron loss in Down syndrome: quantitative comparison with Alzheimer dementia. *Annals of Neurology*, 7, 462-5.

Becker, L. A., Armstrong, D. L., and Chan, F. (1986). Dendritic atrophy in children with Down's syndrome. *Annals of Neurology*, 20, 520-6.

Belza, M. G., and Urich, H., (1986). Cerebral amyloid angiopathy in Down's syndrome. *Clinical Neuropathology*, 6, 257-60.

Bertrand, I., and Koffas, D. (1946). Cas d' idiotie mongolienne adult avec nombreuses plaques séniles et concretions calcaires pallidales. *Revue Neurologie (Paris)*, 78, 338-45.

Blumbergs, P., Beran, R., and Hicks, P. (1981). Myoclonus in Down's syndrome. Association with Alzheimer's disease. *Archives of Neurology*, 38, 453-4.

Burger, P. C., and Vogel, F. S. (1973). The development of the pathologic changes of Alzheimer's disease and senile dementia in patients with Down's syndrome. *American Journal of Pathology*, 73, 457-76.

Casanova, M. F., Walker, L. C., Whitehouse, P. J., and Price, D. L. (1985). Abnormalities of the nucleus basalis in Down's syndrome. *Annals of Neurology*, 18, 310-3.

Chartier-Harlin, M. -C., Crawford, F., Houlden, H., Warren, A., Hughes, D., Fidani, L., *et al.* (1991). Early-onset Alzheimer's disease caused by mutations at codon 717 of the β amyloid precursor protein gene. *Nature*, 353, 844-6.

Coyle, J. T., Oster-Granite, M. L., Reeves, R. H., and Gearhart, J. D. (1988). Down syndrome, Alzheimer disease and the trisomy 16 mouse. *Trends in Neurosciences*, 11, 390-1.

Crapper, D. R., Dalton, A. J., Skopitz, M., Scott, J. W., and Hachinski, V. C. (1975). Alzheimer degeneration in Down's syndrome: electrophysiologic alterations and histopathologic findings. *Archives of Neurology*, 32, 618-23.

Delacourte, A., and Defossez, A. (1986). Alzheimer's disease tau proteins, the promoting factors of microtubule assembly are major components of paired helical filaments. *Journal of the Neurological Sciences*, 76, 173-86.

De La Monte, S. M. (1989). Quantitation of cerebral atrophy in preclinical and end-stage Alzheimer's disease. *Annals of Neurology*, 25, 450-9.

De La Monte, S. M., and Hedley-White, E. T. (1990). Small cerebral hemispheres in adults with Down's syndrome. Contributions of developmental arrest and lesions of Alzheimer's disease. *Journal of Neuropathology and Experimental Neurology*, 49, 509-20.

Edwardson, J. A., Klinowski, J., Oakley, A. E., Perry, R. H., and Candy, J. M. (1986). Aluminosilicates and the aging brain: implications for the pathogenesis of Alzheimer's disease. In *Silicon biochemistry*, (ed. D.Evered and M. O'Connor), Ciba Foundation Symposium No. 121, pp. 160-79. Wiley, Chichester, UK.

Ellis, W. G., McCulloch, J. R., and Corley, C. L. (1974). Presenile dementia in Down's syndrome. Ultrastructural identity with Alzheimer's disease. *Neurology*, 24, 101-6.

Esch, E. S., Keim, P., Beattie, E. C., Blacher, R. W., Culwell, A. R., Oltersdorf, T., *et al.* (1990). Cleavage of amyloid β peptide during constitutive processing of its precursor. *Science*, 248, 1122-4.

Ferrer, I., and Gullotta, F. (1990). Down's syndrome and Alzheimer's disease: dendritic spine counts in the hippocampus. *Acta Neuropathologica*, 79, 680-5.

Flament, S., Delacourte, A., and Mann, D. M. A. (1990). Phosphorylation of tau proteins: a major event during the process of neurofibrillary degeneration. A comparative study between Alzheimer's disease and Down's syndrome. *Brain Research*, 516, 15-9.

Fraser, J., and Mitchell, A. (1876). Kalmuc idiocy: report of a case with autopsy with notes on 62 cases. *Journal of Mental Science*, 22, 161-9.

Frautschy, S. A., Baird, A., and Cole, G. M. (1991). Effects of injected Alzheimer β amyloid cores in rat brain. *Proceedings of the National Academy of Sciences of the United States of America,* **88**, 8362-6.

Gandolfi, A., Horoupian, D. S., and DeTeresa, R. M. (1981). Pathology of the auditory system in trisomies with morphometric and quantitative study of the ventral cochlear nucleus. *Journal of the Neurological Sciences,* **51**, 43-50.

German, D.C., Manaye, K. F., White, C. L., Woodward, D. J., McIntire, D. D., Smith, W. K., *et al.* (1992). Disease specific patterns of locus caeruleus cell loss: Parkinson's disease, Alzheimer's disease and Down's syndrome. *Annals of Neurology,* **32**, 667–76.

Giaccone, G., Tagliavini, F., Linoli, G., Bouras, C., Frigero, L., Frangione, B., *et al.* (1989). Down patients : extracellular preamyloid deposits precede neuritic degeneration and senile plaques. *Neuroscience Letters,* **97**, 232-8.

Gibb, W. R. G., Mountjoy, C. Q., Mann, D. M. A., and Lees, A. J. (1989). The substantia nigra and ventral tegmental area in Alzheimer's disease and Down's syndrome. *Journal of Neurology, Neurosurgery and Psychiatry,* **52**, 193-200.

Glenner, G. G., and Wong, C. W. (1984a). Alzheimer's disease: initial report on the purification and characterisation of a novel cerebrovascular amyloid protein. *Biochemical and Biophysical Research Communications,* **120**, 885-90.

Glenner, G.G., and Wong, C.W. (1984b). Alzheimer's disease and Down's syndrome: sharing of a unique cerebrovascular amyloid fibril protein. *Biochemical and Biophysical Research Communications,* **122**, 1131-5.

Goate, A., Chartier-Harlin, M. -C., Mullan, M., Brown, J., Crawford, F., Fidani, L., *et al.* (1991). Segregation of a missense mutation in the amyloid precursor protein gene with familial Alzheimer's disease. *Nature,* **349**, 704-6.

Godridge, H., Reynolds, G. P., Czudek, C., Calcutt, N. A., and Benton, M. (1987). Alzheimer-like neurotransmitter deficits in adult Down's syndrome brain tissue. *Journal of Neurology, Neurosurgery and Psychiatry,* **50**, 775-8.

Goedert, M., Wischik, C., Crowther, R. A., Walker, J. E., and Klug, A. (1988). Cloning and sequencing of the cDNA encoding a core protein of the paired helical filament of Alzheimer's disease: identification as the microtubule associated protein, tau. *Proceedings of the National Academy of Sciences of the United States of America,* **85**, 4051-5.

Goldgaber, D., Lerman, M. I., MacBride, O. W., Saffioti, U., and Gajdusek, D. C. (1987). Characterization and chromosomal localization of a cDNA encoding brain amyloid of Alzheimer's disease. *Science,* **235**, 877-80.

Haberland, C. (1969). Alzheimer's disease in Down's syndrome: clinical and neuropathological observations. *Acta Neurologica Belgica,* **69**, 369-80.

Hanger, D. P., Brion, J. -P., Gallo, J. -M., Cairns, N. J., Luthert, P. J., and Anderton, B. H. (1991). Tau in Alzheimer's disease and Down's syndrome is insoluble and abnormally phosphorylated. *Biochemical Journal,* **275**, 99-104.

Hardy, J. A., Mullan, M., Chartier-Harlin, M. -C., Brown, J., Goate, A., Rossor, M., *et al.* (1991). Molecular classification of Alzheimer's disease. *Lancet,* **1**, 1342-3.

Hyman, B. T., Van Hoesen, G. W., Beyreuther, K., and Masters, C. L. (1989). A4 amyloid protein immunoreactivity is present in Alzheimer's disease neurofibrillary tangles. *Neuroscience Letters,* **101**, 352-5.

Ihara, Y., Nukina, N., Miura, R., and Ogawara, M. (1986). Phosphorylated tau protein is integrated into paired helical filaments in Alzheimer's disease. *Journal of Biochemistry (Japan),* **99**, 1807-10.

Ikeda, S. -I., Allsop, D., and Glenner, G. G. (1989a). The morphology and distribution of plaque and related deposits in the brains of Alzheimer's disease and control cases: an immunohisto-chemical study using amyloid β protein antibody. *Laboratory Investigation,* **60**, 113-22.

Ikeda, S. -I., Yanagisawa, N., Allsop, D., and Glenner, G. G. (1989b). Evidence of amyloid β protein immunoreactive early plaque lesions in Down's syndrome brains. *Laboratory Investigation,* **61**, 133-7.

Jenkins, E. C., Duncan, C. J., and Wright, C. E. (1983). Atypical Down's syndrome and partial trisomy 21. *Clinical Genetics,* **24**, 97-102.

Jervis, G. A. (1948). Early senile dementia in mongoloid idiocy. *American Journal of Psychiatry,* **105**, 102-6.

Joachim, C. L., Duffy, L. K., Morris, J. H., and Selkoe, D. J. (1988). Protein chemical and immuno-cytochemical studies of meningovascular β amyloid protein in Alzheimer's disease and normal ageing. *Brain Research,* **474**, 100-11.

Joachim, C. L., Morris, J. H., and Selkoe, D. J. (1989). Diffuse amyloid plaques occur commonly in the cerebellum in Alzheimer's disease. *American Journal of Pathology,* **135**, 309-19.

Johnson, S. A., Neill, T., Cordell, B., and Finch, C. E. (1990). Relation of neuronal APP-751/APP-695 mRNA ratio and neuritic plaque density in Alzheimer's disease. *Science,* **248**, 854-7.

Kang, J., Lemaire, H. -G., Unterbeck, A., Salbaum, J. M., Masters, C. L., Grzeschik, K. -H., *et al.* (1987). The precursor of Alzheimer's disease amyloid A4 protein resembles a cell surface receptor. *Nature,* **325**, 733-6.

Kobayashi, K., Emson, P. C., Mountjoy, C. Q., Thornton, S. N., Lawson, D. E. M., and Mann, D. M. A. (1990). Cerebral cortical calbindin D_{28k} and parvalbumin neurones in Down's syndrome. *Neuroscience Letters,* **113**, 17-22.

Kondo, J., Honda, T., Mori, H., Hamada, Y., Miura, R., Ogawara, M., *et al.* (1988). The carboxyl third of tau is tightly bound to paired helical filaments. *Neuron,* **1**, 827-34.

Konig, G., Beyreuther, K., Masters, C. L., Schmitt, H., and Salbaum, J. M. (1989). Pre-A4 RNA distribution in brain areas. *Progress in Clinical and Biological Research,* **317**, 1027-36.

Kosik, K. S., Joachim, C. L., and Selkoe, D. J. (1986). Microtubule associated protein tau is a major antigenic component of paired helical filaments in Alzheimer's disease. *Proceedings of the National Academy of Sciences of the United States of America,* **83**, 4044-8.

Kosik, K. S., Orecchio, L. D., Binder, L. I., Trojanowski, J. Q., Lee, V. M. -Y., and Lee, G. (1988). Epitopes that span the tau molecule are shared with paired helical filaments. *Neuron,* **1**, 817-25.

Lantos, P., Luthert, P. J., Hanger, D., Anderton, B. H., Mullan, M., and Rossor, M. (1992). Familial Alzheimer's disease with the amyloid precursor protein position 717 mutation and sporadic Alzheimer's disease have the same cytoskeletal pathology. *Neuroscience Letters,* **137**, 220-4.

Lennox, G., Lowe, J. S., Morrell, K., Landon, M., and Meyer, R. J. (1988). Ubiquitin is a component of a variety of neurodegenerative disorders. *Neuroscience Letters*, **94**, 211-7.

Lowe, J., Blanchard, A., Morrell, K., Lennox, G., Reynolds, L., Billett, M., *et al.* (1988). Ubiquitin is a common factor in intermediate filament inclusion bodies of diverse type in man including those of Parkinson's disease, Pick's disease and Alzheimer's disease, as well as Rosenthal fibres in cerebellar astrocytomas, cytoplasmic bodies in muscle and Mallory bodies in alcoholic liver disease. *Journal of Pathology,* **155**, 9-15.

McGeer, E. G., Norman, M., Boyes, B., O'Kosky, J., Suzuki, J., and McGeer, P. L. (1985). Acetylcholine and aromatic amine systems in post mortem brain of an infant with Down's syndrome. *Experimental Neurology,* **87**, 557-60.

Malamud, N. (1972). Neuropathology of organic brain syndromes associated with ageing. In *Ageing and the brain: advances in behavioral biology,* (ed. C. M. Gaitz), Vol.3, pp. 63-87. Plenum Press, New York.

Mann, D. M. A. (1985). The neuropathology of Alzheimer's disease: a review with pathogenetic, aetiological and therapeutic considerations. *Mechanisms of Ageing and Development,* **31**, 213-55.

Mann, D. M. A. (1988a). Neuropathological and neurochemical aspects of Alzheimer's disease. In *Handbook of psychopharmacology,* (ed. L. L. Iversen, S. D. Iversen and S. Snyder), Vol. 20, pp. 1-67. Plenum Press, New York.

Mann, D. M. A. (1988b). The pathological association between Down's syndrome and Alzheimer's disease. *Mechanisms of Ageing and Development,* **43**, 99-136.

Mann, D. M. A. (1988c). Calcification of the basal ganglia in Down's syndrome and Alzheimer's disease. *Acta Neuropathologica,* **76**, 595-8.

Mann, D. M. A. (1991). The topographic distribution of brain atrophy in Alzheimer's disease. *Acta Neuropathologica,* **83**, 81-6.

Mann, D. M. A., and Esiri, M. M. (1989). The pattern of acquisition of plaques and tangles in the brains of patients under 50 years of age with Down's syndrome. *Journal of the Neurological Sciences,* 89, 169-79.

Mann, D. M. A., and Jones, D. (1990). Amyloid (A4) protein deposition in the brains of persons with dementing disorders other than Alzheimer's disease and Down's syndrome. *Neuroscience Letters,* 109, 68-75.

Mann, D. M. A., and Yates, P. O. (1986). Neurotransmitter deficits in Alzheimer's disease and in other dementing disorders. *Human Neurobiology,* 5, 147-58.

Mann, D. M. A., Yates, P. O., and Marcyniuk, B. (1984). Alzheimer's presenile dementia, senile dementia of Alzheimer type and Down's syndrome in middle age form an age-related continuum of pathological changes. *Neuropathology and Applied Neurobiology,* 10, 185-207.

Mann, D. M. A., Yates, P. O., and Marcyniuk, B. (1985a). Some morphometric observations on the cerebral cortex and hippocampus in presenile Alzheimer's disease, senile dementia of Alzheimer type and Down's syndrome in middle age. *Journal of the Neurological Sciences,* 69, 139-59.

Mann, D. M. A., Yates, P. O., Marcyniuk, B., and Ravindra, C. R. (1985b). Pathological evidence for neurotransmitter deficits in Down's syndrome of middle age. *Journal of Mental Deficiency Research,* 29, 125-35.

Mann, D. M. A., Yates, P. O., Marcyniuk, B., and Ravindra, C. R. (1986). The topography of plaques and tangles in Down's syndrome patients of different ages. *Neuropathology and Applied Neurobiology,* 12, 447-57.

Mann, D. M. A., Yates, P. O., Marcyniuk, B., and Ravindra, C. R. (1987a). Loss of nerve cells from cortical and subcortical areas in Down's syndrome patients at middle age: quantitative comparisons with younger Down's patients and patients with Alzheimer's disease. *Journal of the Neurological Sciences,* 80, 79-89.

Mann, D. M. A., Yates, P. O., and Marcyniuk, B. (1987b). Dopaminergic neurotransmitter systems with Alzheimer's disease and Down's syndrome at middle age. *Journal of Neurology, Neurosurgery and Psychiatry,* 50, 341-4.

Mann, D. M. A., Tucker, C. M., and Yates, P. O. (1988a). Alzheimer's disease: an olfactory connection? *Mechanisms of Ageing and Development,* 42, 1-15.

Mann, D. M. A., Bonshek, R. E., Marcyniuk, B., Stoddart, R. W., and Torgerson, E. (1988b). Saccharides of senile plaques and neurofibrillary tangles in Alzheimer's disease. *Neuroscience Letters,* 85, 277-82.

Mann, D. M. A., Brown, A. M. T., Prinja, D., Davies, C. A., Beyreuther, K., Masters, C. L., *et al.* (1989a). An analysis of the morphology of senile plaques in Down's syndrome patients of different ages using immunocytochemical and lectin histochemical methods. *Neuropathology and Applied Neurobiology,* 15, 317-29.

Mann, D. M. A., Prinja, D., Davies, C. A., Ihara, Y., Delacourte, A., Defossez, A., *et al.* (1989b). Immunocytochemical profile of neurofibrillary tangles in Down's syndrome patients of different ages. *Journal of the Neurological Sciences,* 92, 247-60.

Mann, D. M. A., Jones, D., Prinja, D., and Purkiss, M. S. (1990a). The prevalence of amyloid (A4) protein deposits within the cerebral and cerebellar cortex in Alzheimer's disease and Down's syndrome. *Acta Neuropathologica,* 80, 318-27.

Mann, D. M. A., Royston, M. C., and Ravindra, C. R. (1990b). Some morphometric observations on the brains of patients with Down's syndrome: their relationship to age and dementia. *Journal of the Neurological Sciences,* 99, 153-64.

Mann, D. M. A., Jones, D., Snowden, J. S., Neary, D., and Hardy, J. (1992a). Pathological changes in the brain of a patient with familial Alzheimer's disease having a missense mutation at codon 717 in the amyloid precursor protein gene. *Neuroscience Letters,* 137, 225-8.

Mann, D. M. A., Jones, D., South, P. W., Snowden, J. S., and Neary, D. (1992b). Deposition of amyloid β protein in non-Alzheimer dementias; evidence for a neuronal origin of parenchymal deposits of β protein in neurodegenerative disease. *Acta Neuropathologica,* 83, 415-9.

Mann, D. M. A., Younis, N., Jones, D., and Stoddart, R. W. (1992c). The time course of pathological events concerned with plaque formation in Down's syndrome with particular reference to the involvement of microglial cells. *Neurodegeneration,* 1, 201–15.

Marcyniuk, B., Mann, D. M. A., and Yates, P. O. (1988). Topography of nerve cell loss from the locus caeruleus in middle aged persons with Down's syndrome. *Journal of the Neurological Sciences*, 83, 15-24.

Masters, C. L., Simms, G., Weinmann, N. A., Multhaup, G., McDonald, B. L., and Beyreuther, K. (1985). Amyloid plaque core protein in Alzheimer's disease and Down's syndrome. *Proceedings of the National Academy of Sciences of the United States of America*, 82, 4245-9.

Melamed, E., Mildworf, B., Sharav, T., Benlenky, L., and Wertman, E. (1987). Regional cerebral blood flow in Down's syndrome. *Annals of Neurology*, 22, 275-8.

Mori, H., Kondo, J., and Ihara, Y. (1987). Ubiquitin is a component of paired helical filaments in Alzheimer's disease. *Science*, 235, 1641-4.

Motte, J., and Williams, R. S. (1989). Age-related changes in the density and morphology of plaques and neurofibrillary tangles in Down's syndrome brains. *Acta Neuropathologica*, 77, 535-46.

Murdoch, J. C., and Adams, H. (1977). Reply to W. Hughes. *British Medical Journal*, 2, 702.

Murdoch, J. C., Rodger, J. C., Rao, S. S., Fletcher, C. D., and Donnigan, M. G. (1977). Down's syndrome: an atheroma-free model. *British Medical Journal*, 2, 226-8.

Murphy, G. M., Eng, L. F., Ellis, W. G., Perry, G., Meissner, L. C., and Tinklenberg, J. R. (1990). Antigenic profile of plaques and neurofibrillary tangles in the amygdala in Down's syndrome: a comparison with Alzheimer's disease. *Brain Research*, 537, 102-8.

Murrell, J., Farlow, M., Ghetti, B., and Benson, M. D. (1991). A mutation in the amyloid precursor protein associated with hereditary Alzheimer's disease. *Science*, 254, 97-9.

Naruse, S., Igarashi, S., Aoki, K., Kaneko, K., Iihara, K., Kubayashi, H., *et al.* (1991). Mis-sense mutation Val→Ile in exon 17 of amyloid precursor protein gene in Japanese familial Alzheimer's disease. *Lancet*, 2, 978-9.

Neumann, N. A. (1967). Langdon Down syndrome and Alzheimer's disease. *Journal of Neuropathology and Experimental Neurology*, 26, 149-50.

Ogomori, K., Kitamoto, T., Tateishi, J., Sato, Y., Suetsugu, M., and Abe, M. (1989). β protein amyloid is widely distributed in the central nervous system of patients with Alzheimer's disease. *American Journal of Pathology*, 134, 243-51.

O'Hara, P. T. (1972). Electronmicroscopical study of the brain in Down's syndrome. *Brain*, 95, 681-4.

Olson, M. I., and Shaw, C. M. (1969). Presenile dementia and Alzheimer's disease in mongolism. *Brain*, 92, 147-56.

Pericak-Vance, M. A., Yamaoka, L. H., Haynes, C. S., Haines, J., Gaskell, P. C., Hung, W. Y., *et al.* (1988). Genetic linkage studies in Alzheimer's disease families. *Experimental Neurology*, 102, 278-9.

Perry, G., Friedman, R., Shaw, G., and Chau, V. (1987). Ubiquitin is detected in neurofibrillary tangles and senile plaque neurites of Alzheimer's disease brains. *Proceedings of the National Academy of Sciences of the United States of America*, 84, 3033-6.

Pogacar, S., and Rubio, A. (1982). Morphological features of Pick's and atypical Alzheimer's disease in Down's syndrome. *Acta Neuropathologica*, 58, 249-54.

Price, D. L., Whitehouse, P. J., Struble, R. G., Coyle, J. T., Clark, A. W., DeLong, M. R., *et al.* (1982). Alzheimer's disease and Down's syndrome. *Annals of the New York Academy of Sciences*, 396, 145-64.

Quon, D., Wang, Y., Catalano, R., Scardina, J. M., Murakami, K., and Cordell, B. (1991). Formation of β amyloid protein in brains of transgenic mice. *Nature*, 352, 239-41.

Rees, S. (1977). The incidence of ultrastructural abnormalities in the cortex of two retarded human brains (Down's syndrome). *Acta Neuropathologica*, 37, 65-8.

Reid, A. H., and Maloney, A. F. J. (1974). Giant cell arteritis and arteriolitis associated with amyloid angiopathy in an elderly mongol. *Acta Neuropathologica*, 27, 131-7.

Reynolds, G. P., and Godridge, H. (1985). Alzheimer-like monoamine deficits in adults with Down's syndrome. *Lancet*, 2, 1368-9.

Reynolds, G. P., and Warner, C. E. J. (1988). Amino acid transmitter deficits in adult Down's syndrome brain tissue. *Neuroscience Letters*, 94, 224-7.

Richards, S. -J., Waters, J. J., Beyreuther, K., Masters, C. L., Wischik, C. M., Sparkman, D. R., *et al.* (1991). Transplants of mouse trisomy 16 hippocampus provide a model of Alzheimer's disease neuropathology. *The European Molecular Biology Organization Journal*, 10, 297-303.

Risberg, J. (1980). Regional cerebral blood flow measurements by 133 Xe inhalation: methodology and application in neuropathology and psychiatry. *Brain and Language*, 9, 9-34.

Robakis, N. K., Wisniewski, H. M., Jenkins, E. C., Devine-Gage, E. A., Houck, G. E., Yao, X. -L., *et al.* (1987). Chromosome 21q21 sublocalization of gene encoding beta-amyloid peptide in cerebral vessels and neuritic (senile) plaques of people with Alzheimer's disease and Down's syndrome. *Lancet*, 1, 384-5.

Ropper, A. H., and Williams, R. S. (1980). Relationship between plaques and tangles and dementia in Down's syndrome. *Neurology*, 30, 639-44.

Ross, M. H., Galaburda, A. M., and Kemper, T. L. (1984). Down's syndrome: is there a decreased population of neurones? *Neurology*, 34, 909-16.

Rumble, B., Retallack, R., Hilbich, C., Simms, G., Multhaup, G., Martins, R., *et al.* (1989). Amyloid A4 protein and its precursor in Down's syndrome and Alzheimer's disease. *New England Journal of Medicine*, 320, 1446-52.

Schapiro, M. B., Haxby, J. V., Grady, C. L., Duara, R., Schlageter, N. L., White, B., *et al.* (1987). Decline in cerebral glucose utilization and cognitive function with ageing in Down's syndrome. *Journal of Neurology, Neurosurgery and Psychiatry*, 50, 766-74.

Schapiro, M. B., Berman, K. F., Friedland, R. P., Weinberger, D. R., and Rapoport, S. I. (1988). Regional blood flow is not reduced in young adult Down's syndrome. *Annals of Neurology*, 24, 310.

Schapiro, M. B., Luxenberg, J. S., Kaye, J. A., Haxby, J. V., Friedland, R. P., and Rapoport, S. I. (1989). Serial quantitative CT analysis of brain morphometrics in adult Down's syndrome at different ages. *Neurology*, 39, 1349-53.

Schapiro, M. B., Grady, C. L., Kumar, A., Herscovitch, P., Haxby, J. V., Moore, A., *et al.* (1990). Regional cerebral glucose metabolism is normal in young adults with Down's syndrome. *Journal of Cerebral Blood Flow and Metabolism*, 10, 199-206.

Schellenberg, G. D., Bird, T. D., Wijsman, E. M., Moore, D. K., Boehnke, M., Bryant, E. M., *et al.* (1988). Absence of linkage of chromosome 21q21 markers to familial Alzheimer's disease. *Science*, 241, 1507-10.

Schochet, S. S., Lampert, P. W., and McCormick, W. F. (1973). Neurofibrillary tangles in patients with Down's syndrome: A light and electron microscope study. *Acta Neuropathologica*, 23, 342-6.

Simpson, M. D., Slater, P., Cross, A. J., Mann, D. M. A., Royston, M. C., Deakin, J. F. W., *et al.* (1989). Reduced D [^{3}H] aspartate binding in Down's syndrome brains. *Brain Research*, 483, 273-8.

Sisodia, S. S., Koo, E. H., Beyreuther, K., Unterbeck, A., and Price, D. L. (1990). Evidence that β amyloid protein in Alzheimer's disease is not derived by normal processing. *Science*, 248, 492-5.

Snow, A. D., Mar, H., Nochlin, D., Sekiguchi, R. T., Kimata, K., Koike, Y., *et al.* (1990). Early accumulation of heparan sulphate in neurones and in the beta-amyloid protein containing lesions of Alzheimer's disease and Down's syndrome. *American Journal of Pathology*, 137, 1253-70.

Solitaire, G. B., and Lamarche, J. B. (1966). Alzheimer's disease and senile dementia as seen in mongoloids: neuropathological observations. *American Journal of Mental Deficiency*, 70, 840-8.

Spargo, E., Luthert, P. J., Anderton, B. H., Bruce, M., Smith, D., and Lantos, P. L. (1990). Antibodies raised against different portions of A4 protein identify a subset of plaques in Down's syndrome. *Neuroscience Letters*, 115, 345-50.

Sparkman, D. R., Hill, C. J., and White, C. L. (1990). Paired helical filaments are not major binding sites for WGA and DBA agglutinins in neurofibrillary tangles of Alzheimer's disease. *Acta Neuropathologica*, 79, 640-6.

Spillantini, M. G., Goedert, M., Jakes, R., and Klug, A. (1990). Topographical relationship between β-amyloid and tau protein epitopes in tangle-bearing cells in Alzheimer disease. *Proceedings of the National Academy of Sciences of the United States of America*, 87, 3952-6.

Struwe, F. (1929). Histopathologische Untersuchungen über Enstehung und Wesen der senilen Plaques. *Zentralblatt für die gesamte Neurologie und Psychiatrie*, **122**, 291-307.

Suenaga, T., Hirano, A., Llena, J. F., Yen, S. -H., and Dickson, D. W. (1990). Modified Bielschowsky staining and immunohistochemical studies on striatal plaques in Alzheimer's disease. *Acta Neuropathologica*, **80**, 280-6.

Suetsuga, M., and Mehraein, P. (1980). Spine distribution along the apical dendrite of the pyramidal neurones in Down's syndrome. A quantitative Golgi study. *Acta Neuropathologica*, **50**, 207-10.

Sylvester, P. E. (1983). The hippocampus in Down's syndrome. *Journal of Mental Deficiency Research*, **27**, 227-36.

Sylvester, P. E. (1986). The anterior commissure in Down's syndrome. *Journal of Mental Deficiency Research,* **30**, 19-26.

Szumanska, G., Vorbrodt, A. W., Mandybur, T. I., and Wisniewski, H. M. (1987). Lectin histochemistry of plaques and tangles in Alzheimer's disease. *Acta Neuropathologica*, **73**, 1-11.

Tabaton, M., Cammarata, S., Mancardi, G., Manetto, V., Autilio-Gambetti, L., Perry, G., and Gambetti, P. (1991). Ultrastructural localization of β-amyloid, tau and ubiquitin epitopes in extracellular neurofibrillary tangles. *Proceedings of the National Academy of Sciences of the United States of America*, **88**, 2098-102.

Tagliavini, F., Giaccone, G., Frangione, B., and Bugiani, O. (1988). Preamyloid deposits in the cerebral cortex of patients with Alzheimer's disease and non-demented individuals. *Neuroscience Letters*, **93**, 191-6.

Takashima, S., and Becker, L. E. (1985). Basal ganglia calcification in Down's syndrome. *Journal of Neurology, Neurosurgery and Psychiatry*, **48**, 61-4.

Takashima, S., Becker, L. E., Armstrong, D. L., and Chan, F. W. (1981). Abnormal neuronal development in the visual cortex of the human fetus and infant with Down's syndrome. *Brain Research*, **225**, 1-21.

Tanzi, R. E., St George-Hyslop, P. H., Haines, J. H., Polinsky, R. J., Nee, L., Foncin, J. -F., *et al.* (1987). The genetic defect in familial Alzheimer's disease is not tightly linked to the amyloid precursor protein gene. *Nature*, **329**, 156-7.

Wegiel, J., and Wisniewski, H. M. (1990). The complex of microglial cells and amyloid star in 3-dimensional reconstruction. *Acta Neuropathologica*, **81**, 116-24.

Whalley, L. J. (1982). The dementia of Down's syndrome and its relevance to aetiological studies of Alzheimer's disease. *Annals of the New York Academy of Sciences*, **396**, 39-53.

Whitson, J. S., Selkoe, D. J., and Cotman, C. W. (1989). Amyloid β protein enhances the survival of hippocampal neurones in culture. *Science*, **243**, 1488-90.

Wischik, C., Novak, M., Thagersen, H. C., Edwards, P. C., Runswick, M. J., Jakes, R., *et al.* (1988). Isolation of a fragment of tau derived from the core of the paired helical filament of Alzheimer's disease. *Proceedings of the National Academy of Sciences of the United States of America*, **85**, 4506-10.

Wisniewski, H. M., and Terry, R. D. (1973). Re-examination of the pathogenesis of the senile plaque. In *Progress in neuropathology,* (ed. H.M. Zimmerman), Vol. 2, pp.1-26. Grune and Stratton, New York.

Wisniewski, H. M., Bancher, C., Barcikowska, M., Wen, G. Y., and Currie, J. (1989). Spectrum of morphological appearance of amyloid deposits in Alzheimer's disease. *Acta Neuropathologica*, **78**, 337-47.

Wisniewski, K. E., Jervis, G. A., Moretz, R. C., and Wisniewski, H. M. (1979). Alzheimer neurofibrillary tangles in diseases other than senile and presenile dementia. *Annals of Neurology*, **7**, 462-5.

Wisniewski, K. E., French, J. H., Rosen, J. F., Kozlowski, P. B., Tenner, M., and Wisniewski, H. M. (1982). Basal ganglia calcification (BGC) in Down's syndrome (DS)—another manifestation of premature ageing. *Annals of the New York Academy of Sciences*, **396**, 179-89.

Wisniewski, K. E., Laure-Kamionowska, M., and Wisniewski, H. M. (1984). Evidence of arrest of neurogenesis and synaptogenesis in brains of patients with Down's syndrome. *New England Journal of Medicine*, **311**, 1187-8.

Wisniewski, K. E., Wisniewski, H. M., and Wen, G. Y. (1985a). Occurrence of neuropathological changes and dementia of Alzheimer's disease in Down's syndrome. *Annals of Neurology*, 17, 278-82.

Wisniewski, K. E., Dalton, A. J., Crapper-McLachlan, D. R., Wen, G. Y., and Wisniewski, H. M. (1985b). Alzheimer's disease in Down's syndrome. Clinicopathologic studies. *Neurology*, 35, 957-61.

Wood, J. G., Mirra, S. S., Pollock, N. J., and Binder, L. I. (1986). Neurofibrillary tangles of Alzheimer's disease share antigenic determinants with the axonal microtubule associated protein tau. *Proceedings of the National Academy of Sciences of the United States of America*, 83, 4040-3.

Yamaguchi, H., Hirai, S., Morimatsu, M., Shoji, M., and Harigaya, Y. (1988). Diffuse type of senile plaque in the brains of Alzheimer type dementia. *Acta Neuropathologica*, 76, 541-9.

Yamaguchi, H., Hirai, S., Morimatsu, M., Shoji, M., and Nakazato, Y. (1989). Diffuse type of senile plaques in the cerebellum of Alzheimer type dementia as detected by β-protein immunostaining. *Acta Neuropathologica*, 77, 314-9.

Yankner, B. A., Dawes, L. R., Fisher, S., Villa-Komaroff, L., Oster-Granite, M. L., and Neve, R. (1989). Neurotoxicity of a fragment of the amyloid precursor associated with Alzheimer's disease. *Science*, 245, 417-20.

Yates, C. M., Simpson, J., Maloney, A. F. T., Gordon, A., and Reid, A. H. (1980). Alzheimer-like cholinergic deficiency in Down's syndrome. *Lancet*, 2, 979.

Yates, C. M., Ritchie, I. M., Simpson, J., Maloney, A. F. J., and Gordon, A. (1981). Noradrenaline in Alzheimer-type dementia and Down's syndrome. *Lancet*, 2, 39-40.

Yates, C. M., Simpson, A., Gordon, A., Maloney, A. F. J., Allison, Y., Ritchie, I. M., *et al.* (1983). Catecholamines and cholinergic enzymes in presenile and senile Alzheimer-type dementia and Down's syndrome. *Brain Research*, 280, 119-26.

Yates, C. M., Fink, G., Bennie, J. G., Gordon, A., Simpson, J., and Eskay, R. L. (1985). Neurotensin immunoreactivity is increased in Down's syndrome but not in Alzheimer type dementia. *Journal of the Neurological Sciences*, 67, 327-35.

Yates, C. M., Simpson, J., and Gordon, A. (1986). Regional brain 5-hydroxytryptamine levels are reduced in senile Down's syndrome as in Alzheimer's disease. *Neuroscience Letters*, 65, 189-92.

III

Diagnosis and management of Alzheimer disease in persons with Down syndrome

6

Alzheimer disease in persons with Down syndrome: diagnostic and management considerations

A. J. Holland, H. Karlinsky, and J. M. Berg

Summary

Retrospective and prospective clinical studies have convincingly shown that a proportion of older adults with Down syndrome develop a progressive dementia consistent with Alzheimer disease. However, establishing this diagnosis clinically is complicated by limitations of current neuropsychological evaluation techniques, by the possible presence of co-existing factors that may mask or mimic the presence of Alzheimer disease, and by diagnostic criteria that are still evolving. Assigning a tenable diagnosis of Alzheimer disease requires careful and comprehensive data assembly, including a medical history, clinical examination, neuropsychological assessment and laboratory investigations. Once the diagnosis is established, effective ongoing management should focus on supporting not only the affected individual (including advocacy for his or her rights) but also the family and professional caregivers. During the course of the illness various medical, psychiatric and psychological interventions can be helpful as can changes in the environment. A wide range of services for persons with Down syndrome who develop Alzheimer disease makes it possible for affected individuals, despite deterioration, to remain in the family home or in community residential settings.

Introduction

Persons with Down syndrome (DS) have a high risk of developing neuropathological changes characteristic of Alzheimer disease (AD) as early as their third to fourth decade of life. In many of them, particularly as age advances, this leads to intellectual and functional decline consistent with the clinical features of AD. With the marked increase in life expectancy of persons who have DS (see Chapter 2), their need for appropriate diagnostic and support services has grown in importance (for a review see Thase 1982a). The extent of this need is highlighted by observations that nearly 15 per cent of males and more than 20 per cent of females with DS are over 55 years of age (Fryers 1986; McGrother and Marshall 1990).

Increased risk of developing AD is not characteristic of disorders other than DS that are associated with mental retardation (Reid *et al.* 1978). This implies that the presence of trisomy 21 and its consequent specific effects on the central nervous system, and not abnormality of brain development in general, greatly increases the likelihood of developing AD. This in turn has led to remarkable research progress regarding the possible aetiological significance of genes on chromosome 21. Current interest is particularly, though not exclusively, focused on the amyloid precursor protein gene (see Chapter 1).

However, despite considerable exploration of the relationship between DS and AD (for a review see Oliver and Holland 1986), the clinical recognition of the dementia of AD in persons with DS, and how to best help them, has received relatively limited attention. This chapter is concerned with these diagnostic and management issues.

Diagnostic considerations

Dementia and Alzheimer disease in the general population

The clinical diagnosis of AD is first dependent on identification of the presence of dementia. Diagnostic criteria for dementia have evolved over the years and although the connotation of the term is still that of a progressive, irreversible syndrome of long duration (i.e. years), present day criteria emphasize clinical manifestations already apparent rather than presumed prognosis. In theory at least, the current conceptualization of dementia includes static or reversible conditions.

Although diagnostic criteria proposed for dementia vary, the presence of multiple acquired cognitive deficits is generally required, with memory impairment usually singled out as a necessary symptom. For example, the widely used DSM-III-R criteria (American Psychiatric Association 1987) for dementia require demonstrable evidence of impairment in short- and long-term memory, as well as at least one of the following: impairment in abstract thinking, impaired judgement, other disturbances of higher cortical function (such as aphasia, apraxia, agnosia or constructional difficulty) and/or personality change (see Table 6.1). The designation of dementia also necessitates that the cognitive deficits reach a certain degree of severity. In DSM-III-R, this is defined as deficits that significantly interfere with work or usual social activities or relationships with others. However, such severity criteria are hardly rigorous, since the sophistication of different individuals' work or social situations varies greatly, a consideration that is particularly relevant in evaluating persons with DS. Other organic brain syndromes, such as delirium (acute confusional state), especially in the elderly, may be misdiagnosed as dementia and must be excluded.

Dementia is a descriptive term that does not in itself imply a specific aetiology. Relatively frequent causes in the population at large include ischaemic vascular disease (multi-infarct dementia), extrapyramidal syndromes (e.g. Parkinson disease), toxic and metabolic encephalopathies (particularly chronic

A. Demonstrable evidence of impairment in short- and long-term memory. Impairment in short-term memory (inability to learn new information) may be indicated by inability to remember three objects after five minutes. Long-term memory impairment (inability to remember information that was known in the past) may be indicated by inability to remember past personal information (e.g., what happened yesterday, birthplace, occupation) or facts of common knowledge (e.g., past Presidents, well-known dates).

B. At least one of the following:
    (1) impairment in abstract thinking, as indicated by inability to find similarities and differences between related words, difficulty in defining words and concepts, and other similar tasks
    (2) impaired judgment, as indicated by inability to make reasonable plans to deal with interpersonal, family, and job-related problems and issues
    (3) other disturbances of higher cortical function, such as aphasia (disorder of language), apraxia (inability to carry out motor activities despite intact comprehension and motor function), agnosia (failure to recognize or identify objects despite intact sensory function), and "constructional difficulty" (e.g., inability to copy three-dimensional figures, assemble blocks, or arrange sticks in specific designs)
    (4) personality change, i.e., alteration or accentuation of premorbid traits

C. The disturbance in A and B significantly interferes with work or usual social activities or relationships with others.

D. Not occurring exclusively during the course of Delirium.

E. Either (1) or (2):
    (1) there is evidence from the history, physical examination, or laboratory tests of a specific organic factor (or factors) judged to be etiologically related to the disturbance
    (2) in the absence of such evidence, an etiologic organic factor can be presumed if the disturbance cannot be accounted for by any nonorganic mental disorder, e.g., Major Depression accounting for cognitive impairment

Criteria for severity of Dementia:
Mild: Although work or social activities are significantly impaired, the capacity for independent living remains, with adequate personal hygiene and relatively intact judgment.
Moderate: Independent living is hazardous, and some degree of supervision is necessary.
Severe: Activities of daily living are so impaired that continual supervision is required, e.g., unable to maintain minimal personal hygiene; largely incoherent or mute.

Table 6.1. DSM-III-R diagnostic criteria for dementia (American Psychiatric Association 1987)

alcohol abuse) and the dementia syndrome of depression. Infections, neoplasms, hydrocephalus, post-traumatic and post-anoxic conditions and subdural haematoma must also be considered in the differential diagnosis. Recently, HIV infection has also been recognized as an important cause of dementia (Catalan 1991). However, AD is widely felt to be the most common cause in the general population (for a review see Cummings and Benson 1992) and is defined in DSM-III-R as a dementia of insidious onset with a uniformly progressive deteriorating course (American Psychiatric Association 1987). Ultimately though, a definitive diagnosis of AD depends on demonstrating its characteristic neuropathological features, which include senile plaques and neurofibrillary tangles. There is no procedure, other than a very rarely justifiable brain biopsy, that can unequivocally yield a diagnosis of AD during life. Nevertheless, diagnostic strategies have been developed, which increase the prospect of correctly diagnosing AD as the cause of observed dementia. These partly rely on excluding other possible causes of dementia (see above) but also depend on the observed pattern of neuropsychological decline. An example of one descriptive model by Cummings and Benson (1992) is provided in Chapter 1. In general, mnemic, visuospatial and language deficits dominate the early and middle stages of AD, with decline occurring in a characteristic fashion. By the end stages of disease, there is profound global cognitive impairment.

The National Institute of Neurological and Communicative Disorders and Stroke (NINCDS) and the Alzheimer's Disease and Related Disorders Association (ARDRA) in the USA have proposed levels of diagnostic certainty for AD (i.e. **definite**, **probable** and **possible**) and have provided criteria for each of these categories (McKhann *et al.* 1984; Chapter 1). In brief, a diagnosis of **definite** AD is made only in those cases of probable AD with histopathological confirmation of the disease. A clinical diagnosis of **probable** AD requires the presence of a progressive dementia and exclusion of other systemic or brain diseases that could account for those deficits. A clinical diagnosis of **possible** AD is assigned if there are atypical variations in the onset, presentation or clinical course of the dementia, or if another illness that may independently affect cognitive functioning is also present.

Clinical observations consistent with Alzheimer disease in persons with Down syndrome

Earlier clinical observations concerning AD in persons with DS primarily were based on institutionalized individuals who may never have had an opportunity to develop sophisticated living skills. The style of life imposed on them was often also relatively undemanding. Intellectual decline in such circumstances, therefore, may not have become evident until it was severe. However, there are relatively early descriptions of changes observed in persons with DS consistent with manifestations of AD (particularly in the later stages) in the general population. For instance, Rollin (1946) described worsened toilet habits, loss of speech and alterations in behaviour, and Jervis (1948) reported profound personality changes in older adults with DS. Many such early studies were retrospective with evidence of decline derived from case notes examined after the subjects concerned had died and been found to have brain pathology characteristic of AD. These reports emphasized changes in behaviour (using terms such as 'unmanageable', 'apathetic' and 'irritable') and loss of skills (for example see Verhaart and Jelgersma 1952; Haberland 1969), but it also became apparent that not all those with AD neuropathology had shown evidence of deterioration in cognition, behaviour or skills prior to death (Olson and Shaw 1969).

More recently, a number of studies (both cross-sectional and prospective) have attempted to clarify the clinical features of AD in DS, as well as discrepancies between clinical and neuropathological findings. Unfortunately, the diagnostic criteria for dementia of AD in DS were not always well-defined. Thase *et al.* (1982), in a study of an institutionalized population, concluded that 45 per cent of individuals with DS aged 45 years or older had a full syndrome of dementia consistent with AD, compared with only five per cent of matched controls. In this investigation, the findings of disorientation, memory impairment and a history of deterioration in functioning were considered evidence for a provisional diagnosis of dementia. In a prospective study of 96 individuals with DS over the age of 35 years, Lai and Williams (1989) defined the presence of dementia as a decline in one or more tests of orientation, memory, verbal and motor skills and level of functioning. The prevalence of dementia rose from eight per cent (2/25 individuals between ages 35 and 49 years) to 75 per cent (6/8 individuals aged over 60 years).

In a study by Schapiro *et al.* (1986) dementia was defined using DSM-III criteria (American Psychiatric Association 1980), modified by taking into account manifestations of dementia of DS reported in previous studies. For example, personality change was considered to be present on the basis of at least one of five possible profiles: (i) 'loss of usual affectionate response to others; (ii) hyperkinetic, impulsive: low frustration level, temper tantrums, irritable, spiteful, obstinate, destructive, demanding, aggressive; (iii) withdrawal: slowness, loss of interest in surroundings, apathy, sudden changes from cheerfulness to sullenness, inattentiveness to vocational tasks, loss of social interests; (iv) automatic movements: rocking, wriggling hands, perseveration, stereotypes (speech, motor); (v) psychiatric changes: hallucinations (auditory, visual), paranoid delusions, depression'.

Evenhuis (1990) also used modified DSM-III criteria, in this instance based on the revised edition, in her detailed clinical descriptions of 14 individuals with DS and dementia. Specifically, affected individuals demonstrated evidence of short- and long-term memory loss as well as at least one of the following: disturbance of spatial or temporal orientation, aphasia, apraxia and personality change. As with the standard DSM-III-R criteria (see Table 6.1), the disturbances (which could be observed in daily living circumstances as opposed to by neuropsychological evaluation only) had to interfere significantly with work or usual social activities or relationships with others and could not occur exclusively during the course of delirium. Evenhuis found that in the moderately retarded cases, early symptoms of dementia consisted mainly of apathy and withdrawal, daytime sleepiness and loss of self-help skills. In those with severe retardation, gait deterioration, myoclonus and seizures were the most common early findings.

The clinical manifestations of AD in persons with DS have been summarized by a number of authors. Burt *et al.* (1992) noted that seizures, personality changes, apathy/inactivity and loss of self-help skills were the most frequently observed findings across studies. Dalton and Crapper-McLachlan (1986) reviewed the psychiatric, medical, neurological, behavioural and functional characteristics of 33 adults with DS who had neuropathologically confirmed AD. Seizures, focal neurological signs and personality changes were the features most often recorded.

Although dementia is not always diagnosed in adults with DS, investigations that have compared older adults with younger adults with DS or with others of similar mental level but without DS, have found that older individuals with the syndrome usually score lower on a range of cognitive tests and behaviour scales and show an increase of abnormal neurological signs (see Chapters 4 and 8). For example, Haxby (1989), using a comprehensive battery of neuropsychological tests, looked for patterns of cognitive deficits that might distinguish cognitive changes in young individuals with DS from such changes associated with either ageing or dementia. Selective patterns of cognitive deficits were found in older persons with DS without dementia, which primarily affected their ability to form new long-term memories and their visuospatial construction. In contrast, the four older individuals in the study diagnosed as suffering from dementia had deficits across all cognitive functions, except for some preserved simple language abilities.

Other studies have failed to show evidence of cognitive decline with increasing age in at least a proportion of adults with DS (Hewitt *et al.* 1985; Devenny *et al.* 1992). Varied methodology and the use of different psychological instruments may account for some of the conflicting results between studies. Nevertheless, there is convincing evidence that some adults with DS develop the characteristic clinical deterioration of AD. However, it is equally clear that dementia does not occur as often as would be expected from the observed high frequency of AD neuropathological changes.

The clinical diagnosis of Alzheimer disease in persons with Down syndrome

Extant clinical diagnostic criteria for AD have been developed for the general population. Their focus is on identifying deficits in cognition. Such deficits are usually present to some degree from childhood in mental retardation syndromes in general, including DS. However, the previously mentioned NINCDS-ARDRA criteria for the clinical diagnosis of probable AD also specify that a progressive worsening of memory and other cognitive functions must be present, while the DSM-III-R diagnostic criteria for 'Primary Degenerative Dementia of the Alzheimer Type' require a dementia with a generally progressive deteriorating course. Hence, decline in comparison with the individual's previous level of function is necessary for the diagnosis of AD, a point of particular importance in assigning a diagnosis of AD in individuals with DS. If there is uncertainty about the maximal level of functioning previously achieved, an initial evaluation is likely to be insufficient for a diagnostic opinion, but does serve to document current cognitive abilities and level of functioning. Longitudinal follow-up would then be necessary to document decline and thus to establish a diagnosis of AD according to DSM-III-R or NINCDS-ADRDA criteria.

There are other considerations that distinguish the diagnostic process for AD in persons with DS from that in others. First, as discussed in this book (see Chapters 4 and 8), the objective measurement of cognitive functions and detection of change remains, perhaps particularly in DS, an ongoing challenge to investigators.

Secondly, there are criteria for AD in the general population that are clearly not applicable to persons with DS. For instance, the NINCDS-ADRDA criteria for probable AD include onset between ages 40 and 90 years, which would arbitrarily exclude individuals with DS who develop AD symptomatology before their fourth decade. Similarly, the DSM-III-R criteria for severity of dementia are based upon the capacity for independent living, already compromised to some extent in many individuals with DS due to underlying mental retardation. In more general terms, Sovner (1986) has discussed in detail the difficulties of evaluating individuals with mental retardation through use of diagnostic criteria, such as those in DSM-III, designed for persons with normal intelligence and psychosocial functioning. Four factors in particular limit the applicability to persons with mental retardation of criteria designed for the general population. These are:

(i) intellectual distortion: diminished ability to think abstractly and often to communicate clearly lead to difficulty in eliciting subjective symptoms;
(ii) psychosocial masking: impoverished social skills and life experiences result in unsophisticated and bland clinical presentations;
(iii) cognitive disintegration: disorganization with emotional stress may occur during a diagnostic interview and produce an atypical or bizarre presentation of a psychiatric illness;
(iv) baseline exaggeration: exacerbation of pre-existing cognitive deficits and maladaptive behaviours creates difficulty in establishing the onset of manifestations of a newly developed disorder.

Thirdly, results from both retrospective and prospective clinical studies emphasize that cognitive changes in DS may not be the most obvious features of AD (see preceding section). Loss of self-care skills or deterioration in vocational performance may be more evident to relatives and other caregivers, but presumably often reflect cognitive decline which, although overshadowed by the above-mentioned functional or vocational observations, would probably be demonstrable upon formal evaluation. In others with DS, it may be the development of non-cognitive signs and symptoms, such as personality or behavioural changes or the onset of dysphoric or psychotic features, that are considered to be the initial manifestations of AD. By contrast, in the general population, these non-cognitive changes tend to be viewed as co-existing phenomena of AD that do not always occur and that when they do, would be insufficient, in themselves, to diagnose AD in the absence of cognitive decline. In DS, however, it may well be that the cognitive decline, although present, is masked by the more dramatic and readily recognizable personality, behavioural or psychiatric changes; alternatively, it may be that these non-cognitive changes precede cognitive decline when AD occurs in persons with DS. In view of this uncertainty and because non-cognitive changes have not been established as universal features of AD in DS, it seems prudent for clinicians to have a high index of suspicion for AD when such changes are noted, but to reserve the diagnosis of the disease until those individuals demonstrate objective cognitive decline and therefore meet current AD diagnostic criteria.

Similar considerations apply when incontinence, seizures or extrapyramidal signs are suspected as initial manifestations of AD in individuals with DS. These features usually do not occur until the late stages of AD in the general population and therefore, particularly in severely and profoundly retarded individuals with DS, could be overshadowing existing but difficult to detect cognitive changes. Again, the most prudent course in these circumstances may be to reserve the diagnosis of AD until current diagnostic criteria are met, but to be highly suspicious for its possible presence.

In view of the above considerations, applying current diagnostic criteria for AD in the general population would probably lead to underdiagnosis or late diagnosis of this disease in persons with DS. A modified approach, such as that used by Schapiro *et al.* (1986), may be more appropriate, particularly greater emphasis on clinical observations than on formal neuropsychological detection criteria. In the future, the development of more sophisticated neuropsychological tests designed specifically for persons with DS may allow previously formulated criteria to be utilized.

A caveat in the above regards is indicated. The clinical heterogeneity of AD in the general population is well-established and clinical subtypes of the disease have been proposed, including those based upon age-of-onset, rapidity of decline, cognitive profiles, extrapyramidal signs, seizures and psychotic symptoms. As aetiological heterogeneity of AD is also now virtually certain, it is unclear whether the observed clinical variability is due to primary aetiological factors or to epistatic modifying events (both genetic and non-genetic). Individuals with DS affected by AD probably share a common causal process, albeit not yet understood. It is possible therefore that a clinical profile of AD in DS will emerge that is distinct from that of AD in the general population, with less emphasis on cognitive changes as a defining feature. Other shared intrinsic characteristics of DS (e.g. abnormal brain cytoarchitecture) may also contribute to determination of distinctive clinical features when AD occurs. Limitations in diagnostic tools suitable for persons with DS may also result in identification of only some clinical features. With the accumulation of additional clinical data, distinct criteria for AD in persons with DS are likely to emerge. Indeed, Sovner (1986) has advocated using a diagnostic framework in the mentally retarded based upon empirically derived criteria.

The diagnostic process when Alzheimer disease is suspected in persons with Down syndrome

As in the general population, the information needed to diagnose AD in persons with DS includes a medical history (necessitating collateral sources of information), detailed clinical examination, neuropsychological assessment and various laboratory investigations.

Medical history
The person with DS and knowledgeable informants should be interviewed about presenting symptoms as well as about past medical, psychiatric, personal

and family history. Ideally, both a relative and a professional caregiver should provide collateral information, including details of daily living skills and work productivity. Since individuals with DS generally will be unable to provide comprehensive or accurate historical information, particularly when new cognitive deficits and loss of insight have occurred, the history from informants is crucial to the assessment process. As professional caregivers may have known the individual being assessed for only a limited period of time, or relatives may be only peripherally involved, the reliability of an informant's history must be established.

In DS, there are two aspects of the medical history that require special emphasis. First, as discussed in the preceding section, the clinical characteristics of AD in DS are less well-defined than, and indeed may differ from, those of the disease in the general population. Results from both retrospective and prospective studies emphasize that in addition to questions focused on changes in intellectual functions (including memory, language and visuospatial abilities), it is also necessary to clarify the presence or absence of behavioural deterioration, psychiatric symptomatology, personality change, loss of self-care skills and alterations in work performance. Whether there have been recent onset of seizures, urinary or bowel incontinence and gait disturbances (secondary to extrapyramidal signs) must also be determined.

Secondly, intellectual and functional decline due to the onset of AD must be distinguished from that due to pre-existing mental retardation. Failure to accurately establish an individual's previous abilities may preclude recognition that particular skills, once present, have now been lost. One purpose of the medical history, therefore, is to establish the pre-morbid level of intellectual and functional abilities of the individual with DS being assessed.

Clinical examination

In an individual with DS suspected to have AD, the purpose of a careful physical examination is twofold. First, other conditions that may cause intellectual or functional decline must be excluded (see next section on differential diagnosis). In particular, clinicians must be aware of conditions over-represented in adults with DS, such as hypothyroidism, the presentation of which may be misdiagnosed as AD. Secondly, potential exacerbating factors of the decline associated with an underlying AD process, such as diminished hearing or sight, must also be identified and treated if at all possible.

Mental status evaluation is also an important component of the clinical examination. Specific inquiry about the presence of mood disturbances (including associated neurovegetative symptoms) and psychotic symptoms must be undertaken, both to exclude primary psychiatric illnesses and to identify potentially distressing and treatable co-existing symptoms of AD. Cognitive screening is, of course, an essential component of the mental status examination and requires knowledge of limitations due to mental retardation as well as of deficits associated with AD. The major cognitive processes impaired in AD should be tested, including orientation to time and place, memory, language skills, visuospatial abilities, attention, praxis, calculation and problem-solving skills. Ideally, the testing should be undertaken in an objective and standardized manner as described below.

Neuropsychological assessment

Neuropsychological testing can provide extremely useful additional information for clarifying the possible presence of AD. At its most basic level, this may simply consist of utilization by the clinician of a standardized screening test, such as the Mini-Mental State Examination (Folstein *et al.* 1975), in order to objectively document cognitive performance. As with the medical history, however, it may not be immediately apparent whether difficulties are related to pre-existing mental retardation or to superimposed dementia consequent on the presence of AD. Furthermore, if test results are at the test floor or are no different than might be expected by chance (as is the case with many of the neuropsychological tests designed for the general population and applied to those with DS), subsequent testing would not be able to detect deterioration.

As discussed in more detail by Crayton and Oliver (Chapter 8), a number of initiatives have attempted to overcome these obstacles. These include constructing tests that attempt to distinguish cognitive deficits associated with the pre-existing mental retardation from those deficits due to AD in persons with DS. In recent years, there has also been considerable success in simplifying various neuropsychological tests targeted at the classic cognitive changes of AD in the general population. For example, a number of investigators (e.g. Thase *et al.* 1982; Dalton and Crapper-McLachlan 1986) have utilized a matching-to-sample task that has quite successfully documented memory decline in a proportion of older individuals with DS.

Laboratory investigations

A number of laboratory tests should be undertaken for all individuals with DS in whom dementia is suspected, regardless of the findings on history, physical and neuropsychological examination. There has been much debate regarding which of these tests should be mandatory when AD is suspected in the general population, and various batteries have been recommended. For example, a conference sponsored by the National Institute on Aging and the National Institute on Health (Consensus Conference 1987) advocated the following investigations: haematological examination, urinalysis, evaluations of levels of serum urea or BUN, glucose, electrolytes, calcium, phosphorous, bilirubin, B_{12} and folic acid, tests for thyroid function, serological testing for syphilis, chest X-ray and electrocardiogram. Computerized tomography (CT) scan of the brain was reserved for those dementias of brief duration, with focal neurological signs or with a history suggestive of a mass lesion. On the other hand, a recent work group in Canada (Clarfield 1989) recommended that only thyroid indices, electrolytes, calcium, glucose and complete blood count are indicated as general routine tests, with all other investigations targeted for selected individuals on the basis of suspicious history and clinical examination. As the yield of laboratory tests in identifying specific causes of dementia in individuals with DS is still uncertain, a cautious approach would be to initiate the more extensive screening battery. As clinical experience accumulates, many of the tests may become reserved for persons with particular historical data and/or clinical findings.

The use of neuroimaging and electrophysiological assessment merits consideration. Although not always well tolerated by individuals with DS, Schapiro and his colleagues (see Chapter 10) have demonstrated that cerebral atrophy is associated with clearly diagnosed dementia in persons with DS, but not in those with cognitive decline who do not meet criteria for dementia. A neuroimaging procedure, either CT or magnetic resonance imaging (MRI), therefore appears to be warranted, both to exclude other causes of dementia and to provide suggestive evidence of AD.

Electrophysiological measurements are now generally not included in the routine diagnostic work-up of suspected AD in the general population. However, in DS, seizures occur in a substantial proportion of those affected with AD and an electro-encephalogram is indicated if seizures are suspected (Lai and Williams 1989). P-300 auditory-evoked potentials have also been studied in relationship to age and cognitive decline in persons with DS. Blackwood *et al.* (1988) measured the latency of the P-300 evoked potential in 89 individuals with DS and found it to increase in those aged 37 years and older; the increase was noted in the 16 individuals who had evidence of dementia. A control group with fragile-X syndrome who did not show dementia had evidence of increased latency from 54 years of age. In a follow-up study two years later, a further 14 per cent of the individuals with DS were dementing and seven out of nine of them showed increased latency (Muir *et al.* 1988). The authors concluded that auditory-evoked potential latency is a useful confirmatory test for the presence of AD in DS and further experience may establish it as a routine component of a diagnostic work-up.

Differential diagnosis of intellectual and functional decline in persons with Down syndrome

As in the general population, various conditions besides AD can cause intellectual and/or functional decline in persons with DS. McCreary *et al.* (Chapter 7) emphasize that thyroid disease, major depression and anticonvulsant drug toxicity, in particular, must be considered in the differential diagnosis. The rationale is readily apparent. There have been numerous reports of increased rates of thyroid disease (primarily hypothyroidism) in persons with DS (e.g. Mani 1988; Dinani and Carpenter 1990). Hypothyroidism can be associated with cognitive decline, particularly mental slowing, as well as with other clinical features (e.g. lack of motivation) that may be present in AD (see case report by Thase 1982b). Hypothyroidism may also exacerbate the cognitive dysfunction of AD in persons with DS (Percy *et al.* 1990).

Depression in the general population can also be associated with significant cognitive changes (so-called pseudodementia, or more preferably, the dementia syndrome of depression) and is an eminently treatable condition. Major depression can occur in persons with DS as well, with detailed case reports highlighting the overlap of manifestations with those of AD, particularly withdrawal and loss of self-care skills (see summary by Pary 1992). Unlike thyroid disease, however, the diagnosis of major depression is not always straightforward and both depression and AD may co-exist (Szymanski and Biederman 1984; Burt *et al.* 1992). A practical recommendation is that potential symptoms of depression should be treated even when the presence of AD is also suspected (Pary 1992).

Many medications taken in excess can cause central nervous system toxicity (Cummings and Benson 1992). McCreary *et al.* (Chapter 7) suggest that toxicity secondary to anticonvulsant medication must be considered and carefully excluded in adults with DS, particularly in view of the high rate of seizure disorders in such adults (Lai and Williams 1989). A careful review of all medications should, of course, regularly take place.

Other conditions common in adults with DS which may cause functional decline and possibly mimic AD, at least to some extent, include sensory impairments (hearing, vision), infections, vitamin deficiencies, malignancies, joint problems and sleep apnoea (Pary 1992; Chapter 4). The purpose of the diagnostic work-up is to identify these conditions if present, with routine investigations augmented by additional tests and different specialist consultations according to historical details and clinical circumstances. Such conditions should be excluded as causes of decline before the diagnosis of AD is assigned.

Management considerations

Since there are no specific means as yet to prevent the neuronal cell death in AD and thus to arrest progress of the disease, the therapeutic approach must be primarily one of 'management' in contrast to definitive 'treatment'. Data are still relatively scanty concerning the management of AD specifically in individuals with DS. However, the considerable similarity in the ways AD affects persons whether they have DS or not suggests that it is appropriate, at least as a starting point, to apply the lessons learned from managing AD in the general population also to persons with DS who develop AD. Furthermore, certain clinical features of AD (e.g. seizures) may occur in DS for reasons other than AD. Established treatment strategies that are successful in these latter circumstances can also be usefully applied when the same clinical features occur in the context of AD.

Informing and supporting the caregivers

When AD occurs in the general population, partners and other family members are almost invariably the main providers of care (Shanas 1979). When AD occurs in individuals with DS, particular factors influence who provides the care and the extent and nature of the stress experienced by families. Although persons with DS may form long-standing stable relationships with a partner, they much more frequently live with other family members or in supported housing. Many reside with parents who are becoming increasingly frail and who are themselves in an age of risk for AD. Parents and other relatives who have looked after a person with DS with litle support over many years now have to face the additional burden of AD. In addition to the stress involved, the increasing needs of adults with DS who develop AD can result in decreased independence of the caregivers and hence a marked change in their lives. Families may have to make difficult decisions about the care of their affected relative. If he or she is being cared for at home by family members, they may no longer be able to cope with the growing burden, yet may feel considerable guilt about transferring the care to local services. In some cases, services may not be available or even be unwilling to take care of someone with the behavioural problems often found in those with dementia.

As a first step in management, it is crucial to educate family members about the characteristics of AD. In addition to personal guidance, an increasing number of publications on the subject is available for the lay public, albeit predominantly focused on AD in the general population. Contact with a local Alzheimer disease or Down syndrome society can be an important source of advice and encouragement, particularly through participation in a caregiver support group. Family caregivers need to know the extent and nature of their affected relative's disability, how it may change from day to day (or even from hour to hour) and what long-term cognitive decline is likely. Such knowledge helps to reduce negative and damaging interpretations of difficult behaviour. For example, a person with AD may seem stubborn for not complying with a particular request when, in fact, he or she may not comprehend the request and be unable to communicate that lack of understanding.

It is equally important to educate and support professional caregivers. In an instructive publication, Newroth and Newroth (1980) described their residential community's experience in caring for two persons with DS who developed AD. These authors highlighted the potential for caregiver staff 'burn-out' and the need for a team approach in sharing responsibilities for an affected person's care. They also noted that staff may feel inadequate if they have little prior experience or training in the care of persons with AD. Additionally, an unwanted professional role change to that of palliative caregiver may lead to emotional unrest. Also anxiety-provoking is a professional caregiver's perception that he or she is no longer 'valued, needed or even recognized' by the affected individual. To deal with these issues, Newroth and Newroth (1980) advocated educational staff counselling, while recognizing that more intense personal counselling may be required. They also emphasized that the needs and morale of other developmentally handicapped residents in the setting must not be overlooked. Since a resident who develops AD will need a good deal of extra staff attention, other residents may feel relatively neglected. Residents must also be educated about any changing health status of a co-resident who is often a close friend as well.

Holden and Woods (1982) have argued that the attitude of support staff and other caregivers is critical in the management of dementia. Interactions with affected individuals must not be devaluing and demeaning to them but instead should encourage choice where possible and maintain individual dignity (Woods 1987), even though this may be difficult when physical problems (e.g.

1. People with dementia have the same human values as anyone else irrespective of their degree of disability or dependence.
2. People with dementia have the same varied human needs as anyone else.
3. People with dementia have the same rights as other citizens.
4. Every person with dementia is an individual.
5. People with dementia have the right to forms of support which do not exploit family or friends.

Table 6.2. Key principles for services for people with dementia (from Marshall 1990)

incontinence) are considerable and disinhibited behaviour is prominent. The stress frequently associated with working with individuals with dementia can easily lead staff to introduce management strategies (e.g. physical methods of restraint) which normally would not have been considered. Understandable development of negative attitudes in caregivers under stress, with resultant deterioration in standards of care, can be at least partially prevented if the services involved establish positive principles of care (see Table 6.2). These principles are pertinent to persons with DS, no less than to others, who develop AD.

Medical, psychiatric and psychological interventions

In persons with DS, as in others, medical involvement does not end with the diagnosis of AD after excluding other possible causes for the observed deterioration. Continuing necessary medical contributions include prompt treatment of concurrent ailments, such as infections and anaemias, as well as regular reviews of sensory impairments and of thyroid and cardiac status, to minimize the possibility that additional treatable disorders will add to the already considerable disability consequent upon the presence of AD. Additionally, particular physical disorders that emerge secondary to the AD process may respond to standard medical treatment. For example, anticonvulsants may be required if seizures or myoclonus occur.

The concurrence of specific psychiatric disorders or manifestations may also respond to intervention. Safe and effective therapies for depressed individuals with DS, albeit without dementia, include psychotherapy, pharmacotherapy and electroconvulsive therapy (Sovner and Hurley 1983; Lazarus *et al.* 1990). Medications may also be indicated for psychotic symptoms, agitation, day–night reversal and poor aggressive impulse control. However, medication usage should be as minimal as possible in demented individuals with DS, particularly with the anticipated high risk of side-effects of centrally active agents. Furthermore, it is first necessary to investigate all possible explanations for the symptom in question. For example, agitated behaviour may be due to an additional physical illness requiring treatment in its own right. If a psycho-active medication is considered necessary, low doses should be used initially, with efficacy and possible side-effects regularly monitored.

Where possibly helpful, non-pharmacological interventions, such as various forms of psychotherapy and behaviour management, are preferable to psychopharmacological medications. In view of the pre-existing mental retardation, traditional psychodynamic psychotherapies are of limited use in persons with DS, but counselling and supportive psychotherapy may help affected individuals (particularly those with greater ability in the early stages of AD) understand that they have an illness that will eventually necessitate assistance in everyday activities and affect work performance (Newroth and Newroth 1980). As in the general population, reality orientation programmes may also be useful, particularly in maintaining orientation to time and place (Holden and Woods 1982).

Other specific psychological strategies have been developed to help minimize the extent of particular disabilities that commonly occur in individuals with AD in the general population, and these strategies are applicable also to affected persons with DS. For example, with progression of AD, individuals may take longer to feed themselves and caregivers may feel obliged to undertake this task. However, if the affected persons are given more time and appropriate verbal and physical prompts, they may be able to satisfactorily manage self-feeding (Melin and Gotestam 1981). Similarly, the skills of bathing or of dressing and undressing may be maintained if caregivers break the task down into small steps and prompt and assist with each step (Rinke *et al.* 1978). These examples have in common the need to assess skills and to understand the different components of the activities necessary to undertake the tasks in question. Completion of each step in the task should be accompanied by appropriate positive reinforcement in the form of praise or simple rewards. Even though such an approach may not result in the learning of new skills, it maximizes retained abilities.

Finally, certain maladaptive behaviours may respond to behaviour management programmes. As affected individuals will generally not be able to comprehend the consequences of their behaviours, the success of such programmes depends on first identifying those environmental circumstances or occurrences that are antecedents to the aberrant behaviours. These environmental cues are then modified in order to extinguish the unwanted behaviours (Cummings and Benson 1992, p. 373).

Changes in the environment

The level of disability when AD occurs is not simply a product of the extent to which the disease has progressed. It also depends on environmental factors which either mitigate against success or help to compensate for the presence of disability. This has been a driving force in the move away from regarding mental retardation as something fixed and unchangeable. Models of social care for those with learning disabilities are also applicable to dementia, but with the recognition in the latter case that intrinsic abilities are deteriorating, so that any given approach will need modification with changing circumstances. Interaction between biologically determined impairment and environmental factors on the level of disability is complex. The relative importance of these components varies throughout the course of the illness and even during a given day.

Understanding the nature of the disability associated with a dementing illness like AD helps in planning the most suitable environment for an affected person. Poor memory and disorientation are inevitable features, but the extent to which these problems may handicap persons in a given residential setting not only depends on the level of impairment but also on the degree to which the environment is helpful (Hanley 1981). For example, living in a residence in which all doors to bedrooms are identical increases the probability that persons with dementia will find themselves in an unfamiliar room and thus become even more confused. Personalized doors can reduce this problem and increase safety. Similarly, clear markings on a bathroom door may reduce or resolve incontinence.

As intellectual deterioration progresses, some household items can become a major concern and may need to be replaced by safer alternatives and the institution of other precautions. For instance, because of a danger of failure to turn off the gas on a cooker, an electric cooker with provision of fire alarms and smoke detectors could reduce the risk of serious accident while still allowing a homely environment. Physical disability also may increase, with unsteadiness, falls and general frailty becoming a problem. In that circumstance, installation of support rails can help to prolong an affected person's independence.

Services for persons with Down syndrome and Alzheimer disease

The much improved life expectancy of persons with DS than in the past, and a markedly increased risk of dementia at an age well below that of the general population, has created the necessity for expansion of established services and for the development of new ones. Usually, the needs of individuals with dementia in the general population are primarily met through psychiatric and social service provisions for the elderly (over 60 or 65 years). This age criterion and the presence of mental retardation in persons with DS have been used as reasons for not meeting, through established 'dementia services', the particular needs of these persons consequent on developing dementia. However, in provisions for the elderly in the general population, it is well-established that a range of specific services are required. The same is true also for those with DS who have dementia. They also require particular services that should include: provision of appropriately trained staff for assessment with, if necessary, availability of in-patient assessment facilities; a range of residential accommodations; and additional support to families who are able to continue to care for the affected person at home, including community nursing and homemaker support, an available and interested general practitioner, specialist advice on the management of problem behaviours, day programmes, respite care, financial help and practical assistance for necessary structural changes to the residence (e.g. a downstairs bedroom, an alarm system, additional toilets). Individuals with DS, if not residing with their families, are increasingly likely to be living in supported housing rather than in large, relatively isolated institutions. It may be possible with additional services to enable these persons to continue to reside in the community despite deterioration (Tinker 1984).

As mentioned earlier, the quality of services primarily depends on the quality of the participating staff and their attitudes. Studies of residential settings for the elderly in general have established some important principles. For instance, engagement of elderly people in meaningful activity depends on organization into smaller living units that can allow involvement in day-to-day activities (Rothwell *et al.* 1983). Persons with dementia can be prompted to engage in these activities and they can be continued for some time after extra staff input has ceased (Melin and Gotestam 1981). Caregivers who are knowledgeable about the nature of dementia will be more conscious of the need for a consistent and predictable environment and will use various techniques to reduce the negative effects of its characteristic symptoms (see section on medical, psychiatric and psychological interventions).

Rights and decision-making

With positive changes in attitude towards the care of persons with mental retardation and/or illness, including those with DS who develop AD, there has been increasing emphasis on how they can be supported in making complex decisions about their own lives. Often, there is no simple legal structure to enable decision-making to take place for adults deemed unable to make decisions for themselves. In the UK, for example, common law provides the framework whereby, in cases of necessity or emergency, appropriate action may be taken, even if the affected person is unable to give informed consent. Good practice would dictate that close family and caregivers be consulted, but they cannot consent on the individual's behalf. These legal questions have often been extensively discussed, for instance in a British Law Commission Consultative Paper (Law Commission 1991). The basic issue concerns the balance between the rights of individuals to self-determination and their rights to protection. Adults with DS when previously functioning well may have established considerable independence. For example, they may have been travelling on public transport independently to and from supported employment, yet recently become lost on some occasions and may even have been returned by police after wandering in the street. The dilemma is to know when to intervene. Caregivers could usefully rehearse the route again with the affected person, help draw a simple map and encourage the person to carry appropriate identification. However, with the progression of AD, a time will eventually come when independent travel is judged to be too risky. The next stage may be the provision of an escort. Nevertheless, the problem remains as to whether more restrictive intervention is needed and, if so, with what legality. Those providing services for adults with DS need to consider these issues. There is a place for legislation that allows a 'framework of care', but that does not become a charter for inappropriate and over-restrictive practices (Greengross 1986). The best insurance against abuse on the one hand, and the provision of an appropriate level of care on the other, is regular consultation and communication with all concerned.

Conclusions

Deterioration in adults with DS requires an explanation. AD can be diagnosed on the basis of longitudinal information about changes in cognitive ability, skills, speech and behaviour. These changes are similar to those observed in the presentation of AD in the general population, but pre-existing deficits in levels of functioning of persons with DS can affect the point at which changes consequent upon AD are recognized. Other causes of cognitive decline and behavioural disturbance common in DS (e.g. hypothyroidism, sensory impairments) must be considered in the differential diagnosis of AD. Although there is no specific treatment for AD, its diagnosis should provide the starting point for developing management strategies to maintain as high a quality of life as possible for the affected person and his or her family and other caregivers. Support for all of them requires the skills and commitment of people from various professional backgrounds. Management should be based on regular assessments of the changing capacities and needs of those affected. All concerned have to be aware of appropriate strategies for maintaining an acceptable quality of life, given the increasing disability associated with the progression of AD. The focus of interventions should be to maintain as good physical and mental health as possible and in other ways also to reduce the effects of the disabilities consequent upon AD. This requires caregivers to be well-informed as to the nature of AD, the use of different psychological and medical strategies and suitable modifications to the living environment.

References

American Psychiatric Association. (1980). *Diagnostic and statistical manual of mental disorders*, (3rd edn). Washington, DC.

American Psychiatric Association. (1987). *Diagnostic and statistical manual of mental disorders*, (3rd edn revised). Washington, DC.

Blackwood, D. H. R., St Clair, D. M., Muir, W. J., Oliver, C. J., and Dickens, P. (1988). The development of Alzheimer's disease in Down's syndrome assessed by auditory event-related potentials. *Journal of Mental Deficiency Research*, **32**, 439-53.

Burt, D. B., Loveland, K. A., and Lewis, K. R. (1992). Depression and the onset of dementia in adults with mental retardation. *American Journal on Mental Retardation*, **96**, 502-11.

Catalan, J. (1991). HIV-associated dementia: review of some conceptual and terminological problems. *International Review of Psychiatry*, **3**, 321-9.

Clarfield, A. M. (1989). *Canadian consensus conference on the assessment of dementia*. Montreal, Canada.

Consensus Conference. (1987). Differential diagnosis of dementing diseases. *Journal of the American Medical Association*, **258**, 3411-6.

Cummings, J. L., and Benson, D. F. (1992). *Dementia: a clinical approach*. Butterworth-Heinemann, Boston.

Dalton, A. J., and Crapper-McLachlan, D. R. (1986). Clinical expression of Alzheimer's disease in Down's syndrome. *Psychiatric Clinics of North America*, **9**, 659-70.

Devenny, D. A., Hill, A. L., Patxot, O., Silverman, W. P., and Wisniewski, K. E. (1992). Ageing in higher functioning adults with Down's syndrome: an interim report in a longitudinal study. *Journal of Intellectual Disability Research*, **36**, 241-50.

Dinani, S., and Carpenter, S. (1990). Down's syndrome and thyroid disorder. *Journal of Mental Deficiency Research*, **34**, 187-93.

Evenhuis, H. M. (1990). The natural history of dementia in Down's syndrome. *Archives of Neurology*, 47, 263-7.

Folstein, M. F., Folstein, S. E., and McHugh, P. R. (1975). 'Mini-Mental State.' A practical method for grading the cognitive state of patients for the clinician. *Journal of Psychiatric Research*, 12, 189-98.

Fryers, T. (1986). Survival in Down's syndrome. *Journal of Mental Deficiency Research*, 30, 101-10.

Greengross, S. (1986). *The law and vulnerable elderly people.* Age Concern, Mitcham, Kent, England.

Haberland, C. (1969). Alzheimer's disease in Down's syndrome: cliniconeurological observations. *Acta Neurologica Belgica*, 69, 369-80.

Hanley, I. (1981). The use of signposts and activity training to modify ward disorientation in elderly patients. *Journal of Behaviour Therapy and Experimental Psychiatry*, 12, 241-7.

Haxby, J. V. (1989). Neuropsychological evaluation of adults with Down's syndrome: patterns of selective impairment in non-demented old adults. *Journal of Mental Deficiency Research*, 33, 193-210.

Hewitt, K. E., Carter, G., and Jancar, J. (1985). Ageing in Down's syndrome. *British Journal of Psychiatry*, 147, 58-62.

Holden, U. P., and Woods, R. T. (1982). *Reality orientation: psychological approaches to the 'confused elderly'.* Churchill Livingstone, Edinburgh.

Jervis. G. A. (1948). Early senile dementia in mongoloid idiocy. *American Journal of Psychiatry*, 105, 102-6.

Lai, F., and Williams, R. S. (1989). A prospective study of Alzheimer disease in Down syndrome. *Archives of Neurology*, 46, 849-53.

Law Commission (1991). *Mentally incapacitated adults and decision-making: an overview.* Consultation Paper No. 119, HMSO, London.

Lazarus, A., Jaffe, R. L., and Dubin, W. R. (1990). Electroconvulsive therapy and major depression in Down's syndrome. *Journal of Clinical Psychiatry*, 51, 422-5.

McGrother, C. W., and Marshall, B. (1990). Recent trends in incidence, morbidity and survival in Down's syndrome. *Journal of Mental Deficiency Research*, 34, 49-57.

McKhann, G., Drachman, D., Folstein, M., Katzman, R., Price, D., and Stadlan, E. M. (1984). Clinical diagnosis of Alzheimer's disease: report of the NINCDS-ADRDA Work Group under the auspices of the Department of Health and Human Services Task Force on Alzheimer's disease. *Neurology*, 34, 939-44.

Mani, C. (1988). Hypothyroidism in Down's syndrome. *British Journal of Psychiatry*, 153, 102-4.

Marshall, M. (1990). *Working with dementia.* Venture Press, Birmingham, UK.

Melin, L., and Gotestam, K. G. (1981). The effects of rearranging ward routines on communication and eating behaviours of psychogeriatric patients. *Journal of Applied Behavioural Analysis*, 14, 47-51.

Muir, W. J., Squire, I., Blackwood, D. H. R., Speight, M. D., St Clair, D. M., Oliver, C., and Dickens, P. (1988). Auditory P300 response in the assessment of Alzheimer's disease in Down's syndrome: a two year follow-up study. *Journal of Mental Deficiency Research*, 32, 455-63.

Newroth, A., and Newroth, S. (1980). *Coping with Alzheimer's disease: a growing concern.* National Institute on Mental Retardation, Downsview, Ontario, Canada.

Oliver, C., and Holland, A. J. (1986). Down's syndrome and Alzheimer's disease: a review. *Psychological Medicine*, 16, 307-22.

Olson, M. I., and Shaw, C. -M. (1969). Presenile dementia and Alzheimer's disease in mongolism. *Brain*, 92, 147-56.

Pary, R. (1992). Differential diagnosis of functional decline in Down's syndrome. *The Habilitative Mental Healthcare Newsletter*, 11, 37-41.

Percy, M. E., Dalton, A. J., Markovic, V. D., Crapper-McLachlan, D. R., Gera, E., Hummel, J. T., *et al.* (1990). Autoimmune thyroditis associated with mild "subclinical" hypothyroidism in adults with Down's syndrome: a comparison of patients with and without Alzheimer's disease. *American Journal of Medical Genetics*, 36, 148-54.

Reid, A. H., Maloney, A. F. J., and Aungle, P. G. (1978). Dementia in ageing mental defectives: a clinical and neuropathological study. *Journal of Mental Deficiency Research*, 22, 233-41.

Rinke, C. L., Williams, J. J., Lloyd, K. E., and Smith-Scott, W. (1978). The effects of prompting and reinforcement on self-bathing by elderly residents of a nursing home. *Behavior Therapy*, 9, 873-81.

Rollin, H. R. (1946). Personality in mongolism with reference to incidents of catatonic psychosis. *American Journal of Mental Deficiency*, 51, 219-33.

Rothwell, N., Britton, P. G., and Woods, R. T. (1983). The effects of group living in a residential home for the elderly. *British Journal of Social Work*, 13, 639-43.

Schapiro, M. B., Haxby, J. V., Grady, C. L., and Rapoport, S. I. (1986). Cerebral glucose utilization, quantitative tomography, and cognitive function in adult Down syndrome. In *The neurobiology of Down syndrome*, (ed. C. J. Epstein), pp. 89-108. Raven Press, New York.

Shanas, E. (1979). The family as a social support system in old age. *Gerontologist*, 19, 169-74.

Sovner, R. (1986). Limiting factors in the use of DSM-III criteria with mentally ill/mentally retarded persons. *Psychopharmacology Bulletin*, 22, 1055-9.

Sovner, R., and Hurley, A. D. (1983). Do the mentally retarded suffer from affective illness? *Archives of General Psychiatry*, 40, 60-7.

Szymanski, L. S., and Biederman, J. (1984). Depression and anorexia nervosa of persons with Down's syndrome. *American Journal of Mental Deficiency*, 89, 246-51.

Thase, M. E. (1982a). Longevity and mortality in Down's syndrome. *Journal of Mental Deficiency Research*, 26, 177-92.

Thase, M. E. (1982b). Reversible dementia in Down's syndrome. *Journal of Mental Deficiency Research*, 26, 111-3.

Thase, M. E., Liss, L., Smeltzer, D., and Maloon, J. (1982). Clinical evaluation of dementia in Down's syndrome: a preliminary report. *Journal of Mental Deficiency Research*, 26, 239-44.

Tinker, A. (1984). *Stay at home: helping elderly people*. HMSO, London.

Verhaart, W. J. C., and Jelgersma, H. C. (1952). Early senile dementia in mongolian idiocy. Description of a case. *Folia Psychiatrica Neerlandica*, 55, 453-9.

Woods, R. (1987). Psychological management of dementia. In *Dementia,* (ed. B. Pitt), pp. 281-95. Churchill Livingstone, Edinburgh.

7

Experiences in an Alzheimer clinic for persons with Down syndrome

B. D. McCreary, J. B. Fotheringham, J. J. A. Holden, H. Ouellette-Kuntz, and D. M. Robertson

Summary

In 1988 a special clinic concerned with Alzheimer disease and Down syndrome was attached to an established psychiatry consulting service for persons with developmental disabilities residing in the Kingston region, Canada. The clinic serves as a tertiary care resource for a general population of about 672,510 (1986 census estimate). The key activities are annual clinical assessments, provision of advice about care and treatment, education of agency staff and health care professionals, and research. By the end of 1991, 202 individuals with Down syndrome had been assessed and consent for post-mortem examination had been obtained in seven (of nine) who died. Hypothyroidism, depression and mental impairment secondary to anticonvulsant drugs were identified as significant factors in the differential diagnosis and treatment of dementia. Alzheimer neuropathology was observed in each post-mortem examination. The average age at death for these subjects was 59 years. Death was 'respiratory' (pneumonia, six cases; pulmonary embolism, one case) and followed a several year interval of dementia. Usually around the mid-point of the dementia the burden of care had increased to the degree that caretakers requested admission of the affected person to a nursing care unit. Continuing longitudinal study of individuals with Down syndrome (including post-mortem examination) with an emphasis on establishing a better understanding of the relative contributions of various factors to mental decline is recommended.

Introduction

This chapter reviews the authors' experience in operating a clinic for ageing persons with Down syndrome (DS). The clinic was established in 1988 as a new component in a long-standing psychiatry consulting service for persons with developmental disabilities in south-eastern Ontario, Canada. The objectives were to improve services to older individuals with DS and to foster genetic, psychiatric and neuropathological research on the association between

DS and Alzheimer disease (AD). The perspective taken in this review is practical and clinical.

First, the literature on dementia in adults with DS that shaped our approach in establishing the clinic is considered. Next, the development and operation of the clinic are briefly described. In the third section, observations made over a three year interval in the clinic are presented, with illustrative case material used to identify some key issues in the differential diagnosis and management of mental change in adults with DS. In the final section, conclusions from our experience and the apparent implications for further work on the association between DS and AD are provided.

Dementia in adults with Down syndrome

The challenges faced by a physician who is consulted about a significant mental or personality change in an older individual with DS involve not only the fundamental distinction between 'mental defect' and 'mental illness' outlined by Penrose (1963), but also the differential diagnosis of a number of mental disorders that can affect these persons. Providing information on prognosis and giving advice that may improve the outlook depend considerably on a competent approach to differential diagnosis. In this section, published observations that influenced our approach in the establishment of the clinic are reviewed.

In his early studies on mental disorders in persons who are mentally retarded, Reid (1972) concluded that mental deficiency modifies the clinical features of an added mental disorder and that diagnostic problems arise particularly amongst the more severely impaired individuals who have little or no powers of verbal communication. Reid and Aungle (1974) proceeded to study the problem of dementia in persons who are mentally retarded. They identified 11 cases of dementia among 155 individuals aged over 45 years living in a Scottish institution. The demented group consisted of three with cerebral arteriosclerosis, six with senile psychosis and two with presenile dementia. In the total sample of 155 persons there were eight with DS—two were demented, one had a paranoid psychosis, another manifested a depressive syndrome and attention-seeking behaviour and the remaining four were not mentally ill. In a follow-up report (Reid *et al.* 1978) further information of interest was provided on the two individuals with DS and dementia: one with an onset of dementia at 54 years in which depression was prominent had died at 59, and the other who died at 53 years had developed seizures at 40 and was 'withdrawn and unsociable' from the age of 50.

Lund (1985b) studied a series of 302 mentally retarded adults (aged 20 years or older) in Aarhus County, Denmark; he found mental disorder in 28 per cent. The rate of affective disorder peaked at six per cent in those aged 45–64 years; the rate for dementia was also six per cent in that age group but peaked at 22 per cent in those over 65 years. DS is not mentioned with respect to affective disorders but four of eight individuals with DS over the age of 45 years had 'the full syndrome of dementia' and there were four persons with DS in a subgroup of seven with a chronic (i.e. 'present in all of them since childhood') psychosis of uncertain type.

Also in 1985, Day reported on mentally handicapped individuals over 40 years who were either long-stay residents in a British institution or who were first admitted there in the seven year period between January, 1976 and March, 1984. Of 357 long-stay cases, 30 per cent had a mental disorder, including 23 with an affective disorder (usually psychotic), and nine were demented. Of the 215 more recent admissions, 20 per cent had a mental disorder, including 19 with affective symptomatology (usually neurotic), and three were demented. In the total of 12 demented individuals, all three with presenile onset had DS.

James (1986) described the psychiatric and behavioural problems in 50 older mentally handicapped individuals in another British institution. Behavioural problems persisting from earlier years were observed in 13 subjects and depression, frequently triggered by actual or threatened loss of close friends or by chronic physical illness, occurred in 11 of them. Subjects who may have had DS were not specified.

These studies suggest that dementia (which might be complicated by affective symptoms and seizures at some stage in the dementing process) and depression are increasingly common in older, retarded individuals, particularly those with DS. The observations contrast with those made on the psychiatric status of 64 younger (aged 8–48 years, average 24 years) retarded patients referred to a psychiatry clinic in an Ontario institution (B. D. McCreary 1979, unpublished observations). In this referred group only one individual with DS was identified; he was aged 15 years and had a psychotic disorder not unlike that described in the Lund study quoted above. By contrast, in 64 controls (i.e. individuals not referred for a psychiatry consultation and identified from the next record file to that of each referred case), there were seven individuals with DS, indicative of either a low prevalence or mild symptoms (such that referral was not initiated) of mental disorders in younger persons with DS.

In a review of the developing knowledge on the association between DS and AD, Oliver and Holland (1986) concluded: 'Although the neuropathological studies are in general agreement about the relationship between the two disorders, the proportion of individuals with Down's syndrome who can be said to dement and have the features of Alzheimer's disease has not been as well established.' Noting the discrepancy between the almost universal finding of Alzheimer neuropathology in older persons with DS and the much lower reported prevalence of dementia (ranging from 24 to 45 per cent), Karlinsky (1986) suggested that sometimes the presence of Alzheimer-type brain changes '...may simply not be clinically significant'. Also in 1986, Dalton and Crapper-McLachlan reviewed the 'clinical expression' of AD in DS and concluded that the existing studies were 'poorly documented'. They called for systematic longitudinal studies to characterize the natural history and course of dementia in persons with DS.

Furthermore, the finding of dementia in individuals with DS does not establish the presence of AD. Mental change associated with hypothyroidism is a particular concern in the differential diagnosis of dementia in these individuals.

For example, Thase (1982) reported on a 38-year-old female with DS referred with a six month history of progressive decline in functional ability and with apathy, lethargy and urinary incontinence. Following a diagnosis of dementia secondary to hypothyroidism, and after four months' treatment with levothyroxine, she had made a full recovery. Kiloh (1986) noted that mental symptoms occur early in the course of hypothyroidism and suggested that if diagnosis is made two years or more after the onset of intellectual decline, there is a risk (even with adequate treatment) of incomplete recovery. These observations are particularly significant given the high rate of hypothyroidism in adults with DS. In 55 adults with the syndrome resident in a British institution, no less than 22 per cent suffered from some degree of hypothyroidism (Mani 1988).

Accordingly, we feel that the assessment of older individuals with DS should involve the recognition not only of the basic intellectual impairment characteristic of the syndrome, but also of the effects of such conditions as seizures, hypothyroidism, mood disorders and AD. Our clinical approach accommodates these various considerations.

The management of older persons with DS who present with mental or personality change, in addition to the matters of differential diagnosis outlined above, also involves consideration of care and treatment. Lund (1986), in calling for 'further development and increased use of elaborate treatment approaches' for all retarded persons with psychiatric disorders, made the following points with respect to the demented sub-group:

- The numbers of retarded persons with these disorders will rapidly grow and constitute a major health problem.
- Medical approaches are directed towards prevention and symptomatic treatment, seeking to minimize the deterioration in abilities by preserving interests and skills.
- Psychotropic medication should be restricted apart from temporary use of hypnotics for night sedation or low doses of neuroleptics for aggressive, restless and paranoid behaviour.
- Anti-epileptic medication is rarely required because 'usually only a few seizures will occur'.

In a separate paper, Lund (1985a) deplored the admission of retarded persons with mental disorders to psychiatric institutions and called for 'special institutions or outpatient clinics' with 'specially trained social psychiatrists and staffs'. Day (1985) noted a failure on the part of health care planners in Britain to appreciate the special requirements of facilities and staff. Reid and Aungle (1974) had earlier recommended that local authority social work departments should make 'appropriate residential provision in the community for many of these elderly people'.

More detailed consideration of management is provided in a 1980 monograph by Newroth and Newroth. They suggested an alternative to institutions in the form of a small community-based facility which can 'change in accordance with the changing needs of its residents even to the point of establishing bed care'. The symptoms and stages of dementia are described along with specific suggestions for care at each stage. Also described is the need for 'a broad-based system of counselling and support for patient, staff, family and the other residents as a major means of problem-solving and coping with human anxiety'.

Thus, by 1988, there was a small but growing literature concerned with care and treatment, but little or no evaluation of the different suggested approaches either in relation to acceptability or cost.

The development and operation of our Down syndrome ageing clinic

Kingston is located in south-eastern Ontario at the juncture of Lake Ontario and the St Lawrence River. It is the site of Queen's University, whose medical school and affiliated teaching hospitals and agencies constitute a tertiary care referral centre for an estimated general population of 672,510 (Canadian census 1986) in the region. The geographic area served is made up of rural and small urban communities, and excludes greater Ottawa, which has its own medical school and referral centre.

Prime responsibility for support of persons with developmental disabilities rests in the social service sector. The provincial Ministry of Community and Social Services directly operates two institutional settings and provides financial support for a variety of community-based agencies across the region. Most of the primary medical care in the region is undertaken by family physicians in both community and institutional settings. However, specialist consultation services, teaching and research relevant to developmental disability are the focus of agreements between the medical school and the ministry referred to above. Faculty appointed to the Division of Developmental Disabilities in the Queen's University Department of Psychiatry (currently three psychiatrists, an epidemiologist and a geneticist), along with a neuropathologist from the Department of Pathology, are the key members of the DS Aging Clinic that was established in 1988 with a special approach for adults with DS.

In developing this new approach, it was decided that each person referred would be assessed annually using a checklist of signs and symptoms shown in Table 7.1. This checklist ensures a comprehensive review of each subject by combining the observations of an informant who knows the subject well and the results of a focused mental status examination by a Clinic psychiatrist. Special laboratory and radiological investigations, as deemed necessary for differential diagnosis by the psychiatrist, are arranged by the subject's primary care physician. Interim psychiatric care is provided by the Clinic psychiatrists in association with family physicians and other professionals in the individual's home community. If a subject dies, consent is sought from the next-of-kin for post-mortem neuropathological examination under the supervision of the Clinic's consultant pathologist.

Table 7.2 indicates the year of enrolment of subjects and certain of their characteristics since the Clinic opened in early 1988 until the end of 1991. All with DS residing in the two institutions previously mentioned, regardless of

Observations of informant (in preceding 12 months)	Observations of examiner
I Self-help skills (including dressing, hygiene, grooming, eating and continence)	I Posture and gait
	II Muscle tone
	III Involuntary movements
II Domestic skills (including setting table, washing dishes and following instructions)	IV Deep tendon reflexes
	V Superficial reflexes (including grasp, snout and palmomental)
III Gait and motor function (including pacing, posture and involuntary movements)	VI Orientation (person, place)
	VII Aphasia (motor, sensory)
	VIII Short-term memory (recall of three items after two minutes)
IV Language (including sensory or motor aphasia)	IX Laterality (hand preferred and distinguishing left versus right)
V Socialization (including interest in others, irritable/aggressive and attention-seeking behaviours)	X Use of pencil to reproduce cross, circle, square and diamond
VI Personality (including energy level and mood variability)	
VII Psychosis (including delusions and hallucinations)	
VIII Memory (including concentration, loss of personal effects and orientation)	
IX Miscellaneous (including appetite, sleep disturbance and mannerisms)	

Note 1: each item is evaluated by informant at one of three levels of magnitude or as 'unknown'.
Note 2: recent changes in health status, use of medications and life situations that may have an impact on adjustment are also recorded.

Table 7.1. Clinical features checklist

age, were enrolled including those who moved to community residential settings in a major de-institutionalization effort in 1989.

Observations

Practical and clinical observations made in the Clinic between 1988 and 1991, supplemented by related information published over the same interval, are presented below under four subheadings. Discussion of the possible significance of these observations is provided in the final section of this chapter.

Early case finding

Early case finding is concerned with the recognition of deterioration in mental functioning in older persons with DS by those involved in their care. The role of family members or agency staff in prompt recognition of mental change is similar to that which applies to dementia in non-DS subjects. Kurz *et al.* (1990) noted that dementia is recognized when mental impairment sufficient to interfere with everyday living is present; in addition to the presence and attention of

Year of first assessment	Gender		Age at entry		Residential setting	
	Male	Female	≤40	>40	Institution	Community
1988	43	30	45	28	53	20
1989	55	41	60	36	47	49
1990	8	7	7	8	0	15
1991	7	11	8	10	3	15
Totals	113	89	120	82	103	99

Table 7.2. Year of enrolment and subject characteristics

an observer, it appears that the level of demand placed on the subject and his or her capacity for compensation are important variables. A vignette serves to illustrate these issues:

> *W was enrolled in the Clinic in April 1988 at 46 years of age. The informant, a residential staff member, reported that W was doing well with no concerns about mental deterioration. At his annual follow-up in 1989 he was reported by residential staff still to be doing well but the physiotherapist, for whom he escorted patients to and from the physiotherapy department, noted that he seemed 'confused and forgetful'. W continues to function well according to informants from the residential area. However, the physiotherapist reports that he continues, on occasion, to lose his way in familiar parts of the institution and that, while he sometimes is agitated and irritable, he often seems blandly unaware of his confusional episodes.*

Thus, the threshold for expressing concern by those in regular contact with ageing persons with DS reflects not only their level of awareness of change but, more importantly, the context of their observations with respect to demands on perception, memory and other intellectual functions. It seems likely that psychometric tests, which involve a structured and concentrated interval of demand, may yield an earlier threshold of recognition than the observations of family or agency staff who relate to the subject in everyday circumstances (Haxby 1989). In some cases, however, agency staff may be better informed than medical advisors in recognizing the need for a special assessment when problems arise. A referral letter concerning a 53-year-old group home resident who had been apathetic, moody and experiencing problems in following complex instructions over the preceding few months is illustrative:

> *'...I am referring this gentleman to you at the request of the Community Living Association of whom I am becoming increasingly weary. They keep making suggestions after conferences regarding my patients... At this time they have requested a referral for C to be assessed for the development of Alzheimer disease. I am not sure that I regard this as an*

*appropriate referral or cost-justifiable or of any significant benefit to C. I
would be grateful for your advice and, if you feel appropriate, your
support in rejecting such requests from the Association.'*

Murdoch and Anderson (1990) described the deficiencies in care when
generically trained physicians must include individuals with syndromes unfamiliar to them, such as DS, in their practice. Beange and Bauman (1990) suggested a special health promotion unit for persons with developmental
disabilities as a possible remedy for such deficiencies. Physicians who have
made developmental disability more than an incidental component of their
practice, such as those attending individuals living in institutional settings, are
better informed on the issues involved.

The DS Aging Clinic team has undertaken a number of educational efforts to
enhance awareness and to facilitate improved care. These include preparation
of an information sheet to supplement information provided by telephone, presentations at lay and professional conferences (e.g. Holden 1989; McCreary *et
al.* 1990) and mail-outs to physicians practising in the region. It has been difficult to evaluate the impact of these educational endeavours but the provision of
good consultative advice on each case reviewed appears to have immediately
obvious effects. For example:

> *B, a 34-year-old group home resident, was referred to the Clinic in the
> fall of 1990. The residential staff had noted depression, crying spells,
> withdrawal and a loss of interest in social activities over the preceding
> few months. She was taken to the local physician for review prior to
> referral to the Clinic. As the physician had referred another person with
> DS a few months previously and had learned about the risk of
> hypothyroidism, he had already diagnosed hypothyroidism and
> commenced treatment with a thyroid supplement by the time B was
> enrolled in the Clinic. Within six weeks of starting the supplement, B's
> mental state had returned to her normal, happy and sociable level. At
> follow-up in the fall of 1991 she remains well on the thyroid treatment.*

Friedman *et al.* (1989) and Pueschel *et al.* (1991) noted the significant risk
that hypothyroidism would be unrecognized in individuals with DS and recommended annual laboratory monitoring of thyroid functioning in primary medical care settings. Thirty-five of 155 (22 per cent) non-demented subjects
attending our Clinic are being treated for hypothyroidism and 18 of 47 (38 per
cent) demented subjects receive thyroid supplement.

Differential diagnosis

In addition to hypothyroidism to which persons with DS are predisposed and
which can mimic the dementia of AD or co-exist with it, two other conditions,
namely depression and anticonvulsant drug toxicity, are important differential
diagnoses.

In the non-DS population, 40–50 per cent of persons with AD show depressive symptoms at some point during their illness (Lauter and Dame 1991). In

contrast, depression is reported less frequently in those with DS. For example, Dalton and Crapper-McLachlan (1986) noted only six references to depression in 33 case reports. Although persons with DS are possibly less prone to become depressed, an acknowledged difficulty in recognizing depression in persons who are mentally retarded (Warren *et al.* 1989; Storm 1990; Clark *et al.* 1991) is a more likely explanation. Three vignettes concerning individuals with DS are provided below to illustrate, respectively, a depression occurring prior to the age of risk for dementia, a depression coinciding with the early development of dementia and a possible atypical depression occurring late in dementia.

B, aged 27, was referred for a psychiatric opinion. He was quite well until the age of 24 when he had two seizures. At that time an EEG revealed 'general slowing' but no specific epileptic features. He took diphenylhydantoin for a year and had no further seizures after its discontinuation. Thyroid function and a CT scan were normal. In the few weeks preceding referral he refused to attend work, feared that he was becoming blind, was suspicious of a friend who lived next door and isolated himself in the home, watching TV well into the night. B had a poor relationship with his father who abused alcohol and was somewhat over-protected by his mother. He was also embarrassed about alopecia but remained withdrawn and isolated even after being provided with a hair-piece. He would not take antidepressant medications despite saying that he felt irritable and sad. The depressive syndrome gradually improved over an 18-month period. At the most recent contact (aged 30 years) he appears to have fully recovered.

Although we had no baseline information for comparison, B never appeared to have cognitive deterioration. Seizures and EEG slowing, however, raise the possibility that his depressive syndrome is an early manifestation of Alzheimer neuropathology. Since a high percentage of non-DS persons with a 'depressive pseudodementia' become demented (Kral 1983), this possibility will be explored in B's annual follow-up assessments.

M was enrolled in the Clinic at the age of 51 in 1989. In 1986 she had a seizure and her EEG revealed a bilateral slow-wave dysrhythmia. Over the next few months she was noted to have crying spells, insomnia and weight loss. Neither her seizures nor her depressive syndrome persisted, however, so medical treatment was not needed. Her thyroid indices were normal. Two years prior to her enrolment, she had problems in following instructions, lost domestic and self-help skills and was often disoriented. The dementia has been slowly progressive. She requires occasional night sedation for insomnia.

Eleven of 155 (seven per cent) non-demented, compared with seven of 47 (14 per cent) demented subjects with records in the Clinic have been treated for depression. For those with dementia, treatment was characteristically given early in the course of the illness. If apathy and social withdrawal, frequent early symptoms usually considered to be reactions to failing cognitive powers, are depressive in origin it is likely that the potential benefits of antidepressant treatment have been under-estimated.

P was enrolled in the Clinic at the age of 59, a few months prior to her death from pneumonia. Previously friendly and sociable, she developed problems in maintaining her vocational performance at the age of 50 and apathy, social withdrawal and seizures at the age of 54. Transferred to a nursing unit at the age of 56, she developed a significant personality change at the age of 58—while mostly apathetic, she had sudden outbursts of irritable, noisy and aggressive behaviour requiring sedation with a major tranquillizer. A post-mortem examination revealed severe Alzheimer neuropathology, including lesions in the mid-brain.

Late in the dementia of some non-DS individuals with AD, Kral (1983) reported an association between a chronic depressed mood, irritability and aggressive outbursts. A possible link between these behaviours and specific neuro-anatomical and neurochemical changes in mid-brain areas involved in mood regulation has also been reported (Torack and Morris 1988; Zubenko and Moossy 1988; Bowen *et al.* 1992). A serotonin-enhancing diet, along with trazodone, was found to be therapeutic for aggressive behaviour in an adult with DS showing signs of AD (Gedye 1991).

Commonly used anticonvulsants may reduce cognitive function to the degree that dementia is suspected. The following vignette illustrates this phenomenon:

P was enrolled in the Clinic at the age of 41 in 1989. At the age of 24 she had developed seizures and her EEG, in addition to general slowing, revealed bilateral temporal foci. Seizure control on diphenylhydantoin and phenobarbital was imperfect. An informant described P as moody, irritable and sometimes confused in completing chores such as setting the table. This picture persisted at each follow-up until late in 1991 when the informant described her as friendly, much less apathetic and more efficient in work activities. In the interim her family physician had discontinued her phenobarbital and reduced the diphenylhydantoin dose. Her serum diphenylhydantoin level, previously in the upper therapeutic-low toxic range, was now in the mid-normal range and her serum folate was normal.

On superficial review P might have been regarded as demented. However, careful review of three successive assessments revealed an absence of deterioration in the judgement of both the informant and the psychiatrist who examined the subject. Evenhuis (1990) reported that anti-epileptic drugs often caused marked lethargy in individuals with DS, even at blood levels in the lower therapeutic range. In such cases carbamazepine or valproic acid may be preferable to diphenylhydantoin or phenobarbital (Committee on Drugs 1985).

Natural history and prognosis

Familiarity with the natural history of mental decline in older persons with DS enables physicians to provide information about care and treatment for family and agency staff. Issues in this regard, such as the duration of the dementia and the nature of the terminal events, are considered below.

Nine subjects died over the 1988–1991 period. Consent for post-mortem examination was granted for seven of them. Table 7.3 summarizes clinical and post-mortem data on these seven cases. In the following discussion, case numbers listed in the table are used for ease of reference.

As expected, Alzheimer neuropathology was observed in all seven instances. For the six subjects in whom dementia was clearly present prior to death, the average age of onset of dementia and of death was 53.8 and 58.8 years respectively. Case 2, the exception, requires separate mention as it was impossible to date confidently the onset of a dementia. He had been admitted to an institution at 45 years of age. He functioned at a profoundly retarded level from the outset of his stay. Informants described a very slow decline in abilities over several years, but we were left with an impression that onset of deterioration may have pre-dated the admission, which occurred when his mother died. In contrast, case 1 demonstrates a more clearly defined interval of dementia, starting with impaired vocational adjustment and progressing slowly over six years in a residential setting and a final three years in a nursing care unit; the subject died of aspiration pneumonia at 59 years. Our overall observations on all the cases closely resemble those reported in prospective studies (Lai and Williams 1989; Evenhuis 1990), but contrast with retrospectively gathered data (summarized by Dalton and Crapper-McLachlan 1986) revealing an average age at death of 50.5 years (range 35–60.3). Presumably sampling issues underlie this discrepancy.

Each of the seven autopsied subjects had pathological evidence of pneumonia except case 5 who had a pulmonary embolism. Apparent 'respiratory deaths' occurred, as well, in the two individuals who died without permission for post-mortem study: one, a profoundly retarded woman with no evidence of dementia, died of pneumonia at 36 years; the other was a 62-year-old woman with a four year history of dementia and a recent deep venous thrombosis of her left leg who died in 'respiratory distress'.

Three cases had been treated for seizures. In one case each the seizures commenced before there was clinical evidence of dementia (case 4), at the onset of dementia (case 3) and after dementia had developed (case 1). Seizure control was easily achieved, without reports of significant side effects. Treatment for depression was undertaken in only one instance (case 3); correspondence with the physician who attended her at the time indicated that it was 'ineffective'. Five of the seven cases were also treated for hypothyroidism, providing further evidence of the high prevalence of this disorder in adults with DS, as previously mentioned. Strokes were recorded in cases 2 and 7. In case 2 there was a large infarct in the area served by the right middle cerebral artery and there was cerebral artery amyloid deposition. Vascular amyloid was also observed in case 5

Table 7.3. Natural history, treatment and autopsy findings

Case number and year of death	Gender	Age at onset of dementia	Clinical features prominent at onset of dementia	Seizures	Treatment for[a] Depression	Hypothyroidism	Age at death	Features of pathologist's report
1. 1989	F	50	Episodes of poor workshop performance.	54	Nil	Nil	59	Aspiration pneumonia. Severe AD (including midbrain lesions).
2. 1989	M	?	No good evidence of onset of dementia. Sudden left hemiplegia and death 1 week later.	Nil	Nil	59	60	Bronchopneumonia. Recent right middle cerebral artery area infarct. Amyloid arteriopathy. AD.
3. 1990	F	57	Apathy and gradual loss of skills.	57	57	'years'	61	Bronchopneumonia. Severe AD (including occasional midbrain lesions).
4. 1990	F	50	Apathy, wandering at night. Gradual loss of skills.	47	Nil	42	57	Viral pneumonia. Lymphoid thyroiditis. AD.
5. 1991	M	56	Apathy, enuretic at times. Marked obsessional traits.	Nil	Nil	56	59	Recent pulmonary embolism and thrombotic occlusion of right coronary artery. Severe AD.
6. 1991	M	51	Difficulty following verbal instructions. Perserverative.	Nil	Nil	51	54	Bronchopneumonia. Endocarditis. Fallot's tetralogy and septic pulmonary emboli. Cirrhosis secondary to hepatitis B infection. AD (including midbrain lesions).
7. 1991	M	59	Gradual loss of skills. Incontinent at times.	Nil	Nil	Nil	63	Bronchopneumonia. Parietal lobe infarct. Fibrosis of liver and passive congestion of spleen. AD.

[a] Age at onset of treatment is listed

but there was no evidence of past or recent infarction. There was post-mortem evidence of cirrhosis in case 6 (secondary to chronic hepatitis B infection) and in case 7 (secondary to a thickened left hepatic duct). The chronic hepatitis in case 6 presumably contributed to the subject's sluggish, passive and perseverative behaviour which complicated attempts to assess his intellectual decline. Similarly, the mental state of case 7 varied considerably in relation to his general health, although the presence of a progressive dementia was never in doubt. Thus, the prognosis in an individual case reflects not only the course of the dementia but the impact of co-existent disorders and their treatment.

Care and treatment

None of the seven individuals represented in Table 7.3 was provided with an evolving residential programme that would preclude the need for a move from a residential to a nursing care setting. Three individuals (cases 1, 4 and 5) were moved from a residential to a nursing home setting at the mid-point of their dementia. Such moves occur when the individual is seen to require mainly nursing care (i.e. has become unsteady, incontinent, confused and often needs to be fed). Case 3 was transferred directly from her home to a nursing home when her mother developed AD; she could then have been supervised in a group home for two to three years but it was thought preferable for her and her mother to live in the same care setting. Cases 2 and 6 were maintained in very supportive residential settings until terminal health problems (i.e. stroke, hepatic failure) prompted admission for acute medical and nursing care. Case 7, living in an institution, had multiple admissions to its infirmary for a variety of health problems (i.e. fractured hip, mild stroke, coccygeal ulcer), with approximately 25 per cent of his care being provided there over the four-year course of the dementia.

Once the dementing process is established those involved with the affected individual must prepare for an interval of care often extending over four to five years. The main burden of care is likely to rest with the family or with agency staff in a residential setting. A role of our Clinic is to provide advice and support directed at maintaining quality of life for the afflicted person and at minimizing the burden of care for the primary caregivers. This role is illustrated in the following vignette:

M was seen first in 1987 in the Clinic. She was aged 43 and had lived in a group home quite successfully for the preceding seven years. Her residential counsellor had noted some changes in the preceding 12 months. These included problems in following complex instruction, reduced concern regarding her dress and personal hygiene and forgetting items of silverware when setting the table. Thyroid function was normal and early AD was suspected. The dementia progressed gradually until the fall of 1991 when M, for the first time, developed seizures. Following a short admission for a neurological review in the local hospital, M's dependency on staff increased dramatically. The agency decided to transfer her to a special support home which she shares with three other persons with developmental disabilities. The residential staff acknowledge that her care requires an approach based on acceptance of her failing adaptation rather than on the 'behaviour management' orientation suitable for the other residents. This acceptance is enhanced by the Clinic psychiatrist who has pointed out M's visual, perceptual, aphasic and apraxic impairments and their impact on her adjustment in the home. Meetings with staff in the home are the most effective method for providing support and education. The family doctor continues to provide primary medical care.

Our present Clinic team does not include an occupational therapist. However, it is clear that experiences of occupational therapists (e.g. Zgola 1990) concerned with AD in the general population are to a considerable degree applicable to individuals like M and the residential staff who care for her.

In addition, the importance of post-mortem examination and of genetic research on the association between DS and AD is discussed with the Clinic clientele's next-of-kin and done only with appropriate informed consent. It is explained that neuropathological study is required to confirm a clinical impression of AD and that an autopsy will also establish the most likely cause of death. Genetic investigations undertaken have been described elsewhere (Holden *et al.* 1991). Sensitivity concerning anxiety about genetic implications is obviously necessary. For example, a rift between a subject's mother who was very interested in the genetic issues, and his sister for whom consideration of these issues was personally traumatic, has followed in the wake of one family study. The need, identified by Alexander *et al.* (1992), to respect the personal dignity and right to privacy of each family member in genetic studies of bipolar affective disorder is equally important in studies on the association between DS and AD.

Conclusions

The observations presented in this chapter allow some conclusions concerning service provision and further research on the association between DS and AD. While these conclusions are related primarily to our endeavours in the Kingston region they may be of some interest in other jurisdictions as well.

From a service perspective, early case finding, differential diagnosis, prognostication and suitable approaches to care and treatment are all important. Assuming a prevalence of DS of about one per 1,000, the arrangements necessary for optimum care must recognize that there are many affected persons. Each one, regardless of age, needs to be reviewed annually by a primary care physician to ensure prompt recognition and treatment of hypothyroidism and other disorders. For those in the age range of 35–40 years, it seems judicious to complete a clinical review, such as that provided in our Clinic, in order to record a basis against which any subsequent mental deterioration can be measured. This assessment could be supplemented by selected psychometric, electrophysiological and radiological examinations, but given the present state of knowledge and the expense of more elaborate investigations, a basic clinical assessment seems sufficient. In contrast, once mental deterioration commences, various special investigations may be very important in the development and monitoring of treatment programmes. It appears unlikely that high quality care can be provided without a special regional clinic; its attachment to a regional psychiatry consulting service makes for efficient use of scarce professional man-power. Families, agency staff and primary care physicians rely on the expert advice and support available through such a clinic. The addition of the services of an occupational therapist, who could provide suggestions for reducing the burden of care faced by family members and agency staff, is desirable. Finally, educational efforts to increase the knowledge level of all those who serve persons with DS must be encouraged.

From a research perspective, there is a continuing need for longitudinal clinical studies, including arrangements for post-mortem examination of those enrolled in such studies. Amongst particular topics that deserve attention is the apparent low prevalence of depression in persons with DS compared with the rates for mood disorder reported in the general population. Further research on this matter requires clinical, psychometric and laboratory observations in living subjects and systematic neuro-anatomical and neurochemical studies after death. The results of such investigations will be of interest to those responsible for the care and treatment of older persons with DS but could also contribute to a broader understanding of mood disorders seen in older members of the general population. As basic research continues on the biological aspects of the DS–AD association, it is important to take into account clinical observations revealing that older persons with DS are predisposed to many conditions that can mimic or complicate the recognition of dementia—the need for a close association between the clinic and the laboratory is obvious.

Acknowledgements

Support for work reported in this chapter was gratefully received from the Ontario Mental Health Foundation and the Ontario Ministry of Community and Social Services and from a research grant (MCSS Research Grants Programme) administered by the Research and Programme Evaluation Unit of that Ministry in cooperation with the Ontario Mental Health Foundation.

References

Alexander, J. R., Lerer, B., and Baron, M. (1992). Ethical issues in genetic linkage studies of psychiatric disorders. *British Journal of Psychiatry*, **160**, 98-102.

Beange, H., and Bauman, A. (1990). Health care for the developmentally disabled: Is it necessary? In *Key issues in mental retardation research*, (ed. W. I. Fraser), pp. 154-62. Routledge, London.

Bowen, D. M., Francis, P. T., Pangalos, M. N., Stephens, P. H., Proctor, A., and Chessell, I. (1992). 'Traditional' pharmacotherapy may succeed in Alzheimer's disease. *Trends in Neurosciences*, **15**, 84-5.

Census of Canada (1986). Pp 80, 121, 206, 217, 229, 283, 350, 373 and 410. Statistics Canada, Ottawa.

Clark, A. K., Reed, J., and Sturmey, P. (1991). Staff perceptions of sadness among people with mental handicaps. *Journal of Mental Deficiency Research*, **35**, 147-53.

Committee on Drugs, American Academy of Pediatrics (1985). Behavioral and cognitive effects of anticonvulsant therapy. *Pediatrics*, **76**, 644-6.

Dalton, A. J., and Crapper-McLachlan, D. R. (1986). Clinical expression of Alzheimer's disease in Down's syndrome. *Psychiatric Clinics of North America*, **9**, 659-70.

Day, K. (1985). Psychiatric disorders in the middle aged and elderly mentally handicapped. *British Journal of Psychiatry*, **147**, 660-7.

Evenhuis, H. M. (1990). Natural history of dementia in Down's syndrome. *Archives of Neurology*, **47**, 263-7.

Friedman, D. L., Kastner, T., Pond, W. S., and Rice O'Brien, D. (1989). Thyroid dysfunction in individuals with Down syndrome. *Archives of Internal Medicine*, **149**, 1990-3.

Gedye, A. (1991). Serotonergic treatment for aggression in a Down's syndrome adult showing signs of Alzheimer's disease. *Journal of Mental Deficiency Research*, **35**, 247-58.

Haxby, J. V. (1989). Neuropsychological evaluation of adults with Down's syndrome: patterns of selective impairment in non-demented old adults. *Journal of Mental Deficiency Research*, **33**, 193-210.

Holden, J. (1989). Down syndrome and Alzheimer dementia: clinical and genetic associations background. In *Preparing for the year 2000*. Proceedings of the 1989 National Conference of the Canadian Down Syndrome Society. University of Victoria, British Columbia, pp. 113-9.

Holden, J., Chalifoux, M., Clairman, C., Dalziel, F., Ditullio, K., Greer, D., *et al.* (1991). Down's syndrome and Alzheimer's dementia: clinical evaluation and genetic association. In *Alzheimer's disease: basic mechanisms, diagnosis and therapeutic strategies*, (ed. K. Iqbal, D.R.C. McLachlan, B. Winblad and H. M. Wisniewski), pp. 435-41. Wiley, New York.

James, D. H. (1986). Psychiatric and behavioural disorders amongst older severely mentally handicapped inpatients. *Journal of Mental Deficiency Research*, **30**, 341-5.

Karlinsky, H. (1986). Alzheimer's disease in Down's syndrome: a review. *Journal of the American Geriatrics Society*, **34**, 728-34.

Kiloh, L. G. (1986). The secondary dementias of middle and later life. *British Medical Bulletin*, **42**, 106-10.

Kral, V. A. (1983). The relationship between senile dementia (Alzheimer type) and depression. *Canadian Journal of Psychiatry*, **28**, 304-6.

Kurz, A., Romero, B., and Lauter, H. (1990). The onset of Alzheimer's disease. A longitudinal case study and a trial of new diagnostic criteria. *Psychiatry,* **53**, 53-61.

Lai, F., and Williams, R. S. (1990). A prospective study of Alzheimer's disease in Down's syndrome. *Archives of Neurology*. **46**, 849-53.

Lauter, H., and Dame. S. (1991). Depressive disorders and dementia: the clinical view. *Acta Psychiatrica Scandinavica*: Supplement, **366**, 40-6.

Lund, J. (1985a). Mentally retarded admitted to psychiatric hospitals in Denmark. *Acta Psychiatrica Scandinavica*, **72**, 202-5.

Lund, J. (1985b). The prevalence of psychiatric morbidity in mentally retarded adults. *Acta Psychiatrica Scandinavica*, **72**, 563-70.

Lund, J. (1986). Treatment of psychiatric morbidity in the mentally retarded adult. *Acta Psychiatrica Scandinavica*, **73**, 429-36.

McCreary, B. D., Holden, J. J. A., Fotheringham, J. B., Burley, J., and Robertson, D. M. (1990). Characterising the association between Down syndrome and Alzheimer's disease. Paper presented at Canadian Psychiatric Association annual meeting, Toronto.

Mani, C. (1988). Hypothyroidism in Down's syndrome. *British Journal of Psychiatry*, **153**, 102-4.

Murdoch, J. C., and Anderson, V. E. (1990). The management of Down's syndrome children and their families in general practice. In *Key issues in mental retardation research*, (ed. W.I. Fraser), pp. 143-53. Routledge, London.

Newroth, A., and Newroth, S. (1980). *Coping with Alzheimer's disease: a growing concern*. National Institute on Mental Retardation, Downsview, Ontario, Canada.

Oliver, C., and Holland, A. (1986). Down's syndrome and Alzheimer's disease: a review. *Psychological Medicine*, **16**, 307-22.

Penrose, L. S. (1963). *The biology of mental defect*, (3rd edn). Sidgwick and Jackson, London.

Pueschel, S. M., Jackson, I. M. D., Giesswein, P., Dean, M. K., and Pezzullo, J. C. (1991). Thyroid function in Down syndrome. *Research in Developmental Disabilities*, **12**, 287-96.

Reid, A. H. (1972). Psychoses in adult mental defectives: I. Manic depressive psychosis; II. Schizophrenic and paranoid psychoses. *British Journal of Psychiatry*, **120**, 205-12 and 213-8.

Reid, A. H., and Aungle, P. G. (1974). Dementia in ageing mental defectives: a clinical psychiatric study. *Journal of Mental Deficiency Research*, **18**, 15-23.

Reid, A. H., Maloney, A. F. J., and Aungle, P. G. (1978). Dementia in ageing mental defectives: a clinical and neuropathological study. *Journal of Mental Deficiency Research*, **22**, 233-41.

Storm, W. (1990). Differential diagnosis and treatment of depressive features in Down's syndrome: a case illustration. *Research in Developmental Disabilities*, **11**, 131-7.

Thase, M. E. (1982). Reversible dementia in Down's syndrome. *Journal of Mental Deficiency Research*, **26**, 111-3.

Torack, R. M., and Morris, J. C. (1988). The association of ventral tegmental area histopathology with adult dementia. *Archives of Neurology*, **45**, 497-501.

Warren, A. C., Holroyd, S., and Folstein, M. F. (1989). Major depression in Down's syndrome. *British Journal of Psychiatry*, **155**, 202-5.

Zgola, J. (1990). Alzheimer's disease and the home: issues in environmental design. *The American Journal of Alzheimer's Care and Related Disorders and Research*, **5**(3), 15-22.

Zubenko, G. S., and Moossy, J. (1988). Major depression in primary dementia: clinical and neuropathologic correlates. *Archives of Neurology*, **45**, 1182-6.

IV

Specific assessment and diagnostic research concerning Alzheimer disease in persons with Down syndrome

8

Assessment of cognitive functioning in persons with Down syndrome who develop Alzheimer disease

L. Crayton and C. Oliver

Summary

The early changes of Alzheimer disease in adults with Down syndrome may not be apparent without neuropsychological assessment. However, the application and interpretation of neuropsychological test procedures utilized to facilitate detection of Alzheimer disease in the general population may not always be suitable for this purpose in persons with Down syndrome. Problems encountered include difficulties in differentiating prior mental retardation from superimposed dementia, floor effects of many of the usual components of a neuropsychological test battery and, in research studies, methodological limitations due to small sample size, absence of appropriately matched controls or cross-sectional design. Nevertheless, results of neuropsychological studies summarized in this chapter provide convincing evidence that as age advances, some, if not all, adults with Down syndrome undergo a pattern of intellectual deterioration consistent with that observed in Alzheimer disease in the general population.

Introduction

Since the early investigations of Jervis (1948), a significantly increased risk of developing Alzheimer disease (AD) in adults with Down syndrome (DS) has frequently been reported. Nearly all persons with DS who die over the age of 35 years show Alzheimer-type neuropathology on post-mortem examination (Ball and Nuttall 1980; Wisniewski *et al.* 1985a; Chapter 5). These pathological changes appear to be the same as those which occur in individuals with AD who do not have DS (Yates *et al.* 1980).

There has been less clarity concerning clinical deterioration typical of AD (e.g. memory loss, aphasia, disorientation) as age advances in adults with DS who presumably have the AD neuropathological changes referred to above. Whilst some studies have reported an increase of specific cognitive impairments with age (e.g. Thase *et al.* 1984), other research (e.g. Silverstein *et al.* 1988) suggests that this is not necessarily the case. This inconsistency may be due to difficulties in recognizing further cognitive impairments in persons with DS who have pre-existing intellectual deficits. Assessment of adults with DS on the basis of caregiver reports and clinical evaluation may miss early changes due to AD. Within the general population, neuropsychological impairments associated with AD involve a wide range of functions (Grady *et al.* 1988). It is therefore necessary to examine this wide range when assessing adults with DS.

Difficulties of neuropsychological assessment for Alzheimer disease in persons with Down syndrome

The difficulties associated with neuropsychological assessment of AD in persons with DS have been noted by several authors (e.g. Malamud 1972; Ellis *et al.* 1974). The application and interpretation of the usual test procedures utilized for detecting AD in the general population may need to be restricted when assessing persons with DS, because of their underlying intellectual impairment. Failure on test items may be due either to this impairment or to more recent dementia. Many cross-sectional age group comparison studies have either ignored this consideration or assumed a similar distribution of degree of mental retardation in all age groups. Others have matched age groups on the basis of IQ to control for the level of mental retardation. However, matching by present IQ alone does not eliminate the possibility that subjects fail the test items either because of their prior mental retardation or because of dementia. Thus the problem of deficits being attributable to dementia or to pre-existing intellectual impairment often goes unresolved.

Further problems in identifying developing cognitive impairments in persons with DS have a more general basis. Individuals with the syndrome typically lead 'sheltered' lives and are often dependent on others to meet their daily needs. Consequently they are not usually subject to substantial intellectual demands or ordinarily expected to accomplish complex tasks. Hence early symptoms of AD such as memory loss and confusion are often not immediately apparent and are difficult to detect from a cursory appraisal of daily functioning. Persons in the general population who develop AD are frequently detected relatively early because a high level of demands in their occupations tends to reveal deteriorating performance. By contrast, persons with DS who develop AD usually maintain personal integrity by continuing to perform routine tasks and to converse socially at a relatively superficial level. Even if difficulties in functioning become apparent, they may be incorrectly attributed to the underlying mental retardation rather than to the developing dementia. Miniszek (1983) has argued that it is the nature of the environment in which the individual lives that masks the presence of cognitive deficits by providing shelter from tasks that might expose the deficit.

In addition, persons with DS are prone to develop many conditions (e.g. hypothyroidism, sensory deficits) that can result in deterioration, so that deterioration due to AD, even if the symptoms become apparent, can easily be attributed to other causes (see Chapter 6). Furthermore, in persons with DS there is commonly a lack of previous information on neuropsychological functioning.

This, combined with great variability on cognitive testing of these persons, makes it difficult to infer previous performance levels and therefore to recognize early signs of deterioration when they occur. The presence of AD is usually diagnosed by a loss of intellectual capabilities severe enough to interfere with occupational or social functioning. Current functioning must therefore be lower than was previously the case. For those who initially have a low level of cognitive, occupational and social functioning, the neuropsychological assessment of AD is particularly difficult.

Neuropsychological assessment of AD in persons with mental retardation often begins with an IQ test. In addition, specific measures of attention, memory, language, motor co-ordination, constructional abilities, social functioning and behavioural skills are frequently undertaken. Whilst standardized psychometric evaluation may provide important information, it is not always satisfactory for use with persons with mental retardation. Hence supplementary new tests, derived from experimental research, are necessary.

Neuropsychological studies

Within the general population, neuropsychological assessment of AD has mainly been cross-sectional in design, comparing affected persons with age-matched normal controls. Several investigations have shown that impaired secondary or recent memory is usually the first neuropsychological deficit to appear, with other cognitive functions such as praxis and language affected later in the course of the disease (e.g. Crapper-McLachlan *et al.* 1984; Haxby *et al.* 1986). In contrast to numerous such cross-sectional studies, there have been few longitudinal examinations of persons with AD. Grady *et al.* (1988) found that memory deficits were the first neuropsychological impairments to occur in AD, followed by problems with attention to complex cognitive sets and abstract reasoning, and thereafter by deficits in language and visuospatial abilities.

Neuropsychological studies of persons with DS have generally shown a similar pattern. Cross-sectional age group investigations of persons with the syndrome have established the presence of neuropsychological changes in a significant proportion of older individuals compared with younger ones and others without DS matched for age and IQ. These changes can occur in a variety of cognitive areas. On some variables critical in the assessment of AD (e.g. orientation, attention span, memory, object naming, praxis), deficits have been most apparent in elderly subjects with DS.

There are, however, several limitations to a cross-sectional age group study design. While providing an accurate assessment of current functioning, inferences concerning observed deficits in older subjects with DS have to be drawn from comparisons with younger ones with the syndrome as well as matched controls. Although unlikely, such an approach does not rule out the possibility that some factor other than AD, albeit currently unknown, specifically and adversely affects older individuals with DS. Studies investigating age-related neuropsychological changes have also been limited by either small sample size or lack of appropriately matched controls. Another criticism is that there is often a lack of comprehensive screening protocols to detect those individuals with dementia of other origins than AD (e.g. hypothyroidism, which is particularly common in DS and can produce marked cognitive impairments in older affected adults with the syndrome). Furthermore, most studies contain few subjects surviving into the fifth decade or beyond, and often have no baseline information on previous functional levels against which to compare current findings.

Cognitive tests that have been used to detect deterioration in adults with DS may not be sufficiently sensitive to the changes which occur. Changes in areas such as memory function are perhaps too subtle to be noted unless tests are specifically designed for this group. Moreover, the difficulties of conducting adequate neuropsychological tests of function in individuals with DS are well known. Assessment of language-based functions and auditory abilities in particular require a degree of co-operation and skills that are absent or minimal in a proportion of adults with DS (see Chapter 9).

The problem of unequivocal neuropathological findings, but limited neuropsychological evidence, of AD in individuals with DS has been addressed in many recent studies by using specific neuropsychological assessments to detect age-related deficits (for a comprehensive review of the early studies see Oliver and Holland 1986).

Owens *et al.* (1971) observed significant differences between a group of 35–50-year-old and a group of 20–25-year-old individuals with DS matched for IQ. The older group gave poorer responses for identification of objects and showed evidence of the presence of focal neurological and frontal release signs. Dalton *et al.* (1974) noted that memory loss in a series of adults with DS aged 44–58 years was similar to that exhibited by persons with AD in the general population. These investigators also compared two groups with DS, aged 39–43 years and 19–23 years respectively, with controls on a simultaneous and delayed visual matching task. The older DS group made more errors than the others. In another study, Wisniewski *et al.* (1978) reported many differences between persons with DS aged above and below 38 years. They found evidence of changes in personality, loss of vocabulary, loss of personal hygiene skills, changes in activities of daily living, recent memory loss, difficulty in object identification and problems with short-term visual retention. Thase *et al.* (1982) found that individuals with DS had greater impairments of orientation, attention span, digit span, visual memory and object identification than controls, with deterioration tending to occur as age advances.

Various studies have also shown deficits in orientation (Wisniewski *et al.* 1978; Thase *et al.* 1982, 1984), object identification (Owens *et al.* 1971; Wisniewski *et al.* 1978; Thase *et al.* 1982), attention (Thase *et al.* 1982) and short-term memory (Dalton *et al.* 1974; Dalton and Crapper 1977; Wisniewski *et al.* 1978). The most consistent finding has been that deficits in neuropsychological functioning are likely to be associated with increasing age in persons with DS.

Recent studies of neuropsychological functioning in Down syndrome

A selection of recent studies that have measured neuropsychological changes associated with AD in persons with DS are summarized in Table 8.1. These studies have demonstrated age-related differences in neuropsychological functioning. Two studies included omnibus measures of overall ability (Stanford-Binet tests and client development evaluation reports) (Fenner *et al.* 1987; Silverstein *et al.* 1988) that provide good indications of general ability, but no information about patterns of neuropsychological performance.

Evaluation of memory functions has shown age-related deficits (e.g. Schapiro *et al.* 1988; Haxby 1989; Lai and Williams 1989). Measurement of memory function has generally concentrated on investigating deficits in short-term memory (e.g. Young and Kramer 1991; Devenny *et al.* 1992). Little attention has been paid to the ability to commit information to long-term memory, an important deficit in very early AD. However, Haxby (1989) found that both non-demented and demented elderly DS groups performed worse on tests of hidden object memory and recognition memory for designs than a young DS group.

Some authors (e.g. Wisniewski and Hill 1985) have felt that identification of neuropsychological deficits indicative of AD may be particularly difficult in persons who have severe or profound mental retardation. However, Wisniewski *et al.* (1985b), using repeated measures, assessed the learning capacities and memory functions of seven individuals who had such marked retardation. They found evidence of impairment of learning capacity, visual memory loss and decreases of occupational and social functioning in five of the seven subjects. Three stages in neuropsychological deterioration were identified. Visual memory loss, impairment of learning capacity and behavioural changes were the initial symptoms marking the first stage. The second stage was characterized by loss of comprehension and communication, impairment of social adaptive skills, bouts of irritability, personality changes and loss of personal hygiene. During the third stage, poor sleep, incontinence, seizures and inability to walk became apparent.

Silverstein *et al.* (1988) compared the effects of age on the adaptive behaviour of individuals with DS and matched controls in three studies. In the first, 413 persons with DS living in institutions (most with profound mental retardation) together with the matched controls were assessed using a client development evaluation form. Five areas were examined: motor development, independent living skills, cognitive competence, social competence and maladaptive behaviour. Some decline in competence of the older DS group was found. In a second study, they used a DS group of 399 persons living within the community and controls to clarify the role played by the subjects' initial level of functioning. No clear evidence of decline in competence among the older DS group was found. They also undertook a longitudinal study using repeated measures on 192 subjects with DS and controls, but again the decline in competence was comparable for those with and without DS in all age groups. Only

Table 8.1. Summary of selected recent studies involving cognitive assessment of persons with Down syndrome

Authors	Design	Subjects	Measures Employed	Findings
Wisniewski *et al.* (1985b)	*Longitudinal* Subjects assessed at approximately one year intervals until death (individual study periods ranged from 2.5 to 9.2 years)	7 (DS) 43.2–57.6 years at onset of AD	• Tests of cognitive function (learning capacities and memory functions) • Staff reports and medical records	• Cognitive decline detected in five subjects (one subject not testable; another failed to learn test) • Signs of 'overall regression' recorded in six subjects
Fenner *et al.* (1987)	**Intellectual functioning** *Longitudinal* • Mean number of assessments per subject was five • Mean test-retest interval was 6.7 years	39 (DS) 20–49 years	• Stanford–Binet Intelligence Scale	• Decline in mental age in less than one-third of total sample and in just over one-third of subjects older than 35 years
	Cross-sectional • Comparison between 5 year age groups	39 (DS) 20–49 years	• Stanford–Binet Intelligence Scale	• Mental age of 45-49 year group significantly lower than that of 40–44 year group
	Behavioural functioning *Cross-sectional* • Comparison between 5 year age groups	39 (DS) 20–49 years	• Modified version of Adaptive Behaviour Scale	• No significant difference in level of self-care skills or degree of behavioural disturbance across age groups

Authors	Design	Subjects	Measures Employed	Findings
Zigman et al. (1987)	*Cross-sectional* • Comparison between 10 year age groups	2,144 (DS) 20–69 years 4,172 (MR controls) mean age = 36.5 years	• Measures of activities of daily living (ADL) and cognitive skills	• Younger subjects with DS performed better on ADL than age-matched controls; older subjects with DS performed worse • Younger subjects with DS similar in cognitive skills to age-matched controls; older subjects with DS performed worse • ADL and cognitive performance of subjects with DS declined much more with age than did the performance of MR controls
Schapiro et al. (1988)	*Cross-sectional* • Test scores in a demented subject with DS compared with scores in non-demented younger subjects with DS	A 47-year-old subject with DS with clinical evidence of dementia Thirteen non-demented DS controls 19–33 years	• Peabody Picture Vocabulary Test (revised) • Block Pattern subtest of Hiskey–Nebraska Test of Learning Aptitude • WISC-R Block Design subtest	• Test scores of general intelligence, visuospatial ability, language and memory function showed poorer performance in the 47-year-old subject than in the younger group
Haxby (1989)	*Cross-sectional* • Comparison of <35 year group (10 subjects) with >35 year group (19 subjects; demented and non-demented)	29 (DS) 19–64 years	• Stanford-Binet Intelligence Scale • Peabody Picture Vocabulary Test	• Four subjects in the >35 year age group classified as demented, compared to none in the <35 year age group

Table 8.1—*continued*

Authors	Design	Subjects	Measures Employed	Findings
Haxby (1989) *contd.*			• DS Mental Status Examination (includes items for recall of personal information, orientation, memory, language, apraxia and visuospatial construction)	• Demented subjects differed significantly from young subjects on measures of overall ability
				• Both non-demented and demented old subjects performed worse on all tests of ability to commit new information to long-term memory; immediate memory spans not significantly different for young and non-demented old subjects
			• Tests of immediate memory spans, ability to commit new information to long-term memory, language function, visuoperceptual discrimination and visuospatial construction	
			• Diagnosis of dementia based on modified DSM-III criteria	• Non-demented old subjects did not differ significantly from young subjects on any test of language function; significant differences between young and demented subjects on comprehension
				• Both non-demented and demented old subjects performed worse than young subjects on a multiple step test of praxis and tests of visuospatial construction
Lai and Williams (1989)	*Longitudinal* • Regular follow-up visits every 6–12 months over eight years	96 (DS) all >35 years	• Tests of orientation, memory and verbal and motor skills	• Forty-nine subjects diagnosed as demented
			• Staff reports and medical records, including objective workshop production	• Average age at dementia onset was 54.2 years (range 43–68 years)
			• Dementia defined as a decline in one or more of the assessed skills	• Dementia in 8 per cent of 35–49 years, 55 per cent of 50–59 years, and 75 per cent of >60 years groups

Authors	Design	Subjects	Measures Employed	Findings
Young and Kramer (1991)	*Cross-sectional* • Comparison of subjects with DS by age	60 (DS) 22–67 years	• Sequenced Inventory of Communication Development • Slosson Intelligence Test • Fairview Self-Help Scale • Psychopathology Inventory for Mentally Retarded Adults • Test of Visual Recognition Memory	• Significant relation between ageing and depressed language comprehension as well as self-help skills • Verbal skills and psychopathology not significantly related to ageing • Trend towards lower mental age and visual memory in older subjects
Devenny et al. (1992)	*Longitudinal and cross-sectional* • All subjects tested at least twice, with assessments spanning a range of no less than three years (mean = 4.1 years) • Comparison of DS <35 year group and >35 year group with MR <35 year group and >35 year group	28 (DS) 27–55 years 18 (MR controls) 27–58 years	• Evaluation of mental status (including tests of orientation, object naming, recitation of alphabet and visuomotor control) • Modified Buschke Memory Test • Visual memory test	• Neither the DS nor MR subjects showed age-associated declines in general or specific functioning during the time period of the study • Orientation to place, naming of objects and motor ability improved at follow-up; younger subjects showed greater improvement for motor ability than older subjects • Comparable improvements in performance on the Buschke memory test in all subject groups • Errors increased with longer delays on the visual memory test in all subject groups

MR=mentally retarded
Note: Studies prior to those in the table are presented in Table 3 of Oliver and Holland (1986).

143

the first study provided evidence of differential effects of age on adaptive behaviour of those who had DS compared with those who did not.

An early sign of AD is subtle loss of language functions. Measurement of language function in DS presents a considerable challenge, as language of persons with the syndrome is often limited before deterioration becomes apparent. Most investigators have described loss of communicative function in general terms or have concentrated on the measurement of anomia (the inability to name objects). Exceptions are the studies of Haxby (1989) who found differential deficits for old demented and young DS groups on tests of comprehension, and of Young and Kramer (1991) who observed age-related deficits on tests of receptive language. The investigation by Young and Kramer (1991) was designed to gather information on decline in neuropsychological functioning, particularly language skills, as a feature of ageing in persons with DS. In comparing young and old persons with the syndrome, they found a significant relationship between age and poor performance on tests of receptive language. Self-help skills also deteriorated with advancing age. There were no significant differences between young and older DS groups on tests of expressive language or in the presence of psychiatric disorders. A trend towards lower scores on a test of visual recognition memory was evident in older subjects.

The age at which decreases in functioning in older adults with DS can be demonstrated may be related to the specificity of the assessment used. Studies that have used behavioural scales have tended to show that the onset of decline in DS begins at a later age than is evident from studies that have directly examined neuropsychological functioning.

Devenny *et al.* (1992) examined whether the early signs of AD could be detected using a variety of psychological measures that are sensitive to subtle changes in memory and overall functioning. All subjects were screened for a previous history of deterioration. All had IQs greater than 35 and were aged 25 years and above. The subjects were tested for orientation to person, place and time, object and colour naming, concentration and visuomotor control. A modified form of the Bushke memory test and a visual memory test based on a shape matching task developed by Dalton and Crapper-McLachlan (1984) were also employed. Four groups were used for comparison (younger and older adults with and without DS). No differences between groups were found on orientation to person and time, colour naming and concentration. All groups showed improvement with the Buschke memory test over time. With the visual memory test, errors increased with longer delays for all subject groups; the younger individuals with DS showed greater decreases in performance at five and 10 second delays of the matching task than did any of the other subject groups. Possibly, group analysis could have masked significant changes for individual subjects. Using a criterion of two standard deviations as indicative of a decrement in performance, three individuals with DS showed decline with the Buschke memory test. With the visual memory test, one person with DS showed a decline at the five second delay. A criticism of the study is that there were few subjects included who were aged 50 years and above. Furthermore, some subjects were tested only twice, whilst others were tested three or four times over the three to five year time period.

In a follow up of the study of Hewitt *et al.* (1985), Fenner *et al.* (1987) investigated changes over time in intellectual functioning of 39 adults with DS. The subjects were assessed on five occasions, on average, over a mean period of 6.7 years. Twelve showed deterioration in mental age, with an age-related trend, as measured by the Stanford–Binet test. A reduced level of skill was evident as subjects reached their late forties: 38 per cent were incontinent, 31 per cent could not wash their hands and faces unaided, 41 per cent were unable to dress without help and behavioural deterioration was present in 44 per cent of the sample. However, use of the Stanford–Binet test has limitations, as many items suppose a degree of formal education rarely possessed by persons who have mental retardation. Also, the estimate of IQ change was based on a comparison between the current and highest previous mental age rather than the most recent one. A cross-sectional design alone will not provide a complete picture of deterioration in individuals with DS and, as no previous measure of behavioural functioning was used, it was not possible to make longitudinal comparison for the behavioural functioning of individuals.

Haxby (1989) utilized an extensive battery of neuropsychological tests designed to examine patterns of age-related neuropsychological changes in older adults with DS. Non-demented elderly persons with DS were compared with young persons and with demented elderly persons with the syndrome. The old DS group considered to be demented showed loss of daily living skills, impaired memory, disorientation, apathy, and reduced speech and verbal comprehension. The test battery used measured overall intelligence, immediate memory span, ability to commit new information to long-term memory, language function, visuoperceptual discrimination and visuospatial construction. Immediate memory span for digit sequences and digit span were tested, as were block tapping span and object pointing span. The measure of ability to commit new information to memory was taken from the DS Mental Status Examination. Language function assessed comprehension of semantic and syntactic aspects, sentence repetition, expressive ability, ability to retrieve and comprehend concrete nouns, and object naming. Visuospatial construction was tested by block design and the Hiskey–Nebraska block patterns test and, for visual perception, picture identification was assessed. Neuropsychological patterns in individuals were examined by constructing test profiles summarizing performance on tests of immediate memory, ability to form new long-term memories, language and visuospatial construction. The non-demented older group demonstrated a significant pattern of diminished new long-term memory and visuospatial construction, with spared immediate memory span and language. Two tests, digit span and ideomotor apraxia, significantly discriminated non-demented from demented DS groups. The neuropsychological profiles reflecting performance on a wide range of tests, namely immediate memory, long-term memory, language and visuospatial construction, also discriminated significantly between these two groups. Of the demented DS group, all scored at the lower end or below the young subject range on all functions, except for two subjects who had language scores within the young range.

Zigman *et al.* (1987), in a study consistent with previous reports, found deterioration in neuropsychological processes and adaptive behaviour among persons with DS. The adaptive skills of 2,144 individuals with DS were compared with those of 4,172 controls with other forms of mental retardation. Information was derived from a Developmental Disabilities Information System (Janicki and Jacobsen 1982). Data on behavioural skills consisted of a series of developmentally sequenced behavioural statements (e.g. uses a spoon to pick up and eat food) containing 80 items divided into eight domains: gross motor development, toileting, dressing/grooming, eating, independent living, language, reading/writing and quantitative. The results were analysed as two major factors, i.e. competency scores for activities of daily living and for cognitive scores. Competence for activities of daily living in persons with DS was found to be related to developmental level and age. The performances of those with DS declined with age to a much greater extent than the performance of the controls. For competence in cognitive skills, there was less interaction between age and diagnosis than was seen for activities of daily living. The most significant decline in cognitive skills in those with DS did not become apparent until 60 years of age, perhaps due to the difficulty of assessing decline in DS when the initial ability level is low.

Lai and Williams (1989) followed up 96 persons with DS aged over 35 years. Subjects were tested every 6–12 months. Neuropsychological assessment included tests of orientation, memory, verbal and motor skills. Decline in one or more of the assessed areas over time occurred in 49 subjects. Three phases of deterioration were identified in those with a higher level of ability, the first signs of deterioration being memory loss, disorientation and reduced verbal output. In less able individuals, the first signs were apathy, inattention and decreased social interaction. The second stage was marked by loss of self-help skills and the third by loss of ambulation, incontinence and the presence of abnormal neurological reflexes. The authors noted that the first signs of AD in adults with DS corresponded to the later signs of the disease identified in the general population. Memory loss, an early sign of AD in the general population, was noted only in a few with mild/moderate mental retardation. In persons with DS, the first evidence of AD was often the need for more assistance with self-help skills.

New assessment procedures used for the early detection of cognitive deficits in persons with Down syndrome

Review of published empirical data and the outline of problems associated with assessment provide a good basis for future research. The challenge is to be able to identify early, and often subtle, cognitive impairments (such as memory loss) in the presence of a varying degree of pre-existing intellectual impairment. In this section, brief case examples are presented to illustrate some of the methods and procedures that help to improve accuracy and sensitivity in assessment.

In the first example, the procedures and tests described were employed in the longitudinal study of Oliver *et al.* (C. Oliver, D. H. R. Blackwood, D. M. St Clair, W. J. Muir, I. Squire, M. D. Speight, and P. Dickens, unpublished observations; see also Blackwood *et al.* 1988; Muir *et al.* 1988) and serve to demonstrate some general principles. Three tests of memory were employed: object and picture memory, memory for position of hidden objects and memory for sentences (from the Stanford–Binet test).

The procedure for object memory was as follows: (i) Subjects were asked to name seven objects (part of a test for anomia as well). Any items not named were discarded. (ii) A number of randomly selected objects were presented (trials of two, three, four and five objects were used) and the subject was again asked to name them. (iii) In full view of the subject, one object was covered with a box and the subject was asked 'What is under the box?'. This procedure was repeated with two other objects. (iv) For the memory test, step (ii) was repeated and then the subject was asked to close his/her eyes whilst the object was covered. (v) The subject was asked to open his/her eyes and asked 'What is under the box?'.

This procedure achieves a number of objectives. Step (i) ensures that the subject knows the name of any object employed. A knife and fork were used and their names were occasionally confused. If this occurred in the memory test proper it could be mistaken for a memory deficit. Even if these items were incorrectly named, this does not necessarily indicate an acquired cognitive deficit; the subject may have always confused the names. Step (ii) is a standardized procedure that ensures that the subject has looked at the objects. Step (iii) ensures understanding of the task without involving the memory component, whilst steps (iv) and (v) are the memory test itself, employing a simple question. Exactly the same procedure was used for the picture memory.

The test for position of hidden objects was similar. In this test, coins were hidden under elevated discs. Initially one coin was used with four discs, then one coin with five discs, and the number of coins and discs were increased to the final trial of two coins with seven discs. In the first stage of the test, the coins were hidden in full view of the subject who was then immediately asked 'Can you give me the (coin)?'. This trial (immediate) acted as a baseline by which to compare 'delay' trials and ensured understanding of the task. The second stage involved the same positioning of the coins and numbers of the discs. This time, after the coins were covered with the discs in full view of the subject, all of the discs were covered by a card. After five seconds the card was removed and the subject was asked 'Can you give me the (coin)?'. The same procedure was repeated with a 30 second delay. Two scores were derived: the number of coins correctly located at the first attempt and the number of errors prior to the coins being located.

The tests for object and picture memory, and position of hidden objects have common features that overcome some of the problems outlined in previous sections. These features are: (i) the tests are procedurally simple to understand and entail little verbal explanation to the subject, thus helping to avoid floor effects that are not due to memory impairment; (ii) the memory components of the tests can be increased systematically without the procedure becoming more difficult to follow. Thus ceiling effects may be avoided. Combined together, these features address the problem of the wide spread of ability noted in persons with DS whilst enhancing test sensitivity.

Another test of memory employed was memory for sentences from the Stanford–Binet test. Subjects were simply asked to 'Say this: (sentence)'. The number of words correctly recalled was then recorded. Again, the procedure is simple but the memory content can be made increasingly taxing by increasing the sentence length.

Figure 8.1 shows the results of these memory tests for two subjects in the Oliver *et al.* (C. Oliver, D. H. R. Blackwood, D. M. St Clair, W. J. Muir, I. Squire, M. D. Speight, and P. Dickens, unpublished observations) longitudinal study. There was a lapse of approximately two years between the first and second assessment in this study. The figure indicates that subject EJ (aged 47 at the first assessment) was scoring lower at the second assessment on the tests of object and picture memory (OPM) and memory for sentences (MS). He also showed a decline in ability to locate correctly the position of hidden objects at 5 second delay (5C) and 30 second delay (30C) and a corresponding increase in errors (5E and 30E). Subject IM (aged 35 at the first assessment) showed a

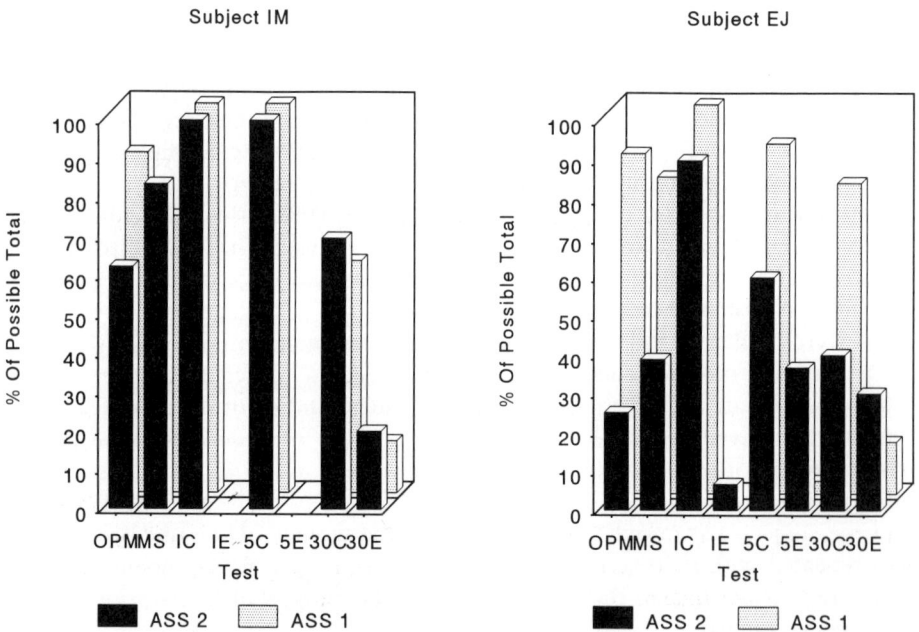

Figure 8.1. Results of memory tests at assessments two years apart for two adults with DS (IM aged 35; EJ aged 47). (OPM=object and picture memory; MS=memory for sentences; IC, 5C and 30 C=respectively, immediate, 5 second delay and 30 second delay correct in Memory for Position Test; IE, 5E and 30E=errors in the same test.)

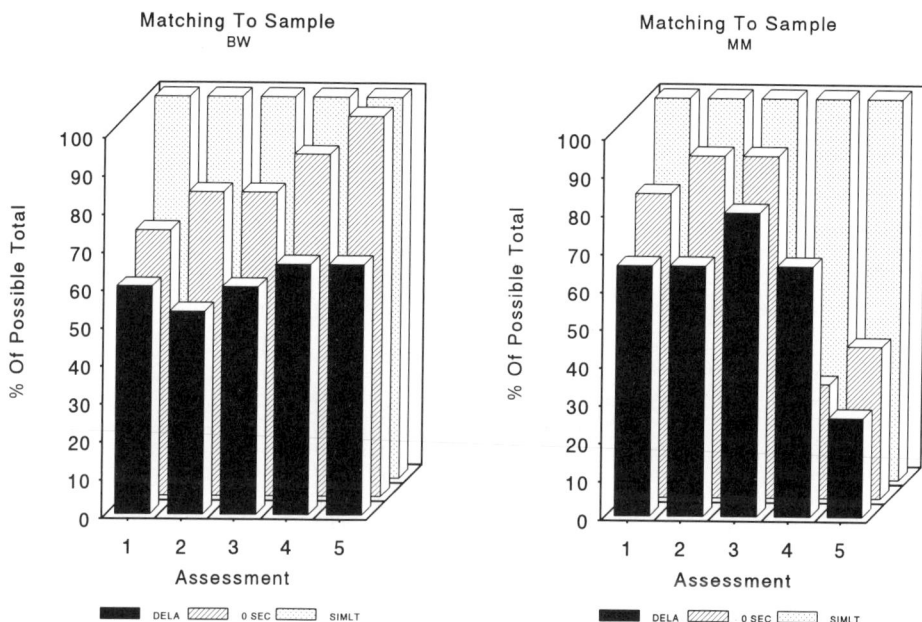

Figure 8.2. Results of computerized matching to sample memory test for two adults with DS (BW aged 37; MM aged 45) assessed at six month intervals. (SIMLT=simultaneous presentation of match and sample; 0 SECS=samples appear as soon as match disappears; DELAY=mean of scores for 4, 8 and 16 second delay between match disappearing and sample appearing.)

moderate decline in object and picture memory but no other indication of increasing memory impairment. It is important to note that neither subject showed any decline in the naming of objects used or demonstrating their use. Similarly on both occasions each subject could name all pictures used, point to them when required and locate the coins correctly when there was no delay. Hence the decline in performance by EJ was almost certainly attributable to an acquired memory deficit.

In a second longitudinal study conducted by the authors with A. J. Holland, a computerized battery of tests devised by Sahakian *et al.* (1988) was employed. Items from this battery have previously been used in the diagnosis of AD in the general population. Three tests of memory are described here to demonstrate their application.

The first was a 'matching to sample' test in which subjects were shown an abstract pattern and asked to match it to one of four patterns presented below it by touching the screen. In the initial stage, match and samples were displayed simultaneously to establish understanding. Delays of 0, 4, 8 and 16 seconds were then introduced, with the match disappearing and the samples then appearing after a delay. Figure 8.2 shows the results of this assessment for two subjects seen at six month intervals over a period of 30 months. Neither subject showed any decline in ability to simultaneously match to sample. (This

Delayed Response

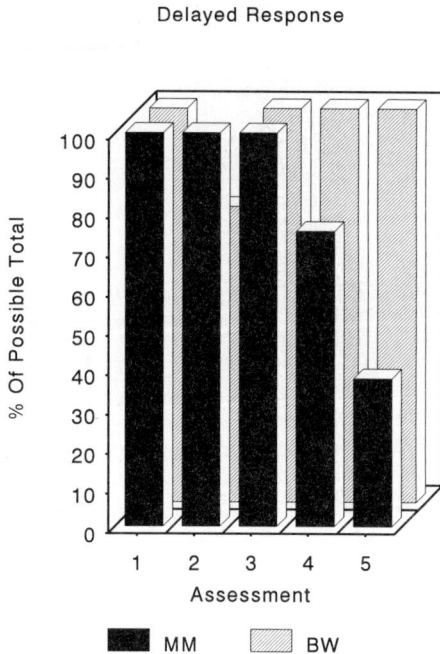

Figure 8.3. Results of computerized delayed response memory test (see text for details) for two adults with DS (BW aged 37; MM aged 45) assessed at six month intervals.

is equivalent to establishing object and picture naming and immediate recall in the tests described above). Subject MM, however, aged 45 at the first assessment, showed some decline in performance between assessments 4 and 5 when there was any delay, including 0 seconds. Subject BW, aged 37 at the first assessment, did not show any deterioration in performance; in fact there appeared to be some increase in score at 0 second delays.

A second test of memory from this battery (delayed response) involved the subject learning to locate patterns 'hidden' in boxes on the screen. Initially the subject was required to locate a previously displayed single pattern in six boxes, five of which were unfilled. The number of patterns that the subject was required to locate was then increased and finally the number of boxes was also increased. The principle behind the test is thus very similar to that described above for the test of memory of position of hidden objects. The results of this test for the two subjects MM and BW are shown in Figure 8.3. BW showed little decline in performance, whilst MM showed a decrease in memory scores at assessments 4 and 5. These results are broadly in agreement with those from the matching to sample test.

Finally, a test of pattern recognition was employed. This test involved showing the subjects 12 disappearing patterns on the screen and then 12 sets of two patterns appearing. The subjects were required to identify the previously observed pattern in each set. Then the whole procedure was repeated. The results of this test for MM and BW are shown in Figure 8.4. Again, these

Pattern Memory

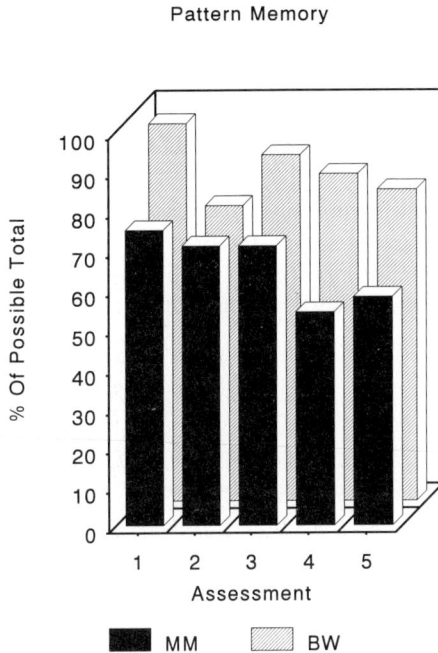

Figure 8.4. Results of computerized pattern recognition and memory test (see text for details) for two adults with DS (BW aged 37; MM aged 45) assessed at six month intervals.

results concur with those from the previous two tests in showing a decline in performance by MM and a limited decrease in score for BW. This test also illustrates a crucial point about the battery used. Although MM appears to have some impairment of memory, her last test result shows a performance of around 55 per cent. The reason for this is that by chance subjects will score 50 per cent correct as the task is one of forced choice. This is important to note because, if initial assessment results fall within chance levels, a floor effect is evident. Additionally, if the test shows poor test–retest reliability and initial assessment results are only slightly above chance levels by a margin within test error, a floor effect is again evident. Hence, initial results from applying the battery described here for the assessment of AD in DS suggest that it may be appropriate for only a limited number of individuals.

Conclusions

Careful neuropsychological testing is necessary to detect and understand the changes that occur in persons with DS who develop AD. Standardized test procedures used for persons in the general population may not be appropriate for those with DS unless they have mild or moderate mental retardation, since performance often reveals floor effects. More sensitive tests of dysfunction developed specifically for individuals with DS are more likely to detect early signs of AD and patterns of deterioration.

Difficulties associated with the assessment of AD in individuals with DS give rise to problems in interpretation of results. However, the neuropsychological differences found between young and old subjects with DS are probably not due to test bias. Studies have consistently shown patterns of deficit that bear some resemblance to patterns of impairment observed in those with AD in the general population.

In essence, evidence from neuropsychological assessment of adults with DS shows clear-cut impairments associated with AD described in detail for a few individuals, significant differences in certain neuropsychological functions for groups of older compared with young individuals, and severe impairments for the oldest individuals.

References

Ball, M. M., and Nuttall, K. (1980). Neurofibrillary tangles, granulovacuolar degeneration and neuron loss of Down's syndrome: quantitative comparison with Alzheimer dementia. *Annals of Neurology*, 17, 278-82.

Blackwood, D. H. R., St Clair, D. M., Muir, W. J., Oliver, C. J., and Dickens, P. (1988). The development of Alzheimer's disease in Down's syndrome assessed by auditory event-related potentials. *Journal of Mental Deficiency Research*, 32, 439-53.

Crapper-McLachlan, D. R., Dalton, A. J., Galin, H., Schlotterer, G., and Daicar, E. (1984). Alzheimer's disease: clinical course and cognitive disturbances. *Acta Neurologica Scandinavica*, 69, 83-90.

Dalton, A. J., and Crapper, D. R. (1977). Down's syndrome and ageing of the brain. In *Research to practice in mental retardation: biomedical aspects,* (ed. P. Mittler), Vol. 3., pp. 391-400. University Park Press, Baltimore.

Dalton, A. J., and Crapper-McLachlan, D. R. (1984). Incidence of memory deterioration in ageing persons with Down's syndrome. In *Perspectives and progress in mental retardation: biomedical aspects,* (ed. J.M. Berg), Vol. 2., pp. 55-62. University Park Press, Baltimore.

Dalton, A. J., Crapper, D. R., and Schlotterer, G. R. (1974). Alzheimer's disease in Down's syndrome: visual retention deficits. *Cortex*, 10, 366-77.

Devenny, D. A., Hill, A. L., Patxot, O., Silverman, W. P., and Wisniewski, K. E. (1992). Ageing in higher functioning adults with Down's syndrome: an interim report in a longitudinal study. *Journal of Intellectual Disability Research*, 36, 241-50.

Ellis, W. G., McCulloch, J. R., and Corley, C. L. (1974). Presenile dementia in Down's syndrome: ultrastructural identity with Alzheimer's disease. *Neurology*, 24, 101-6.

Fenner, M. E., Hewitt, K. E., and Torpy, D. M. (1987). Down's syndrome: intellectual and behavioural functioning during adulthood. *Journal of Mental Deficiency Research*, 31, 241-9.

Grady, C. L., Haxby, J. V., Horwitz, B., Sundaram, M., Berg, G., Schapiro, M., *et al.* (1988). A longitudinal study of the early neuropsychological and cerebral metabolic changes in dementia of the Alzheimer type. *Journal of Clinical and Experimental Neuropsychology*, 10, 576-96.

Haxby, J. V. (1989). Neuropsychological evaluation of adults with Down's syndrome: patterns of selective impairment in non-demented old adults. *Journal of Mental Deficiency Research*, 33, 193-210.

Haxby, J. V., Grady, C. L., Duara, R., Schlageter, N., Berg, G., and Rapoport, S. I. (1986). Neocortical metabolic abnormalities precede nonmemory cognitive deficits in early Alzheimer's type dementia. *Archives of Neurology*, 43, 882-5.

Hewitt, K. E., Carter, G., and Jancar, J. (1985). Ageing in Down's syndrome. *British Journal of Psychiatry*, 147, 58-62.

Janicki, M. P., and Jacobson, J. W. (1982). The character of developmental disabilities in New York State: preliminary observations. *International Journal of Rehabilitation Research*, 5, 191-202.

Jervis, G. A. (1948). Early senile dementia in mongoloid idiocy. *American Journal of Psychiatry*, **105**, 102-6.

Lai, F., and Williams, R. S. (1989). A prospective study of Alzheimer disease in Down syndrome. *Archives of Neurology*, **46**, 849-53.

Malamud, N. (1972). Neuropathology of organic brain syndromes associated with ageing. In *Ageing and the brain,* (ed. C. M. Gaitz), pp. 63-87. Plenum Press, New York.

Miniszek, N. A. (1983). Development of Alzheimer disease in Down syndrome individuals. *American Journal of Mental Deficiency*, **87**, 377-85.

Muir, W. J., Squire, I., Blackwood, D. H. R., Speight, M. D., St Clair, D. M., Oliver, C., and Dickens, P. (1988). Auditory P300 response in the assessment of Alzheimer's disease in Down's syndrome: a two-year follow-up study. *Journal of Mental Deficiency Research*, **32**, 455-63.

Oliver, C., and Holland, A. J. (1986). Down's syndrome and Alzheimer's disease: a review. *Psychological Medicine*, **16**, 307-22.

Owens, D., Dawson, J. C., and Losin, S. (1971). Alzheimer's disease in Down's syndrome. *American Journal of Mental Deficiency*, **75**, 606-12.

Sahakian, B. J., Morris, R. G., Evenden, J. L., Heald, A., Levy, R., Philpot, M., and Robbins, T. W. (1988). A comparative study of visuospatial memory and learning in Alzheimer-type dementia and Parkinson's disease. *Brain*, **111**, 695-718.

Schapiro, M. B., Ball, M. J., Grady, C. L., Haxby, J. V., Kaye, J. A., and Rapoport, S. I. (1988). Dementia in Down's syndrome: cerebral glucose utilization, neuropsychological assessment, and neuropathology. *Neurology*, **38**, 938-42.

Silverstein, A. B., Herbs, D., Miller, T. J., Nasuta, R., Williams, D. L., and White, J. F. (1988). Effects of age on the adaptive behavior of institutionalized and non-institutionalized individuals with Down syndrome. *American Journal of Mental Retardation*, **92**, 455-60.

Thase, M. E., Liss, L., Smeltzer, D., and Maloon, J. (1982). Clinical evaluation of dementia in Down's syndrome: a preliminary report. *Journal of Mental Deficiency Research*, **26**, 239-44.

Thase, M. E., Tigner, R., Smeltzer, D., and Liss, L. (1984). Age-related neuropsychological deficits in Down's syndrome. *Biological Psychiatry*, **19**, 571-85.

Wisniewski, K. E., and Hill, A. L. (1985). Clinical aspects of dementia in developmental disabilities. In *Ageing and developmental disabilities issues and approaches*, (ed. M.P. Janicki and H.M. Wisniewski), pp. 195-210. Paul H. Brookes, Baltimore.

Wisniewski, K. E., Howe, J., Williams, D. G., and Wisniewski, H. M. (1978). Precocious ageing and dementia in patients with Down's syndrome. *Biological Psychiatry*, **13**, 619-27.

Wisniewski, K.E., Wisniewski, H. M., and Wen, Y. (1985a). Occurrence of neuropathological changes and dementia of Alzheimer's disease in Down's syndrome. *Annals of Neurology*, **17**, 278-82.

Wisniewski, K. E., Dalton, A. J., Crapper-McLachlan, D. R., Wen, G. Y., and Wisniewski, H. M. (1985b). Alzheimer's disease in Down's syndrome: clinicopathologic studies. *Neurology*, **35**, 957-61.

Yates, C. M., Simpson, J., Maloney, A. F. J., Gordon, A., and Reid, A. H. (1980). Alzheimer-like cholinergic deficiency in Down syndrome. *Lancet*, **2**, 979.

Young, E. C., and Kramer, B. M. (1991). Characteristics of age related language decline in adults with Down syndrome. *Mental Retardation*, **29**, 75-9.

Zigman, W. B., Schupf, N., Lubin, R. A., and Silverman, W. P. (1987). Premature regression of adults with Down syndrome. *American Journal of Mental Deficiency*, **92**, 161-8.

9

Communication impairments in Alzheimer disease and Down syndrome

I. Campbell-Taylor

Summary

The cognitive deficits of both Alzheimer disease and Down syndrome cause changes in linguistic ability. While these impairments are characteristic for each disorder, there appears to be limited similarity between the two when adults with Down syndrome develop presumed Alzheimer disease. This seems intuitively sensible when it is considered that in individuals with Down syndrome language functioning is never normal, and attempts to compare them with individuals with Alzheimer disease may be inappropriate. There are, however, some areas of possible comparison, most notably in changes in motor speech and in electrophysiological indicators of comprehension that may provide useful clues diagnostically. While the language deficits of Alzheimer disease are now well, if not completely, described, only prospective studies of children and young adults with Down syndrome will allow identification of the specific changes that occur should Alzheimer disease develop.

Introduction

Alzheimer disease (AD) is commonly perceived to be a disease primarily affecting memory. This is indeed the case, but the consequences of the disorder go far beyond simple amnesia. Memory and cognition are inextricably linked. Cognition and language are equally inseparable. They may be considered as representing the two faces of a coin. In many respects they can be observed, examined and described separately but they cannot be forcibly parted without leaving behind relatively worthless metal. Deficits in communication in AD are now well described. They are among the earliest signs and, although subtle, they are distinct, specific and sufficiently characteristic to be diagnostic.

A dementing condition that is presumed to be AD develops in some adults with Down syndrome (DS) (Dalton and Crapper-McLachlan 1984; Cutler 1985; Walz et al. 1986). Early ageing is also characteristic of DS and, while in the population at large increasing age does not cause significant communication impairment, it is not known what effects simple ageing may have on a group of individuals whose linguistic/cognitive abilities are compromised from birth (Zigman *et al.* 1987). Individuals with DS have, as a result of their cognitive deficit, equally characteristic communication impairments without the added pathology of AD. They suffer impairments in the understanding and use of language, spoken and written. Should they develop AD, the linguistic impairments may not even be similar to those found in AD in the general population, although increasing impairments in language with ageing have been noted frequently in adults with DS (Wisniewski *et al.* 1978; Miniszek 1983). The early language deficits of AD may be masked by those due to the effects of DS. In attempts to describe or account for changes in communicative competence in adults with both DS and AD, there is a lack of appropriate tools and methods of assessment, but description and inference may shed some light on commonalities and differences that are related to underlying pathology.

This chapter describes the nature of communication impairments in both AD and DS, outlines some current theories of causation and suggests areas of future research.

Communication deficits in Alzheimer disease

The most frequently cited deficit in AD is in recent memory. Other notable dysfunctions are in lexical-semantic language abilities with diminished vocabulary and naming (Appell *et al.* 1982; Martin and Fedio 1983; Huff *et al.* 1985). Language dysfunction in AD has been likened to both Wernicke's and transcortical sensory aphasia (Appell *et al.* 1982; Murdoch *et al.* 1987). These types of aphasia, following a circumscribed lesion of the left temporal cortex, are characterized by fluent, paraphasic verbal expression, poor auditory comprehension and little or no functional reading or writing (Goodglass and Kaplan 1972). Oral reading in Wernicke's aphasia is, at best, paraphasic. In contrast, oral reading in AD is well-preserved (Cummings *et al.* 1986).

As useful as these comparisons with aphasia may be, to attempt to fit the characteristic changes of AD into aphasic profiles that are, at best, only general descriptions is to perform a procrustean operation in which much valuable information is lost. The exact nature and extent of language impairment in AD has been the subject of much recent research but 'as yet, we do not have a thorough understanding of the range, severity and type of language impairment associated with Alzheimer's disease' (Murdoch *et al.* 1987, p.123). We do have, however, clear descriptions of difficulties in word-finding, naming and spontaneous speech, and of problems with comprehension, use of circumlocutions, paraphasias and relatively well-preserved repetition (Bayles 1982; Murdoch *et al.* 1987).

The pivotal deficit giving rise to the signs described above, according to Bayles and Kaszniak (1987), is a disturbance in the ability to form ideas and a deterioration in the recognition of relationships between concepts. Such deficits underlie the observed semantic dissolution that is said to be the hallmark of AD. This explanation is compatible with the view that rule-governed aspects of language such as syntax and phonology are relatively well-preserved, while conceptual and semantic abilities are severely compromised (Schwartz *et al.* 1979; Nebes *et al.* 1986; Flicker *et al.* 1987).

Naming and semantic function

One of the earliest and most characteristic signs of AD is an inability to name to confrontation (Bayles 1982). Semantic paraphasias (names closely related in meaning) and circumlocutions are common. The problem is one of lexical retrieval, i.e. the inability to select the target name from the individual's store of words. A person with early AD is often able to describe the object and to talk about its function and other attributes such as colour and shape, while being unable to produce a label. Tests of vocabulary commonly provide the subject with a name and ask for a definition or explanation, or ask the subject to point to the appropriate picture after the name is spoken by the examiner. Allowing for auditory comprehension deficits, victims of AD can often perform comparatively well on such tests. They exhibit severely impaired performance on tests of confrontation naming in which they must retrieve the label or refer-ent. The retained ability to describe the object reveals that the naming difficulty is not one of memory but of lexical retrieval.

The extent of the semantic/conceptual problem has been examined in several studies. Some of these have demonstrated that persons with AD retain superor-dinate information about meanings, but that referential details or more precise semantic attributes are eroded. This has been taken to mean that while the semantic representation has not been totally destroyed, it is underspecified and impoverished. Grober *et al.* (1985) have provided evidence to support this assertion. Their subjects were able to pick targets over foils in a forced-choice procedure but were unable to rank the importance of attributes for various ref-erences. These authors interpreted this finding as suggesting that the weighting of attributes in terms of saliency and importance has been altered, resulting in a change in semantic organization but not a loss of the semantic representations. The finding of altered organization, rather than loss, places another perspective on the semantic deficit in AD at variance with the generally held assumption that semantic information is largely eliminated.

Another set of experiments designed to evaluate the extent of the semantic breakdown provides additional support for the possibility of semantic represen-tations still being present. Researchers, who have adopted techniques that do not require the subject deliberately to use semantic information explicitly, have shown that this knowledge is relatively well-preserved. The focus, at present, is to estimate the extent of this preservation using implicit techniques. In one such study, Nebes *et al.* (1984) found sparing of automatic processing of intact semantic knowledge despite the inability to produce this information voluntar-ily.

The distinction between explicit recovery of meaning and involuntary act-ivation effects are also illustrated in studies by Margolin (1988) and Chertkow *et al.* (1989). These studies demonstrated priming effects in which the existence of sufficient information in the semantic system was observed to allow the recognition of a subsequent related word. Margolin (1988) found that words, but not pictures, primed the following word, implying that the strength of association is between words that are closely matched in terms of their internal representations. Pictures and words that are not matched do not show the priming effect. According to Huff *et al.* (1985), if specific semantic information is lost, it will consistently be unavailable to both explicit and implicit recovery.

Moscovitch *et al.* (1986), utilizing implicit testing of memory in AD subjects, have shown that the ability to form and retain new associations is preserved in spite of failing consciously to recollect or to recognize the relevant material. They explained the results by suggesting that their subjects showed a profound memory impairment when explicit reference to a particular episode was required, but that preservation of memory can be demonstrated when explicit recollection is not necessary. The demonstration of unconscious access in both memory and language tests may point to a specific problem in the voluntary recovery of retained information in persons with AD.

Auditory comprehension

As described above, the ability to understand spoken language becomes increasingly impaired in AD. This deficit becomes more marked with more complex input in that single words, short phrases and simple sentences are understood best, but sentences of syntactic complexity with embedded clauses, passive voice and degrees of ambiguity are beyond the comprehension capacity of affected individuals (Flicker *et al.* 1987). Statements about syntactic integrity in AD refer to verbal expression only, in that the output of the individual, at least in the early stages, contains no errors of word order, verb tenses, suffixes and other syntactic markers. Verbal expression becomes simplified and reflects the degree of complexity that can be understood at any given point in the disease process. Auditory comprehension is a function of left temporal lobe integrity so that, in AD, the well-described hippocampal, temporal and parietal pathology is bound to produce increasing problems.

Reading

Reading and reading comprehension provide a controllable method by means of which questions about relative preservation of semantic, syntactic and phonological abilities may be established. Oral reading is well-preserved in AD. That simply means that the individual is able to read aloud with correct phrasing, pronunciation and intonation. Letters, partially obscured words, regular and irregular words and commands can be read by all but the most severely affected individuals (Cummings *et al.* 1986). This intact visual input and verbal output skill should enable detection of effects of semantic preservation without obscuring any pertinent results. There is a lack of sufficient data at present to determine the difference in comprehension between items read aloud and those read silently, although some preliminary observations indicate that reading aloud aids comprehension. Investigations of reading comprehension may prove to be a fruitful source of information about the nature of cognitive/linguistic impairments in AD.

Assessment of reading skills may also facilitate diagnosis. For example, Lytton and Brust (1989) described a 70-year-old man with evidence of strokes in the left frontal and temporal lobes. There was some indication of declining cognitive abilities in the previous year and, after some initial improvement, his gross deterioration over the following two years indicated a probable dementing illness with superimposed cerebral infarctions. Described as having Wernicke's aphasia, this patient in many respects better fits the description of language dysfunction found in AD. He had poor auditory comprehension, circumlocutory confrontation naming and preserved oral reading. This preservation of oral reading argues against a diagnosis of Wernicke's aphasia in which, by definition, there is no functional reading. Although he seemed to have little or no reading comprehension, it is notable that he was able to follow written commands after reading them aloud. It is possible that having the added input modality of his own auditory input, as well as reference to his internal phonological coding, assisted comprehension.

Reading as an adjunct to communication in AD has not been sufficiently explored and may represent the richest source of information for both scientific and therapeutic imperatives.

Late Alzheimer disease

Communication deficits in AD change as the disease progresses. In the earliest stages, while articulation and syntax are intact, naming to confrontation becomes increasingly impaired with semantic paraphasias predominating. Objects are given the labels of related items such as 'chair' for 'table'. As the disease progresses, these related names become further and further removed from the target object until conversation may be incomprehensible. Articulation remains intact until the mid to late stages of disease when additional pathology affects the muscles of articulation. Disrupted articulation in the early stages of a dementing illness is sufficient to negate a diagnosis of AD (Campbell-Taylor 1990).

Spoken language becomes diminished in quantity and what little is produced may be unintelligible because of a combination of motor speech impairment and semantic/syntactic breakdown. In late stages, affected individuals may become mute or aphonic. The former term refers to the retained ability to produce voice and speech, but either the unwillingness to do so or the inability to initiate due to frontal lobe pathology. Aphonia is the inability to produce voice because of vocal cord paralysis or other laryngeal pathology. Mute individuals may be heard to moan, cough or laugh involuntarily, while those who are aphonic are incapable of making sound willingly or otherwise. Differentiation between these conditions is important for diagnostic purposes, true muteness in dementia being indicative of significant frontal lobe pathology.

Synopsis

To summarize, the communication impairments found in the early stages of AD are naming deficits with semantic paraphasias and circumlocutions in which the object may be accurately described in terms of function, shape, colour and other pertinent attributes, but the name resists retrieval from the lexicon. Articulation (phonology), syntax, repetition and oral reading are intact. Auditory comprehension is impaired and may be exacerbated by the decrements in hearing acuity that commonly afflict the older individual. As the disease progresses, deficits become more marked in all domains with the exception of oral reading that remains intact when conversation is no longer possible. Articulation becomes affected as further pathology develops, in particular the parkinsonism that affects a majority of victims of AD should they survive long enough (Chen *et al.* 1991).

Communication deficits in Down syndrome

It is well-recognized that children with DS do not demonstrate the same language ability as normal children of the same chronological age. Whether or not they exhibit unique and characteristic language impairments is still a matter for debate, although there is convincing evidence that some aspects of communicative competence are affected in ways that are peculiar to DS. The evidence is obtained from electrophysiological measures and neuropathological studies discussed later in this chapter.

Delay versus Deviance

The vast body of literature on linguistic competence in developmentally handicapped children and adults, including those with DS, has predominantly concerned itself with the delay versus deviance controversy (Cicchetti and Pogue-Hesse 1982). The major issue is whether or not language is quantitatively or qualitatively different relative to non-verbal skills. The view that differences are of a quantitative nature is consistent with the assumption that development of language is contingent upon development in cognition. According to this view, language skills can never be advanced or delayed relative to non-verbal mental age (Lenneberg *et al.* 1964; Graham and Graham 1971; Mahoney and Snow 1983). Learning rate becomes the primary problem with development characterized as 'slow'. In contrast, the qualitative difference view argues that development is deviant relative to the performance level predicted by mental age (Rohr and Buhr 1978; Haxby 1989). Both positions recognize that cognitive development is essential for language development. The delay position assumes that if adequate cognitive underpinnings are in place, then language will follow (Piaget and Inhelder 1969; McCall 1972; Layton and Sharifi 1978; Miller *et al.* 1978). The deviance position adds that additional conditions are necessary for language to develop over time (Rondal 1978; Kernan 1990).

Rondal (1988) argues that the delay versus deviance argument is conceptually ill-founded. He indicates that both circumstances exist in that slow development in DS is the result of neurological lag, visual and auditory deficits and motor impairments, but that deviances also exist. It should be remembered that memory impairments have also been demonstrated in individuals with DS (Ellis *et al.* 1985). Children with DS use more stereotyped expressions, more gesture to fill in gaps and less sophisticated syntax than do normally developing children of comparable mental age and language development. The former have difficulty with the use of the interrogative form, using intonation rather than inverted word order to ask a question. They are impaired in the use of personal pronouns and sentence complexity is low. Language development in DS appears to plateau at puberty but there is some evidence that adolescents may be assisted to further development by consistent and well-planned intervention programmes.

Auditory comprehension

Hartley (1982) found significant differences in auditory comprehension in children with DS compared with normal controls and with children with developmental delays of other aetiology. Using the Token Test For Children (DiSimoni 1978), he found that children with DS were significantly more impaired except on the part of the test related to spatial tasks. This is consistent with other research that indicates that individuals with DS are selectively impaired in tasks requiring auditory sequential processing and perform better on visual spatial items (Ashman 1982; Carni and French 1984). These difficulties with sequential processing are believed to be the cause of immature syntactic forms that persist in individuals with DS, syntax being dependent on variations in word order. More importantly, sequencing deficits probably underlie the significant auditory comprehension deficits identified in DS, deficits that become even more marked with increasing age (Rohr and Buhr 1978; Kernan 1990; Young and Kramer 1991). Haxby (1989) determined that as adults with DS aged, language abilities became significantly more impaired than other cognitive processes.

Syntax and phonology

Although linguistic function in DS is significantly compromised, children with the syndrome nevertheless acquire normal forms of language. Their linguistic competencies may be of a simplified nature but they use normal phonological and syntactic rules (Naremore and Dever 1975; Layton and Sharifi 1978; Miller 1981). They appear to acquire the rules concerning acceptable word order as well as the rules governing acceptable order of sounds within words according to their native language. They do not produce neologisms or bizarre syntactic structures. In short, they follow the rules of phonology, morphology and syntax, but remain at an earlier level of sophistication than individuals without developmental handicap. Syntactic understanding depends on the ability to deal with information that arrives in specific sequences, word order being critical to accurate syntax in any language. Word order of a complicated nature, with embedded clauses, subordinate clauses separated by phrases and similar constructions, are too complex for the limited sequential processing abilities of individuals with DS. Possible reasons for this are discussed later in the section on neurobiological perspectives.

Naming

For reasons that are, as yet, unclear, children with DS tend to have greater vocabularies than other children with developmental disabilities (Rondal 1978; Young and Kramer 1991). This fact represents some of the best evidence available to support the contention that measures of vocabulary do not reflect linguistic competence. Young and Kramer (1991) found that vocabulary measures remained elevated despite increasing comprehension deficits in adults with DS. They tested 60 non-institutionalized adults aged from 22 to 67 years. They found that auditory comprehension deficits increased significantly with age (P <0.02) but there was no significant relationship between increasing age and expressive language. Similar findings have been reported by Wisniewski *et al.* (1978).

Articulation/motor speech

Children with DS are at significant risk for language problems for reasons that go beyond the obvious cognitive impairment. There is a significantly increased incidence of middle ear pathology that is well known to contribute to language learning difficulties as a result of fluctuating hearing loss (Brandeis and Elsinger 1981). Deficits in muscle tone and motor coordination affect the speech production system, that is, articulation as opposed to language. The pervasive hypotonia and morphological anomalies of the face adversely affect the coordination of movements necessary for precise articulation (Coleman 1973; Sanger 1975; Cicchetti and Sroufe 1976). The environment in which individual children are raised will also differentially contribute to the development of language/learning abilities (Jones 1980; Leuder *et al.* 1981).

Effects of early stimulation

Adults with DS currently being investigated for AD are, for the most part, of a generation that did not have access to early childhood intervention programmes and many have been in institutions since infancy. This is an extremely important point because it has been demonstrated repeatedly that cognitive/linguistic functioning in children with DS is dependent on early, appropriate and consistent stimulation (Rogers 1975; Clunies-Ross 1979; Ludlow and Allen 1979; Price-Williams and Sabsay 1979; Jones 1980; Wiegel-Crump 1981). Close interactions and stimulation by the mother or other consistent caregiver is essential for the development of joint attention, affective responsiveness and referential skills. Mutual gaze and joint attention form the basis of later cognitive/linguistic development (Berger and Cunningham 1981; Gunn *et al.* 1982; Bakeman and Adamson 1984; Landry and Chapieski 1988, 1989, 1990). In addition, affective development is predictive of cognitive processes and is developed almost exclusively by consistent interaction and stimulation by close caregivers (McCall 1972; Cicchetti and Sroufe 1976; Sroufe and Waters 1976).

The development of pragmatic skills, which is the ability to communicate appropriately according to the context or environment, also depends on early stimulation and modelling. Children and adolescents with DS are frequently described as being docile and friendly with an implicit assumption that they are, therefore, pragmatically intact. This is not necessarily the case, since docility and compliance are not always pragmatically appropriate. They simply do not cause unpleasantness. There is some evidence that children and adolescents with DS are not pragmatically appropriate, having difficulty with turn-taking in conversations and in repairing breakdowns in listeners' understanding (Rondal 1988). Adults with DS who were deprived of early stimulation are likely to be so impaired linguistically that attempting to draw comparisons with individuals with AD is untenable. Prospective studies of children and adolescents with DS who are currently well-assessed and enrolled in early intervention programmes may be the most fruitful path to understanding the relationship between AD in the general population and AD in adults with DS.

Similarities and differences: Down syndrome and Alzheimer disease

The most striking similarity in communicative impairment in DS and in AD is in the area of auditory comprehension. In DS, this faculty is always impaired but becomes significantly worse with increasing age. This does not happen in normal ageing. In AD, one of the earliest signs of communicative/cognitive breakdown is a diminution in the ability to understand spoken language. This is not difficult to explain when it is realized that auditory comprehension is dependent on intact temporal lobe function, particularly on the left, and the well-known hippocampal involvement in both AD and DS implies some decrement in left temporal lobe function (McKhann *et al.* 1984; Wisniewski *et al.* 1986).

A reduction in total verbal output is common to both DS and AD, but this is not peculiar to these conditions alone. Diminished amount of speech is also found in depression and, therefore, by itself does not serve to distinguish depression from AD or DS (Darby *et al.* 1984; Nilsonne 1987). The incidence of depression in adults with DS is reported to be significantly high and is frequently found to be related to the elevated rate of hypothyroidism in the syndrome (Sovner and Hurley 1983). Burt *et al.* (1992), in a study of 61 adults with DS (60 with regular trisomy 21 and one with mosaicism) and 43 adults with mental retardation of other aetiology, found that only those with DS showed depression and decline in functioning. They suggested that depression and dementia are related in DS. They used the Vineland scale to obtain measures of functioning from family members and other caregivers and found that scores related to communication function, in particular scores of receptive language, were significantly related to depression scores (P <0.001). The Vineland scale does not address speech and language in fine detail and these results may simply reflect the fact that depression and dementia are related in other conditions as well, including AD in the general population, and that each can produce apparent cognitive, linguistic and motor speech alterations (Darby *et al.* 1984; Gotham *et al.* 1986; Nilsonne 1987). The specific changes found in each have yet to be fully documented.

Articulation (motor speech) is impaired in DS. The pervasive hypotonia characteristic of the syndrome leads to imperfect functioning of the articulators: jaw, lips, tongue, soft palate and facial muscles (Coleman 1973). Within the classic taxonomy of dysarthria, this type can best be described as belonging to the ataxic group (Darley *et al.* 1975). Speech is slow, imprecise and lacking intonational variation. Ataxic or cerebellar dysarthria is the only type of motor speech impairment that can truly be described as 'slurred speech', sounding, as it does, similar to that of an individual who is intoxicated. In DS, ataxic dysarthria is undoubtedly related to the well-described cerebellar hypoplasia found in the syndrome (Jernigan and Bellugi 1990).

In persons with DS who develop AD, articulation changes toward the hypokinetic or extrapyramidal as a result of the parkinsonian features found in the later stages of the disease (Vieregge *et al.* 1991). Extrapyramidal dysarthria in its classic form is characterized by breathiness, short rushes of speech, low intensity and monotone quality. It is doubtful that the classic form would be found in a person who has already demonstrated the motor speech disorder of DS, but it can be safely assumed that with the progression of rigidity and parkinsonian features, the nature and quality of speech and articulation in DS would change. This phenomenon has been clinically observed but remains to be demonstrated by objective as well as perceptual means.

Persons who do not have DS but develop AD have no articulation impairment until the late stages of the disease when extrapyramidal features develop (Chen *et al.* 1991). This parkinsonism of AD is not identical to Parkinson disease, being usually without tremor and having rigidity closer to gegenhalten than to cogwheeling (Tyrell *et al.* 1990). Lai and Williams (1989), in a group of dementing adults with DS, found that only 20 per cent had parkinsonian features. They were, however, using the strict criteria for Parkinson disease, including bradykinesia, tremor and cogwheeling. Eighty per cent of their dementing subjects were described as having 'shuffling gait' and all, at end-stage, were rigid and flexed. It appears that these subjects demonstrated the same extrapyramidal features described in end-stage AD in the general population (Chen *et al.* 1991). Price *et al.* (1982) also found increased muscle tone in adults with DS and AD, a marked change from the pervasive hypotonia associated with the syndrome in childhood. Rigidity always produces articulation changes tending toward the hypokinetic type. In both advanced AD in the general population and in adults with DS and AD, motor speech changes can be expected. These changes are similar to one another, although probably not identical given that the former group had previously normal articulation and the latter did not.

The characteristic naming dysfunction in AD may be difficult to examine in DS or, indeed, may not exist in the same form. An additional possibility is that individuals with DS always have difficulty with confrontation naming. In AD without DS, it becomes increasingly difficult for the affected person to name objects, pictures, people and places. As outlined above, attempts to do so produce either semantic paraphasias or circumlocutions. If, as seems well-established, cognitive as well as social factors determine language competence, then diseases affecting cognition (including memory) will have seriously deleterious effects on individuals who are already cognitively compromised. In DS there is conflicting information regarding linguistic forms used and communicative competence. There is, of course, complete agreement that language in DS is significantly impaired, but some researchers have concluded that linguistic competence is adequate for functional purposes. Others (Price-Williams and Sabsay 1979; Leuder *et al.* 1981) claim that adults with DS lack the linguistic means for expression of their knowledge of the world. Both positions may be true in that children and adolescents with DS have functional language, while the decrement in linguistic ability seen in adults with DS is the first sign of a dementing process.

Syntax

Syntactic production in early AD in the population at large is intact although simplified in relation to the affected person's earlier abilities. In DS, it is always simplified but not deviant. Reference to each individual's previous functioning level is the only means by which deficits in syntax in DS with AD may be identified before any comparison with syntactic function in AD without DS is attempted.

Reading

There are, currently, insufficient data available to allow informed comment on reading ability in DS with AD. Given the retained oral reading abilities in AD, prospective studies of reading, already known to be impaired in DS, appear to be warranted.

Neurobiological perspectives

Prenatal and early postnatal life are marked by exponential growth of neurones in the brain. Most of this maturation occurs prenatally with the exception of the migration of granule cells of the cerebellum and hippocampus (Purves and Lichtman 1985). Many of the neurones generated during development and a majority of the synapses do not survive, but much of the data on trisomy 21, in particular, indicate that in DS the timetable for these developmental processes is altered. It has been suggested that trisomy 21 results in arrested maturation of neurones and synapses.

There are two widely divergent theories of the nature of cognition. One is that all aspects of cognitive function are the product of one superordinate processing system (Anderson 1983). The other, the modular theory, suggests that there are cognitive subsystems, each treating a different module of information processing (Fodor 1983). If, in DS, there is premature arrest of neuronal development, then there should be differential effects on systems that are designed to mature at different times. This appears to be the case, lending support to the modular cognition approach. This can be further supported by what is known of linguistic function. For example, if auditory comprehension depends on intact left temporal lobe functioning, as appears certain from information gained from studies of aphasia in stroke, then neurological deficit in that area will result in comprehension deficit. Inadequate formation of the hippocampus in infants, adolescents and adults with DS has been confirmed as has auditory comprehension impairment (Ross *et al.* 1984; Wisniewski *et al.* 1986). The hippocampal formation is thought to be important for retaining relational and associative information by linking together disparate neocortical areas. Individuals with hippocampal and closely related temporal lobe damage do not lose the ability to speak, but appear to be unable to acquire new words because of a lack of ability to apply associated meaning (Seifert 1983). This has direct implication for the semantic disturbance of AD and should prompt closer inspection of semantic abilities in DS. Lynch (1986) has suggested that the hippocampus is the most complex integrator of information in the brain. If this is so, then some cognitive/linguistic deficits in AD and DS may be explained, at least in part, by improperly functioning or absent neurones (Horwitz *et al.* 1990; Hyman *et al.* 1990).

In addition, marked poverty of the late maturing granule cells of the cerebellum and brainstem has been well-established in DS (Jernigan and Bellugi 1990). The hypotonia of DS may be attributable to this disruption of cerebellar function. Cerebellar and brain stem hypoplasia in DS may account for the sequential-syntactic deficit often reported. Much of the support for this comes from studies of auditory brainstem-evoked response (ABR) measures in DS. The ABR reflects the earliest stages of the afferent pathway and consists of short latency responses. It is a sequence of five waves, labelled I–V with I originating from the acoustic nerve and later waves from successively higher auditory nuclei. The ABR depends on synaptically secure neurones that fire with brief and stable firing latencies (Moore 1983). Individuals with DS have significantly reduced wave intervals and, in addition, fail to habituate to auditory stimuli (Gigli *et al.* 1984; Galbraith 1986). Children with DS process sequential information more slowly than mental age and chronological age-matched subjects (Lincoln *et al.* 1985). The shorter responses in the brainstem may be due to temporal synaptic instability. These individuals may suffer from perceptual variation from trial to trial although the stimulus remains constant. If this occurs, habituation would be diminished, as is the case in DS. Habituation occurs because of a decrease in neural excitability and is important to learning because the individual must learn to select and attend variably to stimuli that contain no new information.

Longer latency evoked potentials, believed to be of cortical origin, are also uniquely deviant in DS. These evoked potentials, auditory and visual, have larger amplitudes and longer latencies than ABRs. In DS, the middle latency potentials, P1 and P2, are magnified in children and young adults (Straumanis *et al.* 1973; Callner *et al.* 1978). Recent electrophysiological studies of the longer evoked potentials in persons with autopsy-proven AD show that P1 is missing in most of them (Buchwald *et al.* 1989; Green *et al.* 1992). If the same potentials are markedly reduced in older adults with DS who develop AD, this may prove to be of diagnostic value.

It has been suggested that in DS there is right hemisphere or atypical dominance for language, a theory derived mainly from dichotic listening studies in DS (Sommers and Starkey 1977; Elliot *et al.* 1987). This is not yet proven but deserves further investigation.

Conclusion

While there are some similarities in communication impairments in both AD in the general population and in DS with AD, the pre-existing abnormal speech and language in DS makes it difficult to draw parallels. The goal at present should be to gather detailed information prospectively on comprehension, naming, syntax, semantics and reading ability in young adults with DS to determine whether these functions decline in similar fashion and progression as they do in AD without DS. DS is said to represent a halt in maturation at an earlier stage of development than in normals (Thase 1982; Goldman-Rakic *et al.* 1983; Cutler 1985; Walz *et al.* 1986; Zigman *et al.* 1987). If ageing begins in DS at about the age of 35 then we may be seeing the cognitive–linguistic manifestations of a disrupted timetable for both maturation and decline. Comparisons with the communicative function of the old-old (85+ years) with AD may shed more light on the matter, since the largely anecdotal reports of these individuals indicate the possibility of changes that differ from those in younger individuals with AD. Significant alterations in speech and language exist in both AD in the general population and DS with AD, and allow potentially useful inferences to be drawn about the nature and causes of each.

References

Anderson, J. D. (1983). *The architecture of cognition.* Harvard University Press, Cambridge, MA.

Appell, J., Kertesz, A., and Fisman, M. (1982). A study of language functioning in Alzheimer's patients. *Brain and Language*, **17**, 73-91.

Ashman, A. F. (1982). Coding, strategic behavior, and language performance of institutionalized mentally retarded young adults. *American Journal of Mental Deficiency*, **86**, 627-36.

Bakeman, R., and Adamson, L. (1984). Coordinating attention to people and objects in mother-infant and peer-infant interaction. *Child Development*, **55**, 1278-89.

Bayles, K. A. (1982). Language function in senile dementia. *Brain and Language*, **16**, 265-80.

Bayles, K. A., and Kaszniak, A. W. (1987). *Communication and cognition in normal aging and dementia.* Little, Brown & Co., Boston.

Berger, J., and Cunningham, C. (1981). The development of eye contact between mother and normal and Down's syndrome infants. *Developmental Psychology*, **17**, 678-89.

Brandeis, P., and Elsinger, D. (1981). The effects of early middle ear pathology on auditory perception and academic achievement. *Journal of Speech and Hearing Disorders*, **46**, 301-7.

Buchwald, J. S., Erwin, R. J., Van Lancker, D., and Cummings, J. L. (1989). Midlatency evoked responses: differential abnormality of P1 in Alzheimer's disease. *Electroencephalography Clinical Neurophysiology*, **74**, 378-84.

Burt, D. B., Loveland, K. A., and Lewis, K. R. (1992). Depression and the onset of dementia in adults with mental retardation. *American Journal on Mental Retardation*, **96**, 502-11.

Callner, D. A., Dustman, R. E., Madsen, J. A., Schenkenberg, T., and Beck, E. C. (1978). Life span changes in the averaged evoked responses of Down's syndrome and nonretarded persons. *American Journal of Mental Deficiency*, **82**, 398-405.

Campbell-Taylor, I. (1990). Motor speech changes. In *Dementia and communication*, (ed. R. Lubinski), pp. 70-83. Decker, Philadelphia.

Carni, E., and French, L. A. (1984). The acquisition of before and after: What develops? *Journal of Experimental Child Psychology*, **37**, 394-403.

Chen, J. -Y., Stern, Y., Sano, M., and Mayeux, R. (1991). Cumulative risks of developing extrapyramidal signs, psychosis or myoclonus in the course of Alzheimer's disease. *Archives of Neurology*, **48**, 1141-3.

Chertkow, H., Bub, D., and Seidenberg, M. (1989). Priming and semantic memory loss in Alzheimer's disease. *Brain and Language*, **36**, 420-46.

Cicchetti, D., and Pogue-Hesse, P. (1982). Possible contributions to the study of organically retarded persons to developmental theory. In *Mental retardation: the developmental-difference controversy*, (ed. E. Zigler and D. Balla), pp. 277-318. Lawrence Erlbaum, Hillsdale, NJ.

Cicchetti, D., and Sroufe, L. A. (1976). The relationship between affective and cognitive development in Down's syndrome infants. *Child Development*, **47**, 920-9.

Clunies-Ross, G. G. (1979). Accelerating the development of Down's syndrome infants and young children. *Journal of Special Education*, **13**, 169-77.

Coleman, M. (1973). *Serotonin in Down's syndrome*. North Holland, Amsterdam.

Cummings, J. L., Houlihan, J. P., and Hill, M. A. (1986). The pattern of reading deterioration in dementia of the Alzheimer type: observations and implications. *Brain and Language*, **29**, 315-23.

Cutler, N. R. (1985). Alzheimer's disease and Down syndrome: new insights. *Annals of Internal Medicine*, **103**, 556-78.

Dalton. A. J., and Crapper-McLachlan, D. R. (1984). Incidence of memory deterioration in persons with Down's syndrome. In *Perspectives and progress in mental retardation, Biomedical Aspects*, (ed. J. M. Berg), Vol. 2, pp. 55-62. University Park Press, Baltimore.

Darby, J. K., Simmons, N., and Berger, P. A. (1984). Speech and voice parameters of depression: a pilot study. *Journal of Communication Disorders*, **17**, 75-85.

Darley, F. L., Aronson, A. E., and Brown, J. R. (1975). *Motor speech disorders*. Saunders, Philadelphia.

DiSimoni, F. G. (1978). *The Token Test for children*. Teaching Resources Corporation, Hingham, MA.

Elliott, D., Weeks, D. J., and Elliott, C. L. (1987). Cerebral specialization in individuals with Down syndrome. *American Journal on Mental Retardation*, **92**, 263-71.

Ellis, N. R., Deacon, J. R., and Woolridge, P. W. (1985). Structural memory deficits of mentally retarded persons. *American Journal of Mental Deficiency*, **89**, 393-402.

Flicker, C., Ferris, S. H., Crook, T., and Bartus, R. T. (1987). Implications of memory and language dysfunction in the naming deficit of senile dementia. *Brain and Language*, **31**, 187-200.

Fodor, J. (1983). *The modularity of mind*. MIT Press, Cambridge, MA.

Galbraith, G. C. (1986). Unique EEG and evoked response patterns in Down syndrome individuals. In *The neurobiology of Down syndrome*, (ed. C. J. Epstein), pp. 109-19. Raven Press, New York.

Gigli, G. L., Ferri, R., Musumeci, S. A., Tomasetti, P., and Bergonzi, P. (1984). Brainstem auditory evoked responses in children with Down's syndrome. In *Perspectives and progress in mental retardation, Biomedical Aspects*, (ed. J. M. Berg), Vol. 2, pp. 277-86. University Park Press, Baltimore.

Goldman-Rakic, P. S., Isseroff, A., Schwartz, M. L., and Bugbee, N. M. (1983). The neurobiology of cognitive development. In *Handbook of child psychology*, (ed. M. M. Haith and J. J. Campos), Vol. 2. (4th edn), pp. 281-344. Wiley, New York.

Goodglass, H., and Kaplan, E. (1972). *The assessment of aphasia and related disorders*. Lea & Febiger, Philadelphia.

Gotham, A. M., Brown, R. G., and Marsden, C. D. (1986). Depression in Parkinson's disease: a quantitative and qualitative analysis. *Journal of Neurology, Neurosurgery and Psychiatry*, 49, 381-9.

Graham, J., and Graham, L. (1971). Language behaviour of the mentally retarded: syntactic characteristics. *American Journal of Mental Deficiency*, 75, 623-9.

Green, J. B., Flagg, L., Freed, D. M., and Schwankhaus, J. D. (1992). The middle latency auditory evoked potential may be abnormal in dementia. *Neurology*, 42, 1034-6.

Grober, E., Buschke, H., Kawas, C., and Fuld, P. (1985). Impaired ranking of semantic attributes in dementia. *Brain and Language*, 26, 276-86.

Gunn, P., Berry, P., and Andrews, R. J. (1982). Looking behavior of Down syndrome infants. *American Journal of Mental Deficiency*, 87, 344-7.

Hartley, X. Y. (1982). Receptive language processing of Down's syndrome children. *Journal of Mental Deficiency Research*, 26, 263-9.

Haxby, J. V. (1989). Neuropsychological evaluation of adults with Down's syndrome: patterns of selective impairment in non-demented old adults. *Journal of Mental Deficiency Research*, 33, 193-210.

Horwitz, B., Schapiro, M. B., Grady, C. L., and Rapoport, S. I. (1990). Cerebral metabolic pattern in young adult Down's syndrome subjects: altered intercorrelations between regional rates of glucose utilization. *Journal of Mental Deficiency Research*, 34, 237-52.

Huff, F. J., Corkin, S., and Growdon, J.H. (1985). Semantic impairment and anomia in Alzheimer's disease. *Brain and Language*, 28, 235-240.

Hyman, B. T., Van Hoessen, G. W., and Damasio, A. R. (1990). Memory related neural systems in Alzheimer's disease. *Neurology*, 40, 1721-30.

Jernigan, T. L., and Bellugi, U. (1990). Anomalous brain morphology on magnetic resonance images in Williams syndrome and Down syndrome. *Archives of Neurology*, 47, 529-33.

Jones, O. H. M. (1980). Prelinguistic communication skills in Down's syndrome and normal infants. In *High risk infants and children: adult and peer interactions*, (ed. T.F. Field), pp. 205-25. Academic Press, New York.

Kernan, K. T. (1990). Comprehension of syntactically indicated sequence by Down's syndrome and other mentally retarded adults. *Journal of Mental Deficiency Research*, 34, 169-78.

Lai, F., and Williams, R. S. (1989). A prospective study of Alzheimer disease in Down syndrome. *Neurology*, 46, 849-53.

Landry, S. H., and Chapieski, M. L. (1988). Visual attention during toy exploration in preterm infants: effects of medical risk and maternal interactions. *Infant Behavior and Development*, 11, 187-204.

Landry, S. H., and Chapieski, M. L. (1989). Joint attention and infant toy exploration: effects of Down syndrome and prematurity. *Child Development*, 60, 103-18.

Landry, S. H., and Chapieski, M. L. (1990). Joint attention of six-month old Down syndrome and preterm infants: 1. Attention to toys and mother. *American Journal on Mental Retardation*, 94, 488-98.

Layton, T. L., and Sharifi, H. (1978). Meaning and structure of Down's syndrome and nonretarded children's spontaneous speech. *American Journal of Mental Deficiency*, 83, 439-45.

Lenneberg, E., Nichols, I., and Rosenberger, E. (1964). *Primitive stages of language development in mongolism*. Research Publications No. 42, Association for Research in Nervous and Mental Disease .

Leuder, I., Fraser, W. I., and Jeeves, M. A. (1981). Social familiarity and communication in Down syndrome. *Journal of Mental Deficiency Research*, 25, 133-42.

Lincoln, A. J., Courchesne, E., Kilman, B. A., and Galambos, R. (1985). Neuropsychological correlates of information-processing by children with Down syndrome. *American Journal of Mental Deficiency*, 89, 403-14.

Ludlow, J. R., and Allen, L. M. (1979). The effect of early intervention and pre-school stimulus on the development of the Down's syndrome child. *Journal of Mental Deficiency Research*, **23**, 29-44.

Lynch, G. (1986). *Synapses, circuits and the beginnings of memory*. MIT Press, Cambridge, MA.

Lytton, W. W., and Brust, J. C. M. (1989). Direct dyslexia: preserved oral reading of real words in Wernicke's aphasia. *Brain*, **112**, 583-94.

McCall, R. B. (1972). Smiling and vocalization in infants as indices of perceptual-cognitive processes. *Merrill-Palmer Quarterly*, **18**, 341-7.

McKhann, G., Drachman, D., Folstein, M., Katzman, R., Price, D., and Stadlan, E. M. (1984). Clinical diagnosis of Alzheimer's disease: Report of the NINCDS-ARDRA Work Group under the auspices of the Department of Health and Human Services Task Force on Alzheimer's disease. *Neurology*, **34**, 939-44.

Mahoney, G., and Snow, K. (1983). The relationship of sensorimotor functioning in children's response to early language training. *Mental Retardation*, **21**, 248-54.

Margolin, D. I. (1988). Lexical priming by pictures and words in aging, stroke and dementia. *Dissertation Abstracts International*.

Martin, A., and Fedio, P. (1983). Word production and comprehension in Alzheimer's disease: the breakdown of semantic knowledge. *Brain and Language*, **19**, 124-41.

Miller, J. (1981). *Analyzing language production in children: experimental procedures*. University Park Press, Baltimore.

Miller, J., Chapman, R., and Bedrosian, J. (1978). The relationship between etiology, cognitive development and language and communicative performance. *The New Zealand Speech Therapists' Journal*, **33**, 2-17.

Miniszek, N. A. (1983). Development of Alzheimer disease in Down syndrome individuals. *American Journal of Mental Deficiency*, **87**, 377-85.

Moore, E.J. (1983). *Bases of auditory brain stem evoked responses*. Grune & Stratton, New York.

Moscovitch, M., Winocur, G., and McLachlan, D. (1986). Memory as assessed by recognition and reading time in normal and memory impaired people with Alzheimer's disease and other neurological disorders. *Journal of Experimental Psychology: General*, **115**, 331-47.

Murdoch, B. E., Chenery, H. J., Wilks, V., and Boyle, R. S. (1987). Language disorders in dementia of the Alzheimer type. *Brain and Language*, **31**, 122-37.

Naremore, R., and Dever, R. (1975). Language performance of educable mentally retarded and normal children at five age levels. *Journal of Speech and Hearing Research*, **18**, 82-96.

Nebes, R. D., Martin, O. S. M., and Horn, L. C. (1984). Sparing of semantic memory in Alzheimer's disease. *Journal of Abnormal Psychology*, **93**, 321-30.

Nebes, R. D., Boller, F., and Holland, A. (1986). Use of semantic context by patients with Alzheimer's disease. *Psychology and Aging*, **1**, 261-9.

Nilsonne, A. (1987). Speech characteristics as indicators of depressive illness. *Acta Psychiatrica Scandinavica*, **77**, 253-63.

Piaget, J., and Inhelder, B. (1969). *The psychology of the child*. Basic Books, New York.

Price, D. L., Whitehouse, P. J., Struble, R. G., Coyle, J. T., Clark, A. W., Delong, M. R., *et al.* (1982). Alzheimer's disease and Down's syndrome. In *Alzheimer's disease, Down's syndrome and ageing*, (ed. F. M. Sinex and C. R. Merrill), pp. 143-63. New York Academy of Sciences.

Price-Williams, D., and Sabsay, S. (1979). Communicative competence among severely retarded persons. *Simiotica*, **26**, 35-63.

Purves, D., and Lichtman, J. W. (1985). *Principles of neural development*. Sinauer, Sunderland, MA.

Rogers, M. G. H. (1975). A study of language skills in severely subnormal children. *Child: Care, Health and Development*, **1**, 113-26.

Rohr, A., and Buhr, D. B. (1978). Etiological differences in patterns of psycholinguistic development of children IQ 30 to 60. *American Journal of Mental Deficiency*, **82**, 549-53.

Rondal, J. A. (1978). Developmental sentence scoring procedure and the delay-difference question in language development of Down's syndrome children. *Mental Retardation*, **16**, 169-71.

Rondal, J. A. (1988). Language development in Down's syndrome: a life-span perspective. *International Journal of Behavioural Development*, **11**, 21-36.

Ross, M. H., Galaburda, A. M., and Kemper, T. L. (1984). Down's syndrome: is there a decreased population of neurons? *Neurology,* **134,** 909-16.

Sanger, R. G. (1975). Facial and oral manifestations of Down's syndrome. In *Down's syndrome (mongolism): research, prevention and management,* (ed. M. Koch and F. de la Cruz), pp. 32-46. Bruner-Mazel, New York.

Schwartz, M., Marin, O. S. M., and Saffran, E. (1979). Dissociation of language function in dementia: a case study. *Brain and Language,* 7, 277-306.

Seifert, W. (1983). *Neurobiology of the hippocampus.* Academic Press, London.

Sommers, R. K., and Starkey, K. L. (1977). Dichotic verbal processing in Down's syndrome children having qualitatively different speech and language skills. *American Journal of Mental Deficiency,* **82,** 44-53.

Sovner, R., and Hurley, A. D. (1983). Do the mentally retarded suffer from affective illness? *Archives of General Psychiatry,* **40,** 61-7.

Sroufe, L. A., and Waters, E. (1976). The ontogenesis of smiling and laughter: a perspective on the organization of development in infancy. *Psychological Review,* **83,** 173-89.

Straumanis, J. J., Shagass, C., and Overton, D. A. (1973). Auditory evoked responses in young adults with Down's syndrome and idiopathic mental retardation. *Biological Psychiatry,* **6,** 75-9.

Thase, M. E. (1982). Reversible dementia in Down's syndrome. *Journal of Mental Deficiency Research,* **26,** 111-3.

Tyrell, P. J., Ibanez, V., Bloomfield, P. M., Leenders, K. L., Frackowiak, R. S., and Rossor, M. N. (1990). Clinical and positron emission studies in the 'extrapyramidal syndrome' of dementia of the Alzheimer type. *Archives of Neurology,* **47,** 1318-23.

Vieregge, P., Ziemens, G., Piosinski, A., Freudenberg, M., and Kompf, D. (1991). Parkinsonian features in advanced Down's syndrome. *Journal Neurologie (Austria),* **33,** 119-24.

Walz, T., Harper, D., and Wilson, J. (1986). The ageing developmentally disabled person: a review. *Gerontological Society of America,* **26,** 622-9.

Wiegel-Crump, C. A. (1981). The development of grammar in Down's syndrome children between the mental ages of 2.0 and 6.11 years. *Education and Training of the Mentally Retarded,* **16,** 24-30.

Wisniewski, K. E., Howe, G., Williams, D. G., and Wisniewski, H. M. (1978). Precocious ageing and dementia in patients with Down's syndrome. *Biological Psychiatry,* **13,** 619-27.

Wisniewski, K. E., Laure-Kamionowska, M., Connell, F., and Wen, G. Y. (1986). Neuronal density and synaptogenesis in the postnatal stage of brain maturation in Down syndrome. In *The neurobiology of Down syndrome,* (ed. C. J. Epstein), pp. 29-44. Raven Press, New York.

Young, E. C., and Kramer, B. M. (1991). Characteristics of age related language decline in adults with Down syndrome. *Mental Retardation,* **29,** 75-9.

Zigman, W. B., Schupf, N., Lubin, R. A., and Silverman, W. P. (1987). Premature regression of adults with Down syndrome. *American Journal of Mental Deficiency,* **92,** 161-8.

10

Neuroimaging in adults with Down syndrome

M. B. Schapiro

Summary

Both structural and functional brain imaging offer advantages over post-mortem studies, including the ability to study patterns of disease early in the illness and to follow the course of the disease and the response to therapy. This chapter summarizes the results of brain imaging studies in adults with Down syndrome. Young adults with the syndrome (mentally retarded but relatively free of Alzheimer neuropathology) have small brains, but no evidence of cerebral atrophy or the selective alteration in cerebral blood flow or metabolism that is noted in Alzheimer disease. Methods of data analysis other than group-mean comparisons suggest that alterations in brain functional associations involving language areas occur. In older persons with Down syndrome both structural and functional brain imaging can distinguish dementia from lesser cognitive decline. Abnormal patterns of brain glucose metabolism and blood flow selectively involving association areas of parietal and temporal neocortices are present in older individuals with Down syndrome who have dementia. Furthermore, progressive brain atrophy is present in demented individuals with Down syndrome. It remains to be determined whether imaging techniques can distinguish lesser cognitive decline from mental retardation in older, non-demented adults with Down syndrome.

Introduction

A human model of both Alzheimer disease (AD) and mental retardation exists that circumvents the problem of aetiological heterogeneity and allows one to study how brain dysfunction occurs due to suboptimal gene expression. Down syndrome (DS) is a genetic disorder in which an extra portion of chromosome 21 leads to mental retardation and many phenotypic abnormalities (see Chapter 2). Furthermore, persons with DS over the age of 40 years demonstrate neuropathological and neurochemical defects post-mortem that are virtually indistinguishable from those found in brains of individuals with AD in the general population, as well as a universal cognitive deterioration and a 20–30 per cent prevalence of dementia (Schapiro *et al.* 1989b).

Because of opportunities available with newer brain imaging techniques, including positron emission tomography (PET) with [^{18}F]fluoro-2-deoxy-D-glucose (^{18}FDG), single photon emission computed tomography (SPECT) and ^{133}xenon inhalation regional cerebral blood flow technique to study brain glucose metabolism as well as blood flow, and quantitative computed assisted tomography (CT) and magnetic resonance imaging (MRI) to study brain anatomy, laboratories throughout the world have developed imaging programmes to address the age changes and mental retardation in adults with DS, in concert with a study of AD.

Subject selection

The relation of DS and AD has been considered in prior reports, but many methodological issues related to subject selection were ignored. For subjects with DS, some of these issues include: type of screening, results of karyotyping (regular trisomy 21 or other chromosomal findings), source of referral (institution or community), reason for referral, mental age, associated medical problems (e.g. congenital heart disease, hypothyroidism) and medications. Similar selection issues apply to control subjects.

To avoid such methodological problems, our laboratory studied only carefully screened, healthy, non-institutionalized subjects with DS, whose mental status and cognitive deficits could be quantified. In our studies referred to subsequently, all subjects (except one to be mentioned separately later) had a trisomy 21 karyotype; they included 20 young adults aged 19–34 years and nine older adults aged 45–63 years. Not all subjects were used in all studies. Screening included a review of the medical and family history, a physical and neurological examination, blood and urine laboratory tests, ECG and chest radiographs. With the exception of thyroid replacement in several subjects and vitamin B12 injections for celiac sprue in one instance, none was taking medication (Schapiro *et al.* 1989b).

In order to distinguish mental retardation from dementia, subjects with DS in our studies were divided into those aged 18–40 years (mentally retarded but presumed to be relatively free of Alzheimer neuropathology) and those over 40 years (with definite Alzheimer neuropathology and possible dementia) (Schapiro *et al.* 1992). This age distinction was based on neuropathological studies that showed that without regard to density, type (senile plaques and/or neurofibrillary tangles), maturity or location, **some** persons with DS less than 40 years of age have some degree of senile plaques and/or possibly neurofibrillary tangles and that **all** over 40 years have some degree of neuropathology, though again the density, type, maturity and location varies (Mann 1988).

Our controls were participants in the National Institute on Aging, Laboratory of Neurosciences Study on Aging and were screened with medical, neurological and laboratory tests. Controls had no history of neurological, psychiatric or systemic medical disorder. None was taking medication for at least two weeks prior to the study.

In our laboratory the diagnosis in DS of dementia consistent with AD was made using modified criteria from the Diagnostic and Statistical Manual of Mental Disorders (American Psychiatric Association 1987), which specified an acquired, progressive loss of intellectual function, such as loss of daily living and vocational skills, impairment of memory and other cognitive functions, and personality change. The diagnosis of dementia was based on review of records, interviews with caregivers or employers, clinical examination and bed-side mental status tests (Schapiro *et al.* 1987a).

Structural brain imaging

Autopsy studies of the human brain may be affected by many parameters and, furthermore, may reflect only the end-stage of a disease (Dekaban and Sadowsky 1978; Schwartz *et al.* 1985; Rapoport 1990). Some of the problems with autopsy studies may be eliminated by using *in vivo* structural imaging techniques, which has the advantage of allowing one to examine patterns of changes early in a disease. In addition, *in vivo* investigations can be used for longitudinal as well as cross-sectional studies, allowing one to examine the course of disease and the response to any therapeutic interventions. *In vivo* structural measures, therefore, have been used to evaluate the dimensions of brain, cerebrospinal fluid (CSF) and cerebral ventricles in ageing and disease.

In vivo structural studies, however, have their own inherent limitations that are often not addressed in reports. Details about the scanning should address the following questions: 'What type of CT/MRI scanner was used?', 'How many different machines were used?', 'If different scanners were used, how were data from them made comparable?', 'What were the resolution and slice thickness of the machine?', 'At what angle were the CT/MRI scans performed and was this uniform for all subjects in all groups?' and 'For MRI scans, what imaging sequences were used?'.

Details about the image analysis also should include information on the following: 'What type of analysis was performed, such as perceptual rating or linear, planimetric or volumetric measures?', 'How were these measures validated?', 'For CT measures, was a constant window of contrasts used (since changes in bone window have been shown to affect measurements, especially linear ratios)?', 'Were the measurements done blinded?', 'How many raters performed these measurements?', 'What was the inter- and/or intra-rater reliability?', 'Were there any differences in reliability among measures?', 'How sensitive and specific are the measures?' Without such details, one often is not able to address the validity or reliability of results nor accurately compare results from different studies.

Young adults with Down syndrome compared with age-matched controls

With these technical issues in mind, as well as the issues related to subject referral and screening, *in vivo* studies have been performed in young adults with DS.
Discrepant results have been noted, particularly in younger subjects, which may
be related to these methodological issues. For example, Ieshima *et al.* (1984),
using linear measures of brain morphometrics, showed that in 10 adults with
DS aged between 20 and 39 years the third ventricle ratio was not significantly
different and the anterior horn slice ratio was smaller in comparison to age-
matched controls with neurological diagnoses (but with normal examinations).
Other ratio measures showed that the posterior fossa was small relative to the
cerebrum in DS. On the other hand, Wisniewski *et al.* (1982), in a study of
basal ganglia calcification, used perceptual ratings and found, in 17 institutionalized, moderately to profoundly mentally retarded adults with DS aged 21–30
years, two subjects with no brain atrophy, and seven with minimal, six with
mild and two with moderate brain atrophy. Controls were not studied. Their
finding of brain atrophy may be due to their subjective CT analysis and to the
use of institutionalized and possibly lower mentally functioning subjects.

Though perceptual ratings, and linear and planimetric measures are easy to
employ, there are disadvantages. In particular, these methods of analysis are
sensitive to the effect of changes in bone window and the influence of changes
in angle orientation of the scan. Also, the methods do not account for the complex shape of the ventricles (DeCarli *et al.* 1990). Some of these problems may
be solved with quantitative volumetric methods.

Therefore, we used quantitative CT to measure volumes of brain structures.
Non-contrast CT imaging was performed using an 8800 CT scanner (General
Electric Corp., Milwaukee, WI), for which full width at half maximum equaled
1 mm. An image processing program allowed determination of the total number of intracranial pixels for the whole slice and the percentages of the total
number of pixels of gray matter, white matter and CSF. In addition, regions of
interest were outlined around the ventricles, using a digitalizing graphics tablet
and a light pen. Intracranial surface area was derived with an automatic edge-
finding procedure. Volumes of intracranial structures were calculated by
adding the number of pixels of the measure of interest, across slices, and multi-
plying by pixel area (0.0064 cm^2) and interslice distance (0.7 cm) (Schwartz *et
al.* 1985; Schapiro *et al.* 1989b).

We found that mean total intracranial volume was significantly less in 18
young adults with DS (12 males, six females) aged 19–34 years than in 16
healthy control males aged 21–35 years; i.e. mean ± SD of $1067 \pm 90 \text{ cm}^3$ versus $1241 \pm 120 \text{ cm}^3$ (unpaired *t*-test, $P < 0.05$). Gray matter plus white matter
volume (in the seven slices from 30–80 mm above the inferior orbital meatal
line) also was significantly less in the subjects with DS ($664 \pm 40 \text{ cm}^3$) than in
the controls ($759 \pm 48 \text{ cm}^3$) ($P < 0.05$). However, after normalization to seven-
slice intracranial volume (to correct for the smaller cranial vault in DS), there
was no longer any difference between the two groups in gray matter plus white

matter volumes (DS 96.3 per cent $\pm$ 2.0 versus controls 97.6 per cent $\pm$ 1.4, $P >$ 0.05). In addition, CSF and ventricular volumes, either directly or after normalization to seven-slice intracranial volume, did not differ between the two groups (Schapiro *et al.* 1987b; Schapiro *et al.* 1992).

CT scanning, however, has limited spatial resolution, partial voluming and spectral shift artefacts, which limit the ability to accurately measure brain volumes, particularly the cerebral cortex and subcortical nuclei (though ventricular spaces can be delineated) (DeCarli *et al.* 1990). MRI scanning has the advantages, compared with CT scanning, of no radioactivity, better spatial resolution (which allows better delineation of brain structures with lower coefficient of variation), no spectral shift artefact near the bone (which allows the examination of peripheral CSF), flexible sequences to allow particular components of the brain (CSF, white matter or gray matter) to be highlighted, choice of image orientation (which permits detailed analysis of particular brain structures of interest) and better imaging of white matter abnormalities in the brain. With the MRI scan, one is now able to obtain measures of peripheral CSF, regional neocortical brain matter and subcortical structures, such as the hippocampus, not well seen with CT scanning. Furthermore, comparison of volumes of brain regions obtained from CT and MRI scanning shows that MRI values are more consistent with volumes derived from autopsy studies (Murphy *et al.* 1992).

Several MRI studies have begun to examine whether atrophy occurs in young DS adults without dementia, as well as to determine if morphological differences can be identified in the DS brain. Pelz *et al.* (1986), examining 0.15 Telsa MRI scans without rating or quantification in several subjects with DS recruited from group homes, noted 'mild atrophic changes in the temporal regions and generous ventricles' in two subjects aged 25 and 26 years; no changes were reported in four subjects aged 29, 32, 35 and 35 years. Scans from controls were not described. Lott *et al.* (1991) quantified coronal MRI sections and showed, in 10 non-demented subjects with DS (ages not specified) that temporal lobe and parahippocampal gyral volumes were increased in comparison to controls; no differences were noted for hippocampal and striatal volumes.

In a quantitative MRI study of Williams syndrome, a mental retardation disorder characterized by short stature, supravalvular aortic stenosis and particular facies, Jernigan and Bellugi (1990) examined three subjects with DS (one male, two females) aged 14–17 years with MRI scans performed on a 1.5 Telsa scanner. Proton density- and T2-weighted axial images were analysed using a computerized approach that classifies pixels into gray matter, white matter, CSF or signal hyperintensities. In comparison to 14 normal control subjects (nine males, five females), the subjects with DS had reduction of both cerebral and cerebellar volumes. No difference was found in the ratio of cerebellar to cerebral volume, suggesting that there was not a disproportionate reduction of one element relative to the other. In addition, the ratio of cerebral volume to supratentorial cranium volume in DS (0.89) did not differ from that in the normal controls (0.92).

Using different scanning methods and data analysis techniques, slightly different results were found by Weis *et al.* (1991). Seven non-institutionalized adults with DS (five males, two females) aged 30–45 years (mean age = 38 years) and seven sex- and age-matched control subjects were studied on a 0.5 Telsa scanner, and T1-weighted images were analysed manually with a trace procedure. In agreement with, and extending the findings of, others using quantitative CT or MRI scanning, Weis *et al.* found that the subjects with DS had a significantly smaller volume of cerebral cortex, white matter and cerebellum compared with the controls; no volume difference was noted in ventricles, thalamus, caudate and lenticular nuclei, and brainstem. The whole brain volume, the sum of the above mentioned structures, also was significantly smaller in the subjects with DS; i.e. mean ± SD of 1081.6 ± 81.0 cm^3 versus 1313.1 ± 146.0 cm^3 in the controls. Also, when regional volumes were normalized to brain volume, no significant difference was noted between the two groups, suggesting that individual brain structures were not decreased or altered in size disproportionately to each other.

Thus, these quantitative CT and MRI studies show that young adults with DS do not have enlarged ventricular volumes but do have decreased brain volumes. However, there are conflicting results as to whether the intracranial volume is decreased in DS. The results of Weis *et al.* (1991), showing no decrease in cranial cavity volume in DS (mean ± SD of 1443.2 ± 99.0 cm^3 compared with 1571.2 ± 231.0 cm^3 in controls), differ from early studies that showed a smaller skull in subjects with DS, using measurements of head circumference (Benda 1969; Zellweger 1977), length and width of the skull (Benda 1969), and an index that expresses three-dimensional skull growth based on linear measures from skull radiographs (Schmid *et al.* 1969). Furthermore, because brain size determines cranial capacity during human development and because the brains of young adults with DS are smaller than those of age-matched controls (reflecting less growth but not atrophy), one might expect the cranial cavity also to be small in DS.

In examining the data of Weis *et al.* (1991) it can be noted that the cranial cavity in DS was smaller than in controls, but not significantly so ($P > 0.05$). This lack of significance may be related to the small number of subjects whom they studied, as well as to the large coefficient of variation in their control group for intracranial volume, suggesting that the control group was heterogeneous. The basis for this increased coefficient of variation in the controls for cranial cavity and not for brain volume is not apparent, as the ratio of brain to cranial volumes in controls is constant in the age range studied (Davis and Wright 1977). In contrast to our study that used only male controls, Weis *et al.* used male and female controls, sex-matched to their DS group. Because postmortem (Dekaban and Sadowsky 1978) and quantitative CT (Kaye *et al.* 1992) studies have shown that men have larger brains than women, the inclusion of both sexes in a control group would increase heterogeneity. Because our statistical analysis showed no difference between male and female subjects with DS, we used only males as a control group to reduce the variance in this group. Using a male control group should increase the likelihood of finding a difference in intracranial and brain volumes between DS and controls, whereas a female control group probably would produce the opposite effect. Another method of analysis may be to compare persons with DS and controls by each sex separately. Alternatively, one must consider that persons with DS are significantly shorter than controls. Although the relation of head size to height is controversial (Dekaban and Sadowsky 1978), a height-matched control group might better answer the question whether having DS causes one to have a proportionately smaller head.

If there are differences in cranial capacity between DS and controls, are they developmental differences or due to atrophy? To answer this, the ratio of brain volume to cranial volume should be determined in each subject, to reduce the variance within groups and to minimize the effect of heterogeneity of cranial volume, as might be seen between sexes. In using both sexes, however, one must be cautious that there are not proportional differences in percentage brain volumes between male and female controls (Kaye *et al.* 1992). Unlike our finding and that of Jernigan and Bellugi (1990) of no proportional decrease in brain size between DS and controls, Weis *et al.* (1991) suggested that the brain is proportionately smaller in DS compared with controls (brain volume/cranial volume: 75.0 per cent versus 83.6 per cent, $P < 0.001$), implying an atrophic process. However, all three groups of investigators noted no difference in the per cent ventricular volume (of intracranial volume) between DS and controls. The reason for the discrepancy between the results of Weis *et al.* and those of the others is not readily apparent. However, it should be noted that the control ratio reported by Weis *et al.* differs from the value of 92.2 ± 1.6 per cent (SEM) reported by Davis and Wright (1977) in their study of a control population and from the value of 92 per cent reported in the MRI study of Jernigan and Bellugi (1990). Furthermore, some of the subjects (number not specified) of Weis *et al.* were older than 40 years, when all with DS show the neuropathology of AD (Mann 1988). No comments were made on the cognitive profile of these DS subjects, including whether or not dementia was absent. Persons with DS over the age of 40 years show a universal cognitive deterioration on standardized neuropsychological tests, more severe in demented than in non-demented cases (Haxby 1989). Dementia occurs in 40 per cent of older adults with DS. Because AD results in neuronal loss, evident on structural imaging as brain atrophy (Schapiro *et al.* 1989b), inclusion of older subjects with DS when relating mental retardation to brain structure would erroneously indicate brain atrophy.

Thus, young adults with DS have small brains, but this probably reflects a small stature and small cranial vault. CT and MRI scanning show no ventricular dilatation, which would be evidence against cerebral atrophy. Therefore, the mental retardation in DS is most likely related to inherent cerebral dysfunction and not to acquired cerebral atrophy.

Demented older adults with Down syndrome compared with non-demented older or young adults with the syndrome

Previous CT and MRI studies in older adults with DS have sought to differentiate dementia from lesser cognitive decline or mental retardation. Two CT studies have looked for **generalized** brain atrophy and ventricular dilatation using perceptual ratings. Dalton and Crapper (1977), in a study of learning and memory, used subjective ratings of CT scans to examine five institutionalized individuals with DS with IQs from 20–40. On the basis of selection criteria, those with prior dementia were effectively excluded. Two subjects, aged 43 and 58 years, showed generalized cortical atrophy, in conjunction with the development of impairment of performance on a delayed match-to-sample test, which was comparable to the degree of atrophy in two DS individuals with advanced AD not in the study. Two other subjects, aged 39 and 41 years, who had no impairment of cognitive performance, showed no cortical atrophy or ventricular enlargement.

Similar CT findings were noted by Lott and Lai (1982a,b) in a study of subjects with DS referred from institutional or community placements for evaluation of mental deterioration. Subjective ratings of CT scans showed severe cerebral atrophy in three subjects aged 53, 60 and 61 years, and moderate atrophy in four subjects aged 49, 55, 60 and 60 years, in conjunction with change in personality and loss of skills of daily living. Furthermore, two of the subjects showed progressive cerebral atrophy on CT scans performed 12 months apart. In comparison, three non-demented age-matched subjects with DS showed no cerebral atrophy. Thus, these two CT studies suggest that cerebral atrophy and ventricular enlargement occur only in persons with DS who show decline in cognitive performance or clinical dementia, but not in those with stable cognitive performance. A third study of the prevalence of dementia in DS (Franceschi *et al.* 1990) also reported that all demented subjects had evidence of brain atrophy, although no other details were provided concerning the brain imaging.

A potential difficulty in morphometric brain studies in demented persons with DS is the lack of proper control groups. Ageing in normal adults is accompanied by increases in ventricular volumes and decreases in brain mass (Schwartz *et al.* 1985; Murphy *et al.* 1992). Furthermore, as indicated above, younger adults with DS have smaller brains than adults without DS. Therefore, to differentiate the effects of ageing from the superimposed effects of AD, one ideally should use both young and old non-demented adults with DS as reference. Therefore, we used both young and old adults with DS in our quantitative CT study of subjects who had two or more CT scans, in order to distinguish dementia from mental retardation.

In cross-sectional studies comparing five non-demented older and 12 young adults with DS, there were no significant differences in mean volume of CSF, gray matter plus white matter or ventricles, directly or after normalizing to seven-slice intracranial volume (one way ANOVA with Bonferroni correction for multiple comparisons, $P > 0.05$). However, significant increases in mean CSF volume and mean third ventricle volumes, with or without normalizing to seven-slice intracranial volume, were shown between three demented older and 12 young subjects with DS, and between three demented and five non-demented older subjects. Furthermore, demented older subjects had mean total ventricular volumes that were 85 per cent and 102 per cent greater than volumes in young and non-demented older groups, respectively. Also, demented older subjects had a mean gray matter plus white matter volume (normalized to seven-slice intracranial volume) that was significantly less than in the two DS groups (Schapiro *et al.* 1989b).

	Young DS (12)	Non-demented old DS (5)	Demented old DS (3)
CATSEG analysis			
CSF volume	−1.32 ± 1.90	2.50 ± 1.81[a]	10.90 ± 4.32[a,b]
GM + WM	−0.17 ± 5.85	−3.28 ± 11.63	−10.41 ± 3.11
Region of interest analysis (Ventricles)			
III	−0.027 ± 0.056	0.115 ± 0.232	−0.131 ± 0.599
L lateral	−0.52 ± 0.51	0.00 ± 0.76	2.18 ± 2.99[a]
R lateral	−0.27 ± 0.58	0.00 ± 0.75	3.91 ± 2.14[a,b]

Number of subjects is shown in parentheses. Mean ± SD (cm^3/yr). GM + WM: gray matter + white matter.
[a]Differs from mean in young DS group by Bonferroni *t*-test ($P<0.05$)
[b]Differs from mean in non-demented old DS group by Bonferroni *t*-test ($P<0.05$)

Table 10.1. Rates of change of cerebrospinal fluid and brain matter volumes for adults with Down syndrome

In longitudinal studies of subjects with DS, CT scans were separated by an average of 43 months in the 12 young subjects, by 20 months in the five non-demented old subjects and by 22 months in the three demented old subjects. A slight increase for rate of change of CSF volume was observed between the non-demented old DS group as opposed to the young DS group, 2.50 ± 1.81 (mean ± SD) versus −1.32 ± 1.90 cm^3/year, respectively [(one-way ANOVA with Bonferroni correction for multiple comparisons, $P < 0.05$) (Table 10.1)]. Greater rates of increase of CSF volume were noted between the demented older DS group and the young DS group, 10.90 ± 4.32 versus −1.32 ± 1.90 cm^3/year, respectively ($P < 0.05$) (Table 10.1). These rates of change in the demented old DS group also were significantly greater than those in the non-demented old DS group. Increased rates of change for lateral ventricles also were noted in demented older subjects compared to non-demented young and older subjects (Schapiro *et al.* 1989b).

Three additional CT studies have attempted to measure regional brain atrophy, as reflected in **regional** enlargement of lateral ventricles or cisterns, in demented subjects with DS. LeMay and Alvarez (1990) used a perceptual rating scale of dilation of the temporal horns of the lateral ventricles to relate temporal lobe atrophy with cognition. They examined 25 institutionalized adults with DS, aged 29–64 years, and assessed mental status on the basis of serial assessment of neurological status, daily living skills and IQ. A significant correlation of age and CT rating was shown, but, without controls, it is not clear how this differed from healthy ageing. However, there was a significant correlation between CT rating and mental status grade, which remained significant even after controlling for age.

Pearlson *et al.* (1990) used perceptual ratings, linear ratios and planimetry to study cortical and subcortical atrophy in CT images obtained by a standardized protocol, from 18 non-institutionalized karyotyped subjects with DS aged 26–70 years (eight subjects <40 years). One of the eight subjects below 40 years and six of 10 subjects older than 40 years were demented, based on deterioration of behaviour and decline in performance on standardized behaviour and cognitive scales. Compared with age-matched, cognitively normal controls, the demented subjects with DS over 40 years tended to have lateral ventricle-to-brain ratio, bicaudate ratio, cortical sulcal atrophy, and to a lesser degree bifrontal ratio, greater than two standard deviations of control values; non-demented subjects tended to have values within two standard deviations. The suprasellar cistern ratio, a measure of mesial temporal lobe atrophy, showed the earliest change and was increased beginning at about the age of 30 years in both the demented and non-demented subjects with DS. This ratio was correlated with the Folstein Mini-Mental State Examination (Folstein *et al.* 1975), suggesting the increase in older subjects with DS was further increased in the demented compared with the non-demented subjects.

Lai and LeMay (1990), using linear measures of the lateral ventricles, sought to determine whether selective regional brain atrophy occurred in 30 demented subjects with DS (mean age 57 years), compared with eight non-demented DS subjects (39 years), 21 normal controls (66 years) and 36 AD subjects without DS (73 years). No significant difference was shown between the non-demented subjects with DS and controls. However, measures of frontal, temporal and occipital ventricular horns were increased in the demented DS group compared with the non-DS AD group. Furthermore, temporal horn ratios were increased more than occipital horn ratios, which were increased compared with frontal horn ratios.

MRI scanning also has been used to determine whether dementia in DS is accompanied by **regional** brain atrophy. Lott *et al.* (1991; see above) showed reduced volumes of hippocampus (46 per cent), parahippocampal gyrus (61 per cent) and striatum (91 per cent) in one trisomy 21 adult with severe dementia in comparison to asymptomatic, non-demented DS controls (ages not noted), whereas Pelz *et al.* (1986; see above) noted 'marked temporal lobe atrophy with a very prominent left sylvian fissure' in a 45-year-old individual with DS (cognitive status was not described).

Thus, cross-sectional and longitudinal studies using quantitative and qualitative CT and MRI can distinguish non-demented and demented older persons with DS. The results suggest that exaggerated increases in CSF and ventricular volumes, and perhaps temporal lobe and generalized cortical atrophy, occur only in older persons with DS who have clinically evident dementia, but not in those without dementia. Rates of ventricular enlargement can distinguish demented from non-demented individuals with DS. Thus, demented older adults with DS have accelerated brain atrophy, suggesting that neuronal loss, in addition to accumulation of senile plaques and neurofibrillary tangles, is critical for dementia in DS. Because accelerated cerebral atrophy occurs in demented but not non-demented older adults with DS, quantitative structural brain imaging can help to distinguish dementia from the lesser (but nevertheless present) cognitive decline in older persons with DS, as well as from uncomplicated mental retardation in young adults with DS.

Differentiating non-demented older adults with Down syndrome from young adults with the syndrome

Dementia can be reliably diagnosed in adults with DS, despite the mental retardation, using standardized diagnostic criteria. However, more sensitive formal neuropsychological tests are needed to detect lesser cognitive decline in non-demented older individuals with DS (Haxby 1989; Schapiro *et al.* 1989b). As noted above, progressive brain atrophy is present in older adults with DS and distinguishes dementia from lesser cognitive decline and mental retardation. However, it remains to be determined whether structural imaging techniques can distinguish lesser cognitive decline (not evident as dementia) in older adults with DS from mental retardation in younger adults with the syndrome.

We evaluated 18 young (aged 19–34 years) and seven older (aged 37–63 years) subjects with DS (Schapiro *et al.* 1987b; see above). Only one of them, aged 63 years, was demented. Slopes of the regression lines for age versus CSF and individual ventricular volumes differed significantly from zero, but did not differ from similar slopes in age-matched controls (Schapiro *et al.* 1987b). Thus, the relationship between age and generalized cortical atrophy is indistinguishable between non-demented subjects with DS (who show some cognitive impairment on formal neuropsychological testing) and healthy controls. Furthermore, in our later cross-sectional and longitudinal quantitative CT study (see above), as well as in other studies (Dalton and Crapper 1977; Lott and Lai 1982a,b; Lai and LeMay 1990), non-demented older subjects with DS did not show progressive generalized cortical atrophy.

As described earlier, several studies measured regional, as opposed to generalized, brain atrophy. Regional measures allow one to investigate brain areas that are known to be affected early in AD. In the CT study of Lai and LeMay (1990; see above), no difference was shown for frontal, temporal and occipital horn ratios between non-demented subjects with DS and controls. However, the mean age of the non-demented DS group was under 40 years, suggesting that some of them had not reached the age when some degree of significant AD neuropathology usually develops (Mann 1988).

On the other hand, Pearlson *et al.* (1990), using CT scanning in persons with DS, showed that the suprasellar ratio was normal in two non-demented subjects under the age of 30 years, elevated in three of five non-demented subjects aged 30-40 years and also elevated in three of four non-demented, as well as in five of six demented, subjects over the age of 40 years. Thus, although the suprasellar ratio may discriminate demented from non-demented older adults with DS, both groups, but not young adults with DS or controls, have an increased suprasellar ratio. Such a finding suggests that mesial temporal lobe atrophy, as reflected by this ratio, occurs prior to the onset of clinically evident dementia in older persons with DS.

Other evidence suggests that the medial temporal region is an early area of dysfunction in AD in persons with or without DS. Neuropsychological studies show that recent memory impairment, attributed to dysfunction of hippocampus and amygdala, is the earliest and most prominent cognitive deficit in AD with and without DS (Haxby 1989). Neuropathological studies also point to early involvement of the medial temporal lobe in AD (Mann and Esiri 1989; Hyman and Mann 1991); the neuropathology in DS is characterized by senile plaques that are initially formed and mature in the amygdala and hippocampus, and by neurofibrillary tangles with paired helical filaments that form in the entorhinal cortex (which has projections with the hippocampus). Thus, neuropsychological, neuropathological and structural imaging studies suggest that the temporal lobe, and in particular the hippocampal formation, is an area to concentrate on for volumetric imaging in early AD, prior to clinically evident dementia and progressive generalized cerebral atrophy.

Synopsis

Structural imaging, particularly quantitative MRI, should provide new insights into the pathogenesis and topography of AD, as well as of mental retardation, in DS. By measuring regional brain volumes, it can be determined whether there are regional variations in onset and progression of AD related to the selective regional vulnerability of the disease (Rapoport 1990). Central and peripheral CSF spaces can be compared to determine whether there are differences in rates of expansion. With the development of better segmentation routines, any differences in the patterns of change with disease can be explored for the different components of brain, such as CSF, white matter and gray matter. Selective involvement of subcortical structures can be examined to determine anatomic correlates for such features as extrapyramidal symptoms or recent memory deficits, and to detect the earliest onset of AD in DS. White matter hyperintensities can be quantified and localized to explore factors that may be related to their pathogenesis, such as amyloid angiopathy evident post-mortem. Finally, since *in vivo* studies can be repeated, the natural history and response to therapy can be determined.

Functional brain imaging

Functional brain imaging makes it possible to examine patterns of neocortical metabolism or blood flow, which may occur early in AD, and to determine whether such patterns persist. These changes can in turn be related to cognitive variables, during the course of disease. With longitudinal studies, rates of decline can be determined and correlated with trajectories of cognitive decline

(Haxby *et al.* 1992). With such advantages, *in vivo* functional imaging has been used to evaluate changes in the brain metabolism and blood flow with age and disease.

Many variables can affect the interpretation of functional brain imaging studies. Health status of the experimental subjects and controls is one such factor. Different research groups often vary in the degree of health screening and of acceptance criteria for both affected subjects and controls. A second factor is the experimental condition employed. Research groups vary in whether studies are performed with eyes open or closed, or ears unplugged or plugged. Phelps *et al.* (1981) showed that the regional cerebral metabolic rate for glucose (rCMRglc) in visual cortex is progressively increased by increasing the complexity of visual stimulation. A third factor is gender. Women tend to have a lower haemoglobin content than men. As cerebral blood flow (CBF) increases when haemoglobin content decreases, CBF measurements will be affected by gender, resulting in greater CBF in studies that employ relatively more women. Gender may also affect measurements of rCMRglc, depending on the type of transmission correction used with PET. Studies that used a calculated correction for attenuation showed increased rCMRglc in young female compared with male adults (Baxter *et al.* 1987; Yoshii *et al.* 1988). However, a calculated attenuation correction may ignore differences in head size and skull thickness between men and women. When a measured attenuation correction is used, no difference in rCMRglc is found between the sexes (Miura *et al.* 1990). Another factor that affects studies is the influence of brain atrophy. Both quantitative CT and MRI have reported brain atrophy with both normal ageing and certain diseases, such as AD. Methods to correct for such changes have been proposed (Herscovitch *et al.* 1986; Schlageter *et al.* 1987), but have had limited use. Yet another factor of uncertain influence is the stress or anxiety of the subject during the procedure.

Other details need to be considered when interpreting functional imaging results, including: in [133]xenon studies, the number of detectors, whether measurements are corrected for differences in pCO_2 and the index used to calculate regional CBF (rCBF); in SPECT and PET studies, scanner characteristics such as number of scanners used, type, resolution, multislice capability and method of attenuation correction; whether scans of subjects are interspaced with control scans to avoid such effects as drift of machine operating characteristics or changes in isotope synthesis method or purity; and whether the analysis of data is done blindly and with acceptable intra- and inter-rater reliability.

Young adults with Down syndrome compared with age-matched controls

In previous studies, no difference in global cerebral oxygen consumption between young adults with DS and age-matched controls was shown using the Kety–Schmidt nitrous oxide saturation technique, an invasive method that measures arterial–cerebral venous differences of a diffusible indicator (Fazekas *et al.* 1958; Lassen *et al.* 1966). The first functional brain imaging studies in DS were performed using [133]xenon inhalation regional cerebral blood flow technique at rest. Xenon is metabolically inert and freely diffusible; its disappearance from a brain region is a function of blood flow to that region. In the three studies performed in DS, similar measurement techniques were used: low concentrations of [133]xenon gas were inhaled for 1 minute and the saturation and desaturation of the tracer were measured using scintillation detection systems.

In one study (Melamed *et al.* 1987), 14 subjects with DS (10 males; four females) aged 16–44 years (12 subjects less than 35 years and two aged 43 and 44 years) whose mental ages ranged from 2.2 to 7.0 years (mean 3.7 ± 0.3) on the Stanford–Binet Intelligence Scale, were compared with 69 normotensive, neurologically intact controls. Measurement of the initial slope index (ISI) was performed with subjects resting with eyes closed and ears plugged. Mean brain CBF was significantly reduced by 19.5 per cent $\pm$ 2.8 (mean $\pm$ SEM). Reductions were evident in 13 of the 14 subjects with DS, including a reduction of 16.9 per cent $\pm$ 3.1 in seven subjects aged 21–29 years, 24.5 per cent $\pm$ 1.5 in four aged 31–35 years, and 10.2 per cent and 12.6 per cent in two aged 43 and 44 years respectively. CBF in the subjects with DS older and younger than 30 years did not differ. The different cortical regions did not differ in the degree of rCBF reductions; the changes in rCBF in DS at all ages were similar in extent and distribution to changes seen in AD.

In contrast, two other studies suggested that rCBF is not abnormal in DS. Risberg (1980) studied 12 subjects with DS older and younger than 40 years, although the exact ages were not noted. No further information was provided, including whether any of the subjects were demented or whether the studies were performed at rest. A normal level and pattern of CBF was measured using the ISI.

We used the [133]xenon inhalation technique to examine 11 healthy, non-institutionalized subjects with DS aged 19–35 years (eight men and three women) and 22 sex-matched, healthy volunteers aged 19–34 years (Schapiro *et al.* 1989a). No subject with DS was demented. At rest (eyes closed in a darkened room with minimal ambient noise), the initial slope (IS) did not differ between groups for any lobar region. The ratio of regional IS (rIS) to hemispheric IS, and right/left and frontal/non-frontal ratios, did not differ between groups (unpaired *t*-test, $P > 0.05$). The reason for the discrepancy between the studies of Melamud *et al.* (1987) and others (Risberg 1980; Schapiro *et al.* 1989a) is not apparent. However, our subjects were less retarded, not institutionalized and otherwise healthy, and we did not include older subjects. Furthermore, the changes that Melamud *et al.* described (uniform decreases in all cortical brain regions) in AD are not consistent with the neuropathological changes measured in life with PET (Rapoport 1990).

Other techniques have been used to study brain blood flow and metabolism in DS. PET allows local tissue physiological processes to be quantified in a non-invasive manner through the use of biologically active compounds that are labelled with positron-emitting isotopes (Mazziotta and Engel 1984). Administration of a positron-emitting glucose analogue ([18]FDG), combined with a system to detect emitted gamma radioactivity and a tracer kinetic model that adheres to principles of chemical and enzyme reaction kinetics, allow the quantification of glucose metabolism at the regional as well as global level (Mazziotta and Engel 1984; Sokoloff 1984). Both cortical and subcortical regions can be studied (subcortical regions are not examined with the [133]xenon inhalation technique).

An initial study with PET and [18]FDG from our laboratory examined four healthy subjects with DS aged 19–27 years and age-matched controls (Schwartz et al. 1983). Brain glucose metabolism was globally elevated in the DS cases. However, a calculated attenuation correction was used that was later discovered to be invalid in DS, due to the smaller cranial capacity and thinner skulls (Schapiro et al. 1990b).

Schapiro et al. (1990b) re-evaluated resting state (eyes patched, ears plugged, no external stimulation) brain metabolism using PET and [18]FDG, to learn whether there are differences in absolute cerebral metabolic rates for glucose between young adults with DS and age-matched controls. For comparing the two groups, scanning was carried out with a multislice PET scanner (Scanditronix PC-1024-7B, Uppsala, Sweden) that used a measured attenuation correction. This is a seven-slice machine with a transverse resolution (full width at half maximum) of 6 mm and an axial resolution of 10 mm. Fourteen healthy, non-institutionalized subjects with DS aged 25–38 years (10 men and four women) and 13 sex-matched, healthy volunteers aged 22–38 years were examined. Dementia was not present in the subjects with DS. There was no significant difference in mean global rCMRglc between the young DS (8.76 ± 0.76 mg/100 g/min, mean ± SD) and control (8.74 ± 1.19 mg/100 g/min) subjects (unpaired t-test, $P > 0.05$). CMRglc was preserved in neocortical association areas as well as in primary motor and sensory cortical areas, and subcortical areas. Reference ratios (rCMRglc/global CMRglc) also did not differ between the groups.

Another non-invasive technique to examine physiological processes in the brain is SPECT. One class of SPECT studies uses rCBF radiolabelled pharmaceuticals in combination with multidetector scanners to determine patterns of rCBF (Bonte et al. 1990). Unlike PET scanning with [18]FDG, however, models are lacking in this SPECT scanning method to quantify data in absolute units (Gemmeli et al. 1990), which would be useful in studies across groups or time. On the other hand, SPECT does not require an on-site cyclotron and is readily available at many hospitals.

SPECT studies in young adults with DS have been limited to date. In one study, a 30-year-old subject with DS did not have a pattern of blood flow seen in dementia (Deisenhammer et al. 1989). Further details were not described. In another study, a 38-year-old man with DS was studied after hospitalization with a mild quadriparesis following a fall (Nakayasu et al. 1991). His intellectual quotient was 25 on the Suzuki–Binet test; there was no history of dementia. [^{123}I]IMP SPECT showed decreased uptake in the posterior regions of both parietal lobes. To what degree such blood flow findings may have been due to the serious fall, with possible associated head trauma, is not clear.

Thus, young adults with DS (less than 40 years and without dementia), despite their mental retardation, do not have consistently altered resting cerebral glucose metabolism or blood flow compared with healthy controls. In addition, reference ratios show no consistent difference in the intrahemispheric distribution of rCMRglc or rCBF in DS compared with controls. Similar findings were noted in DS with different techniques, including [133]xenon inhalation studies, PET with [18]FDG and SPECT. The lack of selective metabolic involvement of neocortex in young adults with DS differs from the pattern in demented older adults with the syndrome (see below).

As noted by Horwitz *et al.* (1988), the neuropathology of developmental disorders may involve 'anomalous structural and functional organization' due to proliferation and migration abnormalities, increased cell packing or synaptic density secondary to failure of normal programmed death, abnormalities of dendritic branching, dendritic spine density or morphology, or changes in synaptic reorganization. In DS, microscopic changes include decreased neuronal density in neocortex and hippocampus, abnormal dendritic spines and branching, altered synaptic density and size, and abnormal electrophysiological properties. Such changes may be reflected in altered patterns of functional associations between brain regions.

Though standard group comparisons did not show differences in rCMRglc between young adults with DS and controls, a prior correlational analysis, using resting glucose metabolic data from the ECAT PET scanner, showed disruption in patterns of brain regional interactions in young adults with DS (Horwitz *et al.* 1990). In this method of analysis, the correlation coefficient is used as a measure of functional association between brain regions, with the magnitude of the correlation coefficient related to the strength of the association. Ten healthy male and four female DS adults without dementia aged 19–34 years were studied; their mental ages were 3.8–15.8 years (mean 6.8) on the Peabody Picture Vocabulary Test (Dunn and Dunn 1981). Compared with age-matched healthy controls, the subjects with DS had quantitatively smaller correlations for pairs of regions within and between frontal and parietal lobes. The thalamus also had smaller correlations with temporal and occipital regions in the DS group. It was hypothesized that these reduced correlations reflect functional disruption of neural circuits responsible for directed attention. In addition, the inferior frontal gyrus, including Broca's area, was particularly involved, consistent with reports of disproportionate language disability in DS (Fowler 1990). However, this study was limited by technical factors of the scanner, including low spatial resolution, use of a calculated attenuation correction factor and the collection of single slices sequentially.

These results were not replicated in another correlational analysis applied to resting glucose metabolic data obtained from a multislice, high resolution Scanditronix PET scanner, which uses a measured attenuation correction (N. P. Azari, personal communication). Thus, different results have been noted for young adults with DS on two different PET scanners, although other correlational results from data obtained on the lower resolution ECAT scanner have been replicated with the higher resolution Scanditronix scanner in other subject groups (Azari *et al.* 1992c). However, there were differences in technique between the two DS studies, including method of correction for attenuation of radiation, spatial resolution, multislice capability, region of interest analysis and subject age (10 of 14 subjects in the Scanditronix PET study were from the earlier ECAT scanner study and were three to five years older for the Scanditronix study).

As a second approach, a method that uses multiple regression and discriminant analysis has been developed that also can assess functional associations among regional PET measures. Additionally, this method can detect individual differences in patterns of rCMRglc, and unlike the correlational analysis which assesses interactions between only two regions, can evaluate functional associations between more than two regions at one time (Azari *et al.* 1992a). Based on language regions highlighted by the initial correlation analysis, a discriminant function was developed using regional interactions involving two language areas, the left inferior premotor and left superior temporal regions. The probability of being classified as DS or control was calculated for each subject and averages were obtained for each group. Application of the discriminant analysis function derived from the standardized residuals of both language areas showed that all the subjects with DS and 88 per cent of the control subjects were correctly classified. The average total probability of correct classification was 0.98 for DS and 0.88 for control subjects. Of the misclassified controls, the probability of misclassification was 0.60 and 0.88 (Azari *et al.* 1992b).

Other studies suggest that language, especially syntax, is developed and represented separately from, and relatively more affected than, other cognitive spheres in DS (Haxby 1989; Fowler 1990). Thus, the results of the correlational and the discriminant function analyses support the hypothesis that functional associations involving brain language areas are disrupted in young adults with DS. The use of these latter methods of data analysis, rather than group-mean comparisons, emphasizes the importance of using alternate methods of data analysis that can identify subtle changes in patterns of metabolism (Azari *et al.* 1992a). On the basis of the results obtained from such analyses, hypotheses can be generated which can be tested in future experiments, involving stimulation and PET.

Demented older adults with Down syndrome compared with non-demented older or young adults with the syndrome

In older adults with DS, brain metabolism and blood flow were studied using the invasive Kety–Schmidt technique (Fazekas *et al.* 1958; Lassen *et al.* 1966). Although not specifically looked for, age-related differences in blood flow and metabolism were not shown and dementia was not examined. In our laboratory, PET with ^{18}FDG was used to determine whether there are age-related differences in brain glucose metabolism in subjects with DS at rest and whether differences in brain metabolism may help to distinguish dementia from mental retardation in adults with DS over 40 years old. These studies were performed with an ECAT II PET tomograph (ORTEC, Life Sciences,

	Young DS (15)	Non-demented old DS (4) (mg/100 g/min)	Demented old DS (3)
Association neocortex			
Parietal	7.14 ± 0.24	6.29 ± 0.51	4.19 ± 0.49[a,b]
Lateral temporal	5.65 ± 0.18	5.07 ± 0.27	3.68 ± 0.33[a,b]
Primary neocortex			
Sensorimotor	7.58 ± 0.25	6.87 ± 0.59	5.21 ± 0.56[a]
Occipital	6.43 ± 0.21	6.30 ± 0.15	4.92 ± 0.64[a]

Number of subjects shown in parentheses. Mean ± SEM.
[a]Differs from mean in young DS group (P<0.05)
[b]Differs from mean in non-demented old DS group (P<0.05)

Table 10.2. Regional cerebral glucose metabolism in older adults with Down syndrome

Oak Ridge, TN), which used a calculated attenuation correction program. Seven serial slices were collected, with transverse and axial resolutions (full width at half maximum) of 17 mm (Schapiro *et al.* 1987a); the procedure was described in detail.

Table 10.2 shows representative absolute values of rCMRglc for young, non-demented old and demented old DS groups. For association as well as for primary neocortices, the non-demented old subjects had glucose metabolic values that were more similar to those in the young subjects. On the other hand, the demented older subjects had significantly lower values of rCMRglc than did the young subjects, with the greatest reductions in the association neocortices (one-way ANOVA with Bonferroni correction for multiple comparisons, $P < 0.05$). The demented older group also had significantly lower values of rCMRglc in association neocortices in comparison to the non-demented older group.

The intrahemispheric distribution in DS of glucose metabolism was examined with the ratio of parietal association cortex (relatively affected area) to sensorimotor primary cortex (relatively spared area). There was no significant difference between 15 young and four non-demented older subjects (mean ± SEM of 0.94 ± 0.01 and 0.92 ± 0.03 respectively). However, the ratio in three demented older subjects (0.80 ± 0.01) was significantly less (without overlap) than in both the young and non-demented older subjects (one-way ANOVA with Bonferroni correction for multiple comparisons, $P < 0.05$). Similar differences were shown for the ratio of temporal association cortex to primary occipital cortex (Schapiro *et al.* 1992).

Such a pattern of glucose metabolism with PET also has been shown in a demented 45-year-old female with DS, without a history of mental retardation and few stigmata of DS, who had a mosaic 46,XX,–21,+t(21;21) translocation trisomy karyotype (Schapiro *et al.* 1990a). Thus, the typical phenotypic expression of DS may not be necessary for the late expression of dementia in DS, suggesting that dementia in DS may involve expression of genes in regions of chromosome 21 other than the 'obligatory' DS region (21q22.2 or 21q22.3), or differential expression of genes on the long arm of chromosome 21 or that only a portion of cells in the brain needs to be trisomic for the development of AD.

SPECT studies also note abnormalities in patterns of rCBF in demented older adults with DS. Reduced uptake in bilateral parietal and right temporal regions was noted with [^{123}I]IMP in a 52-year-old demented subject (Nakayasu *et al.* 1991); reduced uptake in bilateral temporal and parietal regions was also noted with Tc-99m HMPAO in a 46-year-old demented subject (Rae-Grant *et al.* 1991).

Thus, glucose metabolism in DS is preserved in older subjects without clinical dementia, but glucose metabolism and blood flow are reduced in older subjects with clinical dementia. However, in the demented subjects, the reductions are not uniform, with relatively greater involvement of parietal–temporal association neocortices and relative sparing of primary sensorimotor neocortex, cerebellum, thalamus, and caudate and lenticular nuclei. Such a pattern is similar to that seen in persons with AD who do not have DS. As indicated by Rapoport (1990), in AD this regional distribution of metabolic impairment reflects the localization of neuropathology in association regions that are dominated by interconnecting corticocortical projections, rather than in primary neocortex, which are dominated by specific thalamic input (Brun and Gustafson 1976; Pearson *et al.* 1985; Rogers and Morrison 1985; Lewis *et al.* 1987). A similar regional (Ellis *et al.* 1974; Mann 1988) and laminar (Ellis *et al.* 1974; Rafalowska *et al.* 1988) distribution of Alzheimer neuropathology has been described in DS.

Synopsis

Resting functional imaging scans will continue to provide new information on aspects of mental retardation and dementia in DS, especially with the development of alternate and more powerful methods of data analysis and technological improvements. One area that is relatively unexplored in DS and where improvements in functional brain imaging methods may help is the early detection of AD. At present, older adults with DS who have cognitive impairment but without clinical dementia cannot be identified with resting functional brain imaging techniques. Higher resolution scanners, which reduce partial voluming and increase recovery coefficients, and superimposition of PET and MRI images, which permits better identification of regions of interest, will allow the imaging of the hippocampal and other smaller subcortical structures that appear to be involved early in AD in adults with DS. Furthermore, application of alternate methods of data analysis, searching for differences in metabolic or flow patterns, will allow detection of subtle changes in brain organization known to occur early in AD. Through the early identification of AD, one will be able to chart the development of later neocortical changes and relate these changes to other measures, such as neuropsychological ones.

Another area of opportunity in functional brain imaging involves the use of stimulation paradigms, such as with cognitive and psychophysical stimuli. Prior studies in healthy volunteers have shown that rCBF changes associated with specific cognitive operations are more enhanced in regions associated with those cognitive functions. Similar paradigms can be devised for DS, even tasks that do not require subject compliance. Furthermore, once a task is established, the effect of drug modulation on rCBF during the task can be examined. What brain areas and age groups to probe can be decided by the research question. For the early detection of AD in non-demented older persons with DS, one may probe areas affected early in the disease, such as hippocampal regions, as well as regions known to be prominently affected later, such as the neocortical association regions. In demented subjects, one can stimulate areas with abnormalities of rCBF and rCMRglc (both with and without preserved cognitive functions localized to these areas) to learn about mechanisms of functional failure and the reversibility of abnormalities in rCBF or rCMRglc, as well as the effect of the disease severity on the brain's ability to modify rCBF or rCMRglc in response to stimuli. In young adults with DS, one might design tasks to determine whether they use different brain pathways than controls to perform similar tasks, providing insights as to whether there is reorganization of the brain in DS. As the only example in DS, Berman *et al.* (1988), using the [133]xenon inhalation technique, showed that young adults with DS can increase prefrontal rCBF similar to controls, despite poor performance on a task thought to activate this region.

Thus, both resting and activation studies will allow further assessment of mechanisms of mental retardation and dementia. Other advances, such as newer isotopes and receptor studies, should provide additional tools for the exploration of the brain in DS.

Conclusions

As an example of how structural and functional imaging may help to understand the pathogenesis of a disease, a hypothesis has been generated to help explain the observation that, despite the universal presence of some type and degree of neuropathological changes of AD after age 40 years in DS, only 20–30 per cent of older persons with DS are demented, despite the likelihood, given their poor performance on standardized neuropsychological tests, that all had some neuropathological changes of AD.

It is suggested that a cognitive decline in older adults with DS occurs in two stages that can be separated by as much as 20 years. First, one finds absence of a progressive change in CSF and ventricular volumes on quantitative CT and MRI scanning, and retention of normal patterns of brain metabolism on functional brain imaging in a non-demented older DS group as compared with a young DS group. There is a reduction in cognitive performance on standardized neuropsychological tests, perhaps reflective of poorer processing skills. The reduction coincides with the accumulation of marked numbers of senile plaques, but neurofibrillary tangle accumulation, cell loss and atrophy in the neocortex have not yet occurred to any extent.

In the second stage of cognitive decline, as seen in a demented older DS group, progressive brain atrophy on quantitative CT and MRI is suggestive of such cell death, while decreased brain metabolism and blood flow (particularly in association neocortex) is suggestive of either dysfunctional or dead cells. There is additional loss of overlearned behaviours, leading to deterioration in social, occupational and adaptive skills, and a characteristic dementia. Concurrently, it is suggested that accumulations of neurofibrillary tangles and accelerated cell loss become evident, particularly in the phylogenetically newer neocortical association regions. A similar temporal sequence of pathological changes has been reported in premorbidly normal subjects who were prospectively studied with cognitive tests prior to death: cognitively impaired yet nondemented subjects had senile plaque accumulation only; demented subjects with the diagnosis of AD had accumulations of both senile plaques and neurofibrillary tangles (Crystal *et al.* 1988; Katzman *et al.* 1988).

In another study from our laboratory, we noted that in subjects with AD, densities of neurofibrillary tangles, but not of senile plaques, in the neocortex post-mortem correlated with the extent of reduction of regional brain metabolism prior to death. Furthermore, neocortical association areas, in general, had the lowest metabolic rates and the highest neurofibrillary tangle counts, emphasizing their selective regional vulnerability in AD (DeCarli *et al.* 1992). This latter study suggests that decreased brain metabolism in AD reflects dysfunctional cells (with neurofibrillary tangles a marker of this dysfunction) in the neocortex in AD.

Thus, evidence suggests that progressive brain atrophy and reduced brain metabolism occur only in demented old persons with DS, and that quantitative CT and MRI scanning and functional brain imaging can help to distinguish dementia from lesser cognitive decline in older persons with DS.

References

American Psychiatric Association (1987). *Diagnostic and statistical manual of mental disorders.* (3rd edn, revised). American Psychiatric Association, Washington, DC.

Azari, N. P., Pietrini, P., Grady, C. L., Pettigrew, K. D., Horwitz, B., Rapoport, S. I., *et al.* (1992a). Early detection of Alzheimer's disease using a discriminant analysis applied to regional glucose metabolic data obtained by positron emission tomography. *Journal of Neuroimaging*, 2, 55.

Azari, N. P., Horwitz, B., Pettigrew, K. D., Haxby, J. V., Giacometti, K. R., and Schapiro, M. B. (1992b). Disruption of brain regional functional interactions involving language areas in young adults with Down syndrome. *Annals of Neurology*, 32, 475.

Azari, N. P., Rapoport, S. I., Salerno, J. A., Grady, C. L., Gonzalez-Aviles, A., Schapiro, M. B., *et al.* (1992c). Interregional correlations of resting cerebral glucose metabolism in old and young women. *Brain Research*, 589, 279-90.

Baxter, L. R., Jr, Mazziotta, J. C., Phelps, M. E., Selin, C. E., Guze, B. H., and Fairbanks, L. (1987). Cerebral glucose metabolic rates in normal human females versus normal males. *Psychiatry Research*, 21, 237-45.

Benda, C. E. (1969). *Down's syndrome*, (2nd edn). Grune and Stratton, New York.

Berman, K. F., Schapiro, M. B., Friedland, R. P., Rapoport, S. I., and Weinberger, D. R. (1988). Regional cortical blood flow during cognitive activation in Down syndrome. *Society for Neuroscience Abstracts*, 14, 1012.

Bonte, F. J., Hom, J., Tintner, R., and Weiner, M. F. (1990). Single photon tomography in Alzheimer's disease and the dementias. *Seminars in Nuclear Medicine*, 20, 342-52.

Brun, A., and Gustafson, L. (1976). Distribution of cerebral degeneration in Alzheimer's disease. A clinico-pathological study. *Archiv für Psychiatrie und Nervenkrankheiten*, 223, 15-33.

Crystal, H., Dickson, D., Fuld, P., Masur, D., Scott, R., Mehler, M., et al. (1988). Clinico-pathologic studies in dementia: nondemented subjects with pathologically confirmed Alzheimer's disease. *Neurology*, 38, 1682-7.

Dalton, A. J., and Crapper, D. R. (1977). Down's syndrome and aging of the brain. In *Research to practice in mental retardation*, (ed. P. Mittler), Vol. 3, pp. 391-400. University Park Press, Baltimore.

Davis, J. M., and Wright, E. A. (1977). A new method for measuring cranial cavity volume and its application to the assessment of cerebral atrophy at autopsy. *Neuropathology and Applied Neurobiology*, 3, 341-58.

DeCarli, C., Kaye, J. A., Horwitz, B., and Rapoport, S. I. (1990). Critical analysis of the use of computer-assisted transverse axial tomography to study human brain in aging and dementia of the Alzheimer type. *Neurology*, 40, 872-83.

DeCarli, C., Atack, J. R., Ball, M. J., Kaye, J. A., Grady, C. L., Fewster, P., et al. (1992). Post-mortem regional neurofibrillary tangle densities but not senile plaque densities are related to regional cerebral metabolic rates for glucose during life in Alzheimer's disease patients. *Neurodegeneration*, 1, 11-20.

Deisenhammer, E., Reisecker, F., Leblhuber, F., Holl, K., Markut, H., Trenkler, J., et al. (1989). Single-photon-emissions-computertomographie bei der differentialdiagnose der Demenz. *Deutsche Medizinische Wochenschrift*, 114, 1639-44.

Dekaban, A. S., and Sadowsky, D. (1978). Changes in brain weights during the span of human life: relation of brain weights to body heights and body weights. *Annals of Neurology*, 4, 345-56.

Dunn, L. M., and Dunn, L. M. (1981). *Peabody Picture Vocabulary Test—Revised*. American Guidance Service, Circle Pines, New Mexico.

Ellis, W. G., McCulloch, J. R., and Corley, C. L. (1974). Presenile dementia in Down's syndrome: ultrastructural identity with Alzheimer's disease. *Neurology*, 24, 101-6.

Fazekas, J. F., Ehrmantraut, W. R,. Shea, J. G., and Kleh, J. (1958). Cerebral hemodynamics and metabolism in mental deficiency. *Neurology*, 8, 558-60.

Folstein, M. F., Folstein, S. E., and McHugh, P. R. (1975). 'Mini-Mental State': a practical method for grading the mental state of patients for the clinician. *Journal of Psychiatry Research*, 12, 189-98.

Fowler, A. E. (1990). Language abilities in children with Down syndrome: evidence for a specific syntactic delay. In *Children with Down syndrome*, (ed. D. Cicchetti and M. Beeghly), pp. 302-28. Cambridge University Press, New York.

Franceschi, M., Comola, M., Piattoni, F., Gualandri, W., and Canal, N. (1990). Prevalence of dementia in adult patients with trisomy 21. *American Journal of Medical Genetics* Supplement, 7, 306-8.

Gemmell, H. G., Evans, N. T. S., Besson, J. A. O., Roeda, D., Davidson, J., Dodd, M. G., et al. (1990). Regional cerebral blood flow imaging: a quantatitive comparison of technetium-99m-HMPAO SPECT and $C^{15}O_2$ PET. *Journal of Nuclear Medicine*, 31, 1595-600.

Haxby, J. V. (1989). Neuropsychological evaluation of adults with Down's syndrome: patterns of selective impairment in non-demented old adults. *Journal of Mental Deficiency Research*, 33, 193-210.

Haxby, J. V., Raffaele, K., Gillette, J., Schapiro, M. B., and Rapoport, S. I. (1992). Individual trajectories of cognitive decline in patients with dementia of the Alzheimer type. *Journal of Clinical and Experimental Neuropsychology*, 14, 575-92.

Herscovitch, P., Auchus, A. P., Gado, M., Chi, D., and Raichle, M. E. (1986). Correction of positron emission tomography data for cerebral atrophy. *Journal of Cerebral Blood Flow and Metabolism*, 6, 120-4.

Horwitz, B., Rumsey, J. M., Grady, C. L., and Rapoport, S. I. (1988). The cerebral metabolic landscape in autism. Intercorrelations of regional glucose utilization. *Archives of Neurology*, 45, 749-55.

Horwitz, B., Schapiro, M. B., Grady, C. L., and Rapoport, S. I. (1990). Cerebral metabolic pattern in young adult Down's syndrome subjects: altered intercorrelations between regional rates of glucose utilization. *Journal of Mental Deficiency Research*, 34, 237-52.

Hyman, B. T., and Mann, D. M. A. (1991). Alzheimer-type pathological changes in Down's syndrome individuals of various ages. In *Alzheimer's disease: basic mechanisms, diagnosis and therapeutic strategies* (ed. K. Iqbal, D. R. C. McLachlan, B. Winblad and H. M. Wisniewski), pp. 105-13. Wiley, Chichester.

Ieshima, A., Kisa, T., Yoshino, K., Takashima, S., and Takeshita, K. (1984). A morphometric CT study of Down's syndrome showing small posterior fossa and calcification of basal ganglia. *Neuroradiology*, 26, 493-8.

Jernigan, T. L., and Bellugi, U. (1990). Anomalous brain morphology on magnetic resonance images in Williams syndrome and Down syndrome. *Archives of Neurology*, 47, 529-33.

Katzman, R., Terry, R., DeTeresa, R., Brown, T., Davies, P., Fuld, P., *et al.* (1988). Clinical, pathological, and neurochemical changes in dementia: a subgroup with preserved mental status and numerous neocortical plaques. *Annals of Neurology*, 23, 138-44.

Kaye, J. A., DeCarli, C., Luxenberg, J. S., and Rapoport, S. I. (1992). The significance of age-related enlargement of the cerebral ventricles in healthy men and women measured by quantitative computed X-ray tomography. *Journal of the American Geriatrics Society*, 40, 225-31.

Lai, F., and LeMay, M. (1990). Changes in regional lateral ventricular size in Alzheimer disease with and without Down syndrome. *Fifth Annual Scientific Poster Session*. Massachusetts Alzheimer's Disease Research Center.

Lassen, N. A., Christensen, S., Hoedt-Rasmussen, K., and Stewart, B. M. (1966). Cerebral oxygen consumption in Down's syndrome. *Archives of Neurology*, 15, 595-602.

LeMay, M., and Alvarez, N. (1990). The relationship between enlargement of the temporal horns of the lateral ventricles and dementia in aging patients with Down syndrome. *Neuroradiology*, 32, 104-7.

Lewis, D. A., Campbell, M. J., Terry, R. D., and Morrison, J. H. (1987). Laminar and regional distributions of neurofibrillary tangles and neuritic plaques in Alzheimer's disease: a quantitative study of visual and auditory cortices. *The Journal of Neuroscience*, 7, 1799-808.

Lott, I. T., and Lai, F. L. (1982a). Dementia in Down syndrome. *Annals of Neurology*, 12, 210.

Lott, I. T., and Lai, F. L. (1982b). Dementia in Down's syndrome: observations from a neurology clinic. *Applied Research in Mental Retardation*, 3, 233-9.

Lott, I. T., Kesslak, P., and Nalcioglu, O. (1991). Down syndrome and Alzheimer's disease: quantitative magnetic resonance imaging of brain. *Annals of Neurology*, 30, 288.

Mann, D. M. A. (1988). The pathological association between Down syndrome and Alzheimer disease. *Mechanisms of Ageing and Development*, 43, 99-136.

Mann, D. M. A., and Esiri, M. M. (1989). The pattern of acquisition of plaques and tangles in the brains of patients under 50 years of age with Down's syndrome. *Journal of the Neurological Sciences*, 89, 169-79.

Mazziotta, J. C., and Engel, J., Jr (1984). The use and impact of positron computed tomography scanning in epilepsy. *Epilepsia*, 25(suppl. 2), S86-104.

Melamed, E., Mildworf, B., Sharav, T., Belenky, L., and Wertman, E. (1987). Regional cerebral blood flow in Down's syndrome. *Annals of Neurology*, 22, 275-8.

Miura, S. A., Schapiro, M. B., Grady, C. L., Kumar, A., Salerno, J. A., Kozachuk, W. E., *et al.* (1990). Effect of gender on glucose utilization rates in healthy humans: a positron emission tomography study. *Journal of Neuroscience Research*, 27, 500-4.

Murphy, D. G. M., DeCarli, C., Schapiro, M. B., Rapoport, S. I., and Horwitz, B. (1992). Age-related differences in volumes of subcortical nuclei, brain matter, and cerebrospinal fluid in healthy men as measured with magnetic resonance imaging. *Archives of Neurology*, 49, 839-48.

Nakayasu, H., Araga, S., Takahashi, K., Otsuki, K., and Murata, M. (1991). Two cases of adult Down's syndrome presenting parietal low uptake in ^{123}I-IMP-SPECT. *Rinsho Shinkeigaku*, 31, 557-60.

Pearlson, G. D., Warren, A. C., Starkstein, S. E., Aylward, E. H., Kumar, A. J., Chase, G. A., *et al.* (1990). Brain atrophy in 18 patients with Down syndrome: a CT study. *American Journal of Neuroradiology*, 11, 811-6.

Pearson, R. C. A., Esiri, M. M., Hiorns, R. W., Wilcock, G. K., and Powell, T. P. S. (1985). Anatomical correlates of the distribution of the pathological changes in the neocortex in Alzheimer's disease. *Proceedings of the National Academy of Sciences of the United States of America*, **82**, 4531-4.

Pelz, D. M., Karlik, S. J., Fox, A. J., and Vinuela, F. (1986). Magnetic resonance imaging in Down's syndrome. *The Canadian Journal of Neurological Sciences*, **13**, 566-9.

Phelps, M. E., Mazziotta, J. C., Kuhl, D. E., Nuwer, M., Packwood, J., Metter, J., *et al.* (1981). Tomographic mapping of human cerebral metabolism: visual stimulation and deprivation. *Neurology*, **31**, 517-29.

Rae-Grant, A. D., Barbour, P. J., Sirotta, P., and Gross, P. (1991). Alzheimer's disease in Down's syndrome with SPECT. *Clinical Nuclear Medicine*, **16**, 509-10.

Rafalowska, J., Barcikowska, M., Wen, G. Y., and Wisniewski, H. M. (1988). Laminar distribution of neuritic plaques in normal aging, Alzheimer's disease and Down's syndrome. *Acta Neuropathologica*, **77**, 21-5.

Rapoport, S. I. (1990). Topography of Alzheimer's disease: involvement of association neocortices and connected regions; pathological, metabolic and cognitive correlations; relation to evolution. In *Imaging, cerebral topography and Alzheimer's disease*, (ed. S. I. Rapoport, H. Petit, D. Leys and Y. Christen), pp. 1-17. Springer-Verlag, Berlin.

Risberg, J. (1980). Regional cerebral blood flow measurements by 133 Xe-inhalation: methodology and applications in neuropsychology and psychiatry. *Brain and Language*, **9**, 9-34.

Rogers, J., and Morrison, J.H. (1985). Quantitative morphology and regional and laminar distributions of senile plaques in Alzheimer's disease. *The Journal of Neuroscience*, **5**, 2801-8.

Schapiro, M. B., Haxby, J. V., Grady, C. L., Duara, R., Schlageter, N. L., White, B., *et al.* (1987a). Decline in cerebral glucose utilisation and cognitive function with aging in Down's syndrome. *Journal of Neurology, Neurosurgery, and Psychiatry*, **50**, 766-74.

Schapiro, M. B., Creasey, H., Schwartz, M., Haxby, J. V., White, B., Moore, A., *et al.* (1987b). Quantitative CT analysis of brain morphometry in adult Down's syndrome at different ages. *Neurology*, **37**, 1424-7.

Schapiro, M. B., Berman, K. F., Friedland, R. P., Weinberger, D. R., and Rapoport, S. I. (1989a). Regional cerebral blood flow is not decreased in young adults with Down's syndrome. *Brain Dysfunction*, **2**, 310-5.

Schapiro, M. B., Luxenberg, J. S., Kaye, J. A., Haxby, J. V., Friedland, R. P., and Rapoport, S. I. (1989b). Serial quantitative CT analysis of brain morphometrics in adult Down's syndrome at different ages. *Neurology*, **39**, 1349-53.

Schapiro, M. B., Kumar, A., White, B., Fox, D., Grady, C. L., Haxby, J. V., *et al.* (1990a). Dementia without mental retardation in mosaic translocation Down syndrome. *Brain Dysfunction*, **3**, 165-74.

Schapiro, M. B., Grady, C. L., Kumar, A., Herscovitch, P., Haxby, J. V., Moore, A. M., *et al.* (1990b). Regional cerebral glucose metabolism is normal in young adults with Down's syndrome. *Journal of Cerebral Blood Flow and Metabolism*, **10**, 199-206.

Schapiro, M. B., Haxby, J. V., and Grady, C. L. (1992). The nature of mental retardation and dementia in Down syndrome: a study with PET, CT and neuropsychology. *Neurobiology of Aging*, **13**, 723-34.

Schlageter, N. L., Horwitz, B., Creasey, H., Carson, R., Duara, R., Berg, G. W., *et al.* (1987). Relation of measured brain glucose utilization and cerebral atrophy in man. *Journal of Neurology, Neurosurgery, and Psychiatry*, **50**, 779-85.

Schmid, F., Duren, R., and Ahmadi, K. (1969). Das Mongolismus-syndrom: die mongoloide dyszephalie. *Fortschritte der Medizin*, **87**, 1252-6.

Schwartz, M., Duara, R., Haxby, J., Grady, C., White, B. J., Kessler, R. M., *et al.* (1983). Down's syndrome in adults: brain metabolism. *Science*, **221**, 781-3.

Schwartz, M., Creasey, H., Grady, C. L., DeLeo, J. M., Frederickson, H. A., Cutler, N. R., *et al.* (1985). Computed tomographic analysis of brain morphometrics in 30 healthy men, aged 21-82 years. *Annals of Neurology*, **17**, 146-57.

Sokoloff, L. (1984). Modeling metabolic processes in the brain in vivo. *Annals of Neurology*, **15** (suppl.), S1-11.

Weis, S., Weber, G., Neuhold, A., and Rett, A. (1991). Down syndrome: MR quantification of brain structures and comparison with normal control subjects. *American Journal of Neuroradiology*, **12**, 1207-11.

Wisniewski, K. E., French, J. H., Rosen, J. F., Kozlowski, P. B., Tenner, M., and Wisniewski, H. M. (1982). Basal ganglia calcification (BCG) in Down's syndrome (DS) - another manifestation of premature aging. *Annals of the New York Academy of Sciences*, **396**, 179-89.

Yoshii, F., Barker, W. W., Chang, J. Y., Loewenstein, D., Apicella, A., Smith, D., *et al.* (1988). Sensitivity of cerebral glucose metabolism to age, gender, brain volume, brain atrophy, and cerebrovascular risk factors. *Journal of Cerebral Blood Flow and Metabolism*, **8**, 654-61.

Zellweger, H. (1977). Down syndrome. In *Handbook of clinical neurology*, (ed. P.J. Vinken and G.W. Bruyn), Vol. 31, pp. 367-469. North Holland Publishing, New York.

11

Peripheral biological markers as confirmatory or predictive tests for Alzheimer disease in the general population and in Down syndrome

M. E. Percy

Summary

Considerable progress has been made in developing tests, based on the analysis of peripheral biological markers, for confirming the presence of Alzheimer disease (confirmatory tests) or for identifying individuals at high risk of becoming affected with the disease (predictive tests). Many of the reports describing aberrant biological markers in persons with Alzheimer disease are summarized in tabular form in this chapter. Certain studies that appear to have immediate application to confirmatory testing for Alzheimer disease or that provide novel insights about its pathogenesis or possible aetiology are highlighted. Controversial issues related to the concept of using DNA mutations as potential predictive (including prenatal) diagnostic tests for familial Alzheimer disease and related disorders are identified. The special problem of testing for the disease in persons with Down syndrome is discussed.

Introduction

There is no definitive test for Alzheimer disease (AD) in living individuals. Diagnosis of the disease during life is largely based at present on exclusion of other causes of dementia (see Chapter 6). Although a firm diagnosis of AD can usually be made by histopathological analysis of post-mortem brain tissue, neuropathological diagnosis in cases of late-onset dementia may not be clear-cut, as some elderly individuals without dementia have brain changes that qualitatively resemble those in AD (Regland and Gottfries 1992). Also complicating the neuropathological diagnosis is evidence that there may well be different subtypes of AD (Yankner and Mesulam 1991), and that AD may occur in 'mixed' form along with histopathological features of one or more other neurodegenerative diseases (Mirra *et al.* 1992). Correlative clinical and histopathological studies indicate that, for persons older than 65 years, a designation of probable AD based on ante-mortem examination may be correct in only about 70 per cent of cases (Yankner and Mesulam 1991).

Since the late 1970s there have been many reports suggesting that the Alzheimer process is reflected in peripheral tissues as well as in brain (Tables 11.1–11.4). Hence, there have been attempts to develop confirmatory tests for AD based on the analysis of peripheral biological markers. If such tests could help to distinguish AD from other types of dementia, advantages would include improved monitoring of progression of the disease, and perhaps its more effective medical treatment and overall management.

In 1991, the same missense mutation at codon 717 of the amyloid precursor protein (APP) locus was identified in two families with an early-onset form of AD (Goate *et al.* 1991). Different types of missense mutations in the APP locus have now been identified in at least 11 families with early-onset AD (see Chapter 1). Linkage analysis has established that persons in these families who carry a mutant APP allele are at much increased risk of developing AD. By determining which persons in such families do or do not carry the mutant APP allele, predictive testing for AD in these persons becomes possible.

In this chapter, an account is provided of progress that has been made in understanding the peripheral manifestations of AD in the general population and in Down syndrome (DS), and of advances in developing confirmatory and predictive tests for AD.

Confirmatory testing for Alzheimer disease

Experimental approach and design

In order to have clinical applicability, a confirmatory test for AD must be relatively simple to perform, cost-effective, reproducible, have a high degree of sensitivity and specificity, and not be traumatic to the patient. Also, the biological samples for analysis must be stable to storage. Special precautions, such as the addition of protease inhibitors to blood or cerebrospinal fluid (CSF) samples or the use of particular fixatives for preserving tissues, may be necessary.

Once a biological test is thought to have clinical potential on the basis of pilot studies, it should be applied in a blind fashion to large series of males and females with probable AD, healthy sex- and age-matched control individuals and persons with other types of dementia. There is merit in four groupings of affected cases: those with early- or late-onset familial AD, and those with early- or late-onset sporadic AD. Factors leading to avoidable variability in test results (e.g. deterioration of the biological sample after its collection) should be borne in mind. Furthermore, the particular stage of development of the disease, diurnal variation in the biological variable being tested, diet, drugs or medications, extent of physical activity or the presence of infections, may also affect test results. Ideally, the clinical diagnosis of AD should be validated by eventual post-mortem neuropathological examination.

It is possible statistically to combine the results of more than one AD test, or of a series of measurements made with one type of test, into a single risk estimate that a person has AD. For example, the results of brain imaging, tests of cognitive function and the measurement of a peripheral biological marker may all be combined to arrive at a risk estimate. This requires the same tests to be applied to the reference AD and control subjects. As previous studies of other genetic diseases have shown, if one or more tests that provide(s) even partial discrimination between known affected and unaffected individuals is applied to a number of relatives in a large family with an inherited disorder, and the biological data are incorporated with the pedigree information using discriminant analysis, the risk estimates for certain persons in the pedigree may become greatly clarified (Percy *et al.* 1981, 1988).

Type of tissue for analysis

Virtually all accessible tissues have been utilized for confirmatory testing in individuals suspected to have AD (see Tables 11.1–11.4). The potential advantage of a CSF-based test is that, since AD is primarily a central nervous system (CNS) disorder, aberrant biological processes in the CNS might be directly reflected in the CSF, particularly as the CSF is largely (albeit, as indicated below, not completely) sequestered from the peripheral circulation by the blood brain barrier. The main disadvantage is that the drawing of CSF is a traumatic procedure not routinely done outside of a hospital setting. Obviously, blood-, skin- or urine-based tests are less invasive, better tolerated and would not require hospitalization for the samples to be obtained.

The blood brain barrier is now known to be weak or virtually absent at several key points in the nervous system. Even where it seems to be intact, activated T-cells can penetrate to provide 'routine surveillance' of the brain parenchyma (Rogers *et al.* 1988). Moreover, there is accumulating evidence for a neuroendocrine (hypothalamus–pituitary) regulation of immune function and, conversely, that immunocompetent cells may regulate the function of nerve cells through shared systems of receptors for various neurotransmitters and immune regulators (Besodovsky and del Ray 1990). These observations suggest the existence of immunological and endocrinological abnormalities in various psychiatric or neurological diseases (Fudenberg *et al.* 1984; Singh and Fudenberg 1986; Ferrero *et al.* 1991), and provide a rationale for attempting to develop blood-based confirmatory tests for AD.

Although the analysis of skin biopsies might constitute a diagnostic technique for AD that is minimally invasive, there are difficulties associated with the analysis of cultured skin fibroblasts for the study of primary phenomena. The main problem is that cultured fibroblasts have a limited life span (Goldstein 1990). Chromosomal changes, loss of telomeric DNA (Harley 1991) and decreased DNA methylation (Catania and Fairweather 1991) are all associated with senescence. Unless the efficiency of colony formation in the primary culture and the doubling time for each culture can be determined, it is not possible to match cultures for division number, and it becomes difficult to exclude the possibility that an observed phenomenon is simply the reflection of an 'old' culture. The accumulation of mutations in mitochondrial DNA may also be associated with cellular senescence (Wallace 1992).

Summaries are provided in Tables 11.1–11.4 of many peripheral phenomena that have been reported to be associated with AD in the general population.

Some possibly associated with AD manifestations in adults with DS are given in Table 11.5. Most of the phenomena have been described only once, often in small series of cases, and therefore remain unconfirmed. Some exceptions are manifestations of an aberrant hypothalamus–pituitary axis, the presence of acute phase proteins and altered levels of various APP species in plasma and CSF, increased levels of certain brain markers in CSF, increased red cell choline, altered platelet function, altered immune functions of peripheral blood lymphocytes and functional changes of their muscarinic receptors. In no case has a peripheral marker been unequivocally diagnostic of AD.

Peripheral biological markers that show particular promise in confirmatory testing for Alzheimer disease

APP derivatives in CSF and plasma

Measurement of APP and its derivatives, by means of immunoassays, in CSF has been of special interest because of the involvement of APP in plaque formation in AD. A number of independent studies have shown that AD is associated with decreased levels of APP in the CSF (Henriksson *et al.* 1991; Farlow *et al.* 1992; Read *et al.* 1992; Van Nostrand *et al.* 1992). Erickson (1992) reported that the mean APP level in a small series of individuals with AD was about 3.8-fold lower than in age-matched healthy controls or in persons with non-Alzheimer type dementia, although there was some overlap between the groups. Low APP levels have also recently been noted in persons with early-onset AD associated with an APP mutation (Farlow *et al.* 1992). In another study (Read *et al.* 1992), CSF measurements of APP in a number of affected persons were compared with subsequent post-mortem pathological findings. The lowest levels of APP were found in the individuals who had the highest number of amyloid plaques in their brains; amyloid deposits were particularly prominent in the cerebral blood vessels of these individuals, suggesting that vascular amyloid deposition may be highly correlated with the CSF findings.

A variant of the APP test has been developed for application to plasma (Bush *et al.* 1992a). In this approach, the four different circulating species of APP are quantified on western blots after enrichment by heparin–Sepharose chromatography. Levels of the largest APP derivative were reported to be substantially higher in moderately to severely demented persons with AD than in aged control subjects or in those with other neurodegenerative diseases. Because a plasma-based APP assay would have greater clinical applicability than a CSF-based assay, the value of the western blotting procedure as a confirmatory test of AD is being further assessed.

Alzheimer disease associated protein and neuropil thread protein in CSF

Ghanbari *et al.* (1991, 1992) have developed two new tests for application to brain homogenates which appear to provide excellent discrimination between AD-affected and control individuals. The monoclonal antibody Alz-50 recognizes a protein marker (A68) in brain homogenates that is highly selective for AD; A68 is normally expressed in neurones during fetal and early post-natal development (Wolozin *et al.* 1988). However, as Alz-50 cross-reacts with normal brain components, a sandwich enzyme immunoassay was developed that detects AD-associated proteins (ADAP) with very high specificity. ADAP has three major Alz-50 reactive subunits, including A68. The appearance of ADAP precedes the formation of plaques and NFT. Neuronal thread protein (NTP) is a recently characterized 21 kDa protein that is over-expressed in brains with AD pathology; this protein may also have potential as a discriminating biochemical brain marker for AD.

CSF measurements of ADAP and NTP also appear to discriminate well between AD and other types of dementia. Advances in methods for measuring these parameters are encouraging. A chemiluminescent immunoassay with atomole sensitivity for measuring ADAP levels and a totally automated microparticle enzyme immunoassay for measuring NTP have now been developed and should be practical for large-scale clinical application (Ghanbari *et al.* 1992).

Nerve growth factor receptor in urine

A completely different diagnostic approach for AD involves the measurement of urine levels of a fragment of nerve growth factor (NGF) receptor (Lindner *et al.* 1992). NGF receptors are located on the cholinergic basal forebrain neurones that degenerate in AD. As a normal part of all membrane turnover, the extracellular portion of the receptor is clipped off and excreted in urine. Through the use of an enzyme-linked immunoabsorbent assay, urine levels of truncated NGF receptor were found to be dramatically elevated in mildly demented individuals and decreased to low levels in severely demented ones. This simple, non-invasive test has obvious clinical merits and its value in detecting the presence of AD should be further investigated.

Serum markers of an acute phase reaction

Biological markers that are not very specific for AD, but may be indicators of disease progression or regression, include measurements of plasma or serum α_1-antichymotrypsin (ACT) (Giometto *et al.* 1988; Matsubara *et al.* 1990; Altstiel *et al.* 1992) and cortisol (Davis *et al.* 1986; Nappi *et al.* 1990). The rationale is that amyloid β-protein is not the only major component of 'senile' plaques in AD; ACT, a serine protease inhibitor, is also present in large amounts (Abraham *et al.* 1988, 1990). ACT is known to play a special role in maintaining peripheral physiological homeostasis. It is one of a group of proteins collectively known as 'acute phase proteins', the expression of which is altered in response to peripheral infection, inflammation or immunologically-related disease (Heinrich *et al.* 1990). In the periphery, the acute phase reaction is mediated by the cytokines interleukin 1 and 6 (IL-1 and IL-6), which are produced by monocytes, epithelial and endothelial cells. Because of the presence of ACT in senile plaques, Abraham *et al.* (1988) first proposed that AD was associated with an acute phase reaction in brain. This hypothesis has been further extended and it has been postulated that the deposition of amyloid β-protein in AD is also part of this acute phase reaction (Vandenabeele and Fiers 1991). If this is so, some type of brain injury (e.g. exposure to anaesthetic, a viral infection or autoimmune attack) might cause local production of IL-6. This in turn might stimulate microglial cells, astrocytes and cells of the choroid plexus

Abnormality	Reference
Immunoglobulins	
Increased IgG and IgA in cognitively impaired elderly women	Cohen and Eisdorfer (1980)
Significant correlation between serum Ig levels and tests of intelligence and performance	Eisdorfer and Cohen (1980)
Antibodies	
To brain	Nandy (1978)
To rat neurones	Watts *et al.* (1981)
To human neurofilament antigens	Bahmanyar *et al.* (1983)
To rat cholinergic neurones and choline acetyltransferase	Fillit *et al.* (1985)
To a 200 kDa antigen in cell bodies and axons of Torpedo	Chapman *et al.* (1988)
To human brain tissue in AD but not DS	Singh and Fudenberg (1986)
To blood vessels	Fillit *et al.* (1987)
To human neurofibrillary tangles and brain tissue	Gaskin *et al.* (1987)
To brain reactive antibodies in AD and DS	Kumar *et al.* (1988)
To a cholinergic subpopulation of rat brain synaptosomes	Bradford *et al.* (1989)
To amyloid β-protein	Mönning *et al.* (1991)
Antihistone Ab titre correlated with severity of dementia in early-onset but not late-onset AD	Mecocci *et al.* (1992)
Hormones	
Altered secretory pattern of prolactin, thyroid-stimulating hormone, growth hormone and vasopressin	Davis *et al.* (1986)
Decreased plasma estrogen stimulated neurophysin and delayed response to estrogen challenge	Christie *et al.* (1990)
The hypothalamus-pituitary axis	
High prevalence of escape from dexamethasone suppression	Greenwald *et al.* (1986); Nappi *et al.* (1990)
Increased plasma cortisol	Davis *et al.* (1986)
Increased plasma cortisol with abnormal amplitude of DOPA oscillations	Nappi *et al.* (1990)
Increased corticotrophin-releasing factor	Nappi *et al.* (1990)
Acute phase reaction	
Increased serum acute phase proteins in persons with presenile AD (α_1-antichymotrypsin, cerulo-plasmin, C3, C4, properdin factor B)	Giometto *et al.* (1988)
Increased α_1-antichymotrypsin	Matsubara *et al.* (1990); Furby *et al.* (1991)
Increased C-reactive protein, α_1-antichymotrypsin and tumour necrosis factor	Altstiel *et al.* (1992)
Neurotransmitters/neuropeptides	
Decreased neuropeptide Y	Alom *et al.* (1990)
Decreased norepinephrine	Nappi *et al.* (1990)
Vitamins and minerals	
Increased aluminium, cadmium, mercury and selenium; lower iron and manganese; iron, zinc and calcium levels correlated with memory and cognitive function; iron, manganese and strontium related to changes in behaviour	Basun *et al.* (1991)

Table 11.1. Reported serum or plasma abnormalities in Alzheimer disease

Abnormality	Reference
Amyloid precursor protein (APP) species	
Altered profile of APP species	Whyte *et al.* (1992)
Increased level of 130 kDa APP	Bush *et al.* (1992a)
Decreased levels of amyloid β-protein antigen	Shinoda *et al.* (1992)

Table 11.1. *Continued.*

(which maintain homeostasis of the CSF) to produce IL-1 and more IL-6. These cytokines might then upregulate APP and ACT expression in various target cells and elicit other reactions characteristic of the peripheral acute phase response, including increased secretion of ACTH and glucocorticoid production, leukocytosis and activation of the classical complement pathway. Interestingly, both APP and ACT contain a consensus sequence in their promoter regions corresponding to that found in the promoter region of most identified acute phase proteins (Tsuchiya *et al.* 1987). The APP promoter also contains a heat shock element suggesting that at least one of its functions is protective (Salbaum *et al.* 1988). The demonstration of elevated IL-6 and IL-1 biosynthesis in particular cell populations of AD versus control brains, using tissue *in situ* hybridization, would constitute strong support for the acute phase reaction hypothesis (Vandenabeele and Fiers 1991). Increased IL-1 production in AD brains, possibly due to the associated gliosis, has already been demonstrated using immunostaining (Griffin *et al.* 1989).

Alterations in levels of certain acute phase proteins have been found both in CSF and serum or plasma of individuals with AD (see Tables 11.1 and 11.2). Elevated plasma cortisol and elevated corticotrophin-releasing factor (CRF), characteristic of acute phase reactions, also have been observed in AD. Normally, dexamethasone suppresses cortisol levels (Greenwald *et al.* 1986). Persons with AD, however, show a high prevalence of escape from dexamethasone suppression, suggesting that they have a defect in regulation of the hypothalamus–pituitary axis (Nappi *et al.* 1990). The reason for this abnormality is not understood; it may be of particular significance in AD, as high circulating cortisol levels are known to cause hippocampal neurone death (Deshmukh and Deshmukh 1990).

Studies of peripheral biological markers that may provide new insight into the pathophysiology of Alzheimer disease

Mutations in mitochondrial or genomic DNA coding for mitochondrial proteins

The hypothesis has recently been advanced that mutations in mitochondrial DNA may be a cause of ageing and of different types of neurodegenerative diseases (reviewed by Wallace 1992). The finding of defects in mitochondrial function in platelets, muscle and fibroblasts (see Tables 11.3 and 11.4) in persons with AD will spur a search to determine whether mutations or alterations in the DNA coding for mitochondrial proteins play a causal role in AD.

Abnormality	Reference
Antibodies	
To rat cholinergic neurones and human thyroglobulin	McRae-Deguerce *et al.* (1988)
To rat cholinergic neurones	Dahlström *et al.* (1990)
To amoeboid microglial cells in the developing rat CNS	McRae *et al.* (1991)
To amyloid β-protein	Mönning *et al.* (1991)
To paired helical filaments	Wang *et al.* (1991)
Hormones	
Reduced pre-albumin (transthyretin), the thyroid hormone carrier	Riisoen (1988)
The hypothalamus-pituitary axis	
Increased corticotropin-releasing factor	Nappi *et al.* (1990)
Acute phase reaction	
Increased α_1-antichymotrypsin	Matsubara *et al.* (1990)
Neurotransmitters/neuropeptides	
Decreased acetylcholine and methylhydroxyphenyl glycol	Davis *et al.* (1986)
Decreased norepinephrine	Nappi *et al.* (1990)
Reduced neurotransmitter markers (cholinesterase, homovanillic acid, 5-hydroxyindoleacetic acid and somatostatin-like immunoreactivity)	Alhainen *et al.* (1992)
Vitamins and minerals	
Decreased cadmium, calcium and increased copper; calcium levels correlated with memory and cognitive function; aluminium levels related to behaviour changes	Basun *et al.* (1991)
Amyloid precursor protein (APP) species	
Increased β-APP 751/770	Kitaguchi *et al.* (1990)
Slightly decreased β-APP 695 in living AD persons; no difference from controls in post-mortem samples	Henriksson *et al.* (1991)
Decreased amyloid β-protein	Van Nostrand *et al.* (1992)
Decreased APP in four patients with mild to moderate biopsy-confirmed AD; levels correlated inversely with the severity of vascular amyloid deposition	Read *et al.* (1992)
Brain markers of AD	
Increased ubiquitin (in advanced AD)	Kudo *et al.* (1992)
Increased AD-associated protein and neuropil thread protein	Ghaubari *et al.* (1992)
Increased neuropil thread protein	De La Monte *et al.* (1992)
Increased ganglioside GM1, lower sulphatide in AD than in vascular dementia	Blennow and Wallin (1992)
Decreased sulphated glycosaminoglycan	Willmer *et al.* (1992)
Enzymes	
Decreased acetylcholinesterase	Shen *et al.* (1992)
Anomalous form of acetylcholine esterase	Jobst *et al.* (1992)

Table 11.2. Reported cerebrospinal fluid abnormalities in Alzheimer disease

Abnormality	Reference
Platelets	
Increased monoamine oxidase	Smith *et al.* (1982)
Decreased phosphofructokinase activity	Ksiezak-Reding *et al.* (1983)
Increased membrane fluidity	Zubenko *et al.* (1987)
Altered binding to α_2-adrenoceptors	Adunsky *et al.* (1989)
Cytochrome oxidase deficiency	Parker *et al.* (1990)
Red blood cells	
Altered physical state of membrane proteins	Markesbery *et al.* (1980)
Increased choline	Friedman *et al.* (1981); Blass *et al.* (1985)
Decreased cholinesterase	Chipperfield *et al.* (1981)
Increased Na-Li counter-transport	Diamond *et al.* (1983)
Impaired ouabain binding	McHarg *et al.* (1983)
Increased choline efflux	Butterfield *et al.* (1985)
Abnormal erythrocyte deformability and choline transfer	Rapin *et al.* (1992)
Lymphocytes	
Increased acentric chromosome fragments	Nordenson *et al.* (1980); Moorhead and Heyman (1983)
Increased mutagen sensitivity of chromosomes	Fishman *et al.* (1984)
Increased suppressor cells	Miller *et al.* (1981)
Decreased killer T-cells	Kraus (1983)
Reduced T-cell suppressor activity	Skias *et al.* (1985)
Functional changes in muscarinic receptors	Adem *et al.* (1986); Ferrero *et al.* (1991)
Reduced acetylcholinesterase in senile dementia	Bartha *et al.* (1987)
Immunologic dysfunction in AD and older persons with DS	Singh *et al.* (1987)
Reduced T-cell mitogenic response	Nijhuis *et al.* (1991)
Evidence for activated humoral immunity and a decreased cellular immunity	Ikeda *et al.* (1991a)
Evidence for a possible immune reaction in peripheral blood lymphocytes paralleling that in AD brain (see also McGeer *et al.* (1991))	Ikeda *et al.* (1991b)
Lymphoblasts	
Impaired microtubular assembly in lymphoblasts	Krawczun *et al.* (1991)
Increased sensitivity to X-rays	Robbins *et al.* (1985)
Neutrophils/polymorphonuclear leucocytes	
Altered production of toxic oxygen species in activated peripheral blood neutrophils	Licastro *et al.* (1992b)
Granulocytes	
Decreased motility	Jarvik *et al.* (1982)

Table 11.3. Reported blood cell abnormalities in Alzheimer disease

The development of autoimmune mouse models for Alzheimer disease
The sera of some individuals with AD contain antibodies that bind specifically to the heavy molecular weight neurofilament (NF-H) of mammalian and

Abnormality	Reference
Skin, subcutaneous tissue and intestine	
Increased amyloid β-protein	Joachim *et al.* (1989)
Skin biopsy	
Immune complexes with amyloid β-protein	Heinonen *et al.* (1992)
Fibroblasts	
Microtubular defect	Andria-Waltenbaugh and Puck (1977)
Decreased phosphofructokinase activity	Sorbi and Blass (1983)
Decreased interferon response	Mowshowitz *et al.* (1983)
Decreased free calcium and cell spreading	Peterson *et al.* (1986)
Higher CO_2 and lactate production from glucose; lower CO_2 production from glutamine	Sims *et al.* (1987)
Decreased secretion of cholinergic differentiation factor	Kessler (1987)
Decreased adhesiveness	Uéda *et al.* (1989)
Deficiency of mitochondrial α-ketoglutarate dehydrogenase in familial AD	Sheu *et al.* (1992)
Altered β-adrenergic receptor stimulated, G-protein-mediated cAMP formation in familial and sporadic AD	Huang and Gibson (1992)
Atypical transketolase pattern in some familial and sporadic AD patients	Sorbi *et al.* (1992)
Muscle	
Reduced mitochondrial respiratory activity	Trounce *et al.* (1989)
Increased oxidative activity using biochemical assays; reduced CoQ_{10} level	Mariani *et al.* (1991)
Nasal epithelial tissue	
Increased phosphorylated neurofilament staining	Talamo *et al.* (1989)
Urine	
Greatly increased levels of truncated nerve growth factor receptor in mildly demented and decreased levels in severely demented persons with AD	Lindner *et al.* (1992)
Adrenal medulla	
Inclusion bodies	Averback (1983)

Table 11.4. Reported abnormalities in other tissues in Alzheimer disease

Torpedo cholinergic neurones. This has prompted the development of an experimental immunological model of memory impairment for AD (Michaelson *et al.* 1992). Rats immunized for prolonged periods with Torpedo NF-H developed antibodies against this antigen that cross-reacted with their own NF-H. The immunized rats were found to have an enhanced hyperthermic response to the muscarinic agonist oxothenoline and their short-term working memory was impaired compared to control animals. The impairment in short-term memory was reversed by treatment with the acetylcholinesterase inhibitor physostigmine. This autoimmune animal model may replicate some immunologically induced pathogenic processes in AD. An obvious question is whether an analogous mouse model could be developed by immunization with human amyloid β-protein, perhaps rendered more immunogenic by aggregation or the attachment of a hapten.

Possible deficiencies of certain minerals and vitamins

There is some evidence that alterations in the levels of certain minerals and vitamins may contribute to the pathogenesis of AD. In one group of persons with AD, the serum levels of iron, zinc and calcium correlated with measurements of memory and cognitive functions; in CSF, only the calcium level correlated with these functions (Basun *et al.* 1991). However, this study did not investigate the possibility that the apparent correlation resulted from the age-related variability of both the levels of the minerals and the indices of memory and cognitive function. Zinc is of particular interest as it is a co-factor of many different enzymes and proteins, and has long been known to be important in immunoregulation. With selenium, it also has been shown to affect the metabolism of thyroid hormones (Editorial 1992). The level of zinc is very high in the hippocampus, suggesting that it may play a special role in the CNS (Constantinidis 1991). For instance, in one clinical study of patients with different neurodegenerative diseases, including AD, low serum zinc levels correlated with slowed brain wave patterns. In this study, the intravenous administration of zinc rapidly normalized the brain wave patterns (Fuenfgeld 1992). That vitamin and mineral deficiency may contribute to the pathogenesis of AD is further supported by reports that administration of CoQ_{10}, vitamin B6 and iron had beneficial effects on memory and cognitive function (Imagawa 1991), and regional blood flow (Imagawa 1992) in persons with early- and late-onset AD.

Predictive testing for Alzheimer disease and related disorders

Currently, the normal functions of APP are not known, and it is unclear whether the deposition of amyloid β-protein is a cause or consequence in the Alzheimer process. However, the consistent presence of amyloid β-protein in the brains of persons with AD suggests that APP metabolism plays a central role in this disorder. It has been proposed that there are different aetiologies for AD, but only a single pathogenic process consisting of a cascade that involves APP 'mismetabolism', which leads to amyloid β-protein deposition, tau phosphorylation/tangle formation and cell death (Hardy and Allsop 1991).

Different types of missense mutations in the APP locus on chromosome 21q have now been found in at least 11 different families with early-onset AD (see Chapter 1). These findings have stimulated large-scale screening of persons with AD for APP mutations, but the yield has been extremely low.

Another disorder characterized by amyloid deposition—hereditary cerebral haemorrhage with amyloidosis, Dutch type (HCHWA-Dutch type)—is associated with a point mutation in the APP gene at codon 692 (Levy *et al.* 1990). A similar disease, HCHWA-Icelandic type, is associated with a mutation in the gene coding for cystatin C, a cysteine protease inhibitor (Ghiso *et al.* 1986). Familial Gerstmann–Sträussler syndrome and Creutzfeldt–Jakob disease, conditions that sometimes clinically mimic AD, are associated with mutations in the prion locus on the long arm of chromosome 20 (Brown *et al.* 1992). Petersen *et al.* (1992) have observed that familial thalamic dementia and fatal familial insomnia are diseases with an identical mutation in the prion gene. In all of these conditions, there is a strong association between the presence of the 'mutant' allele and the disease. However, it is unclear whether the particular mutant allele is always predictive that the disease in question will develop. In one carefully examined family with Gerstmann–Sträussler syndrome, some members with the mutation appear to have escaped the disease (Korczyn and Chapman 1992). Other families with identified mutations need to be similarly screened to help determine whether the mutant alleles are pathogenic *per se*, or whether interaction with an environmental agent (e.g. a common virus such as herpes simplex virus 1, Jamieson *et al.* 1992) or other genes are necessary for disease expression.

In families with such mutations, it is now feasible to screen individual members for the presence or absence of the mutation concerned and hence to determine whether they are at high risk of developing the associated disease. Prenatal testing for the presence or absence of the mutation is also technically possible, although not necessarily desirable. As predictive testing for AD and related disorders raises complex ethical, psychological and legal issues, lessons learned from already established predictive testing programmes for Huntingon's disease will provide valuable guidelines for such testing for AD and other adult-onset disorders. Furthermore, because the mutations referred to are rare and the search for new ones is costly, protocols need to be developed for deciding which families should be screened, which family members should be initially tested and which settings should undertake the screening. Because DNA sequencing is tedious and expensive, alternative molecular biological technologies are being developed to ascertain DNA polymorphisms with single base changes and other types of mutations (e.g. Hartmann *et al.* 1992; Podlisny *et al.* 1992; Poduslo and Decker 1992).

Most cases of AD do not appear to be linked to chromosome 21. Thus, besides known APP mutations, additional genetic or environmental factors must be causally relevant. Candidate genes for AD now being investigated include a number of loci on chromosomes 14 (Schellenberg *et al.* 1992) and 19 (Roses *et al.* 1992); one chromosome 19 locus encodes a protein with an overall structure and amino acid sequence similar to APP (Wasco *et al.* 1992). Other candidate genes for AD are proteases that process or degrade amyloid (Abraham *et al.* 1992) and proteins that may bind either to APP or to regulatory sites that govern its transcription or translation.

The special problem of testing for Alzheimer disease in persons with Down syndrome

A puzzling feature is that brain changes resembling those characteristic of AD have been found in virtually all persons with regular trisomy 21 DS who are over the age of 40 years (see Chapter 5). Yet as many as 45 per cent of those in the age group 40–52 years appear not to show clinical manifestations of AD (Franceschi *et al.* 1990). Co-ordinated genetic, other biological and neuropathological studies will help to address this discrepancy. These investigations may lead to better methods for confirmatory diagnosis of AD in DS and possibly also to predictive testing for AD in certain rare cases of DS caused by partial trisomy 21.

Genetic approaches

Genotype–phenotype correlations of individuals with partial trisomy of the long arm of chromosome 21 are being used to molecularly map regions of the chromosome that are linked to characteristic features of DS (Korenberg *et al.* 1990). Long-term follow-up of these individuals ultimately may also enable mapping of chromosome 21 region(s) connected with AD-like brain changes and dementia. It would be particularly significant if certain individuals with partial trisomy 21 do not develop characteristic AD neuropathology. If an AD region of chromosome 21 that is distinct from the 'critical' DS region can be identified, then genotype analysis could be used to predict whether a person with DS due to partial trisomy 21 had a relatively low or high risk of developing AD.

Study of genotype–phenotype correlations in individuals with DS who have an unbalanced Robertsonian translocation involving chromosome 21 also will contribute valuable information. Because Robertsonian translocation chromosomes are formed by 'centromeric fusion', a portion of the p arm, possibly including the pericentromeric region, in one or both of the participating chromosomes may be deleted. Thus, in these cases, phenotypic or other changes could be due to partial trisomy of chromosome 21 and/or to partial monosomy of the other acrocentric chromosome that participated in the translocation.

Other biological approaches

Studies of confirmatory peripheral biological markers of AD in DS first require accurate classification of participants as having or not having clinical manifestations of AD. Because of the variability in the degree of underlying mental retardation and decreasing auditory and visual acuity that occurs with advancing age in DS, this classification is a challenging diagnostic task (Hewitt *et al.* 1985; Chapter 6). Currently, the most rigorous clinical diagnostic classification involves the application of NINCDS-ADRDA criteria (McKhann *et al.* 1984). Secondly, investigations should include a comparison of the putative biological marker of AD in persons with DS with and without signs of dementia, as well as in sex- and age-matched controls. Appropriate statistical methodology should be employed to discriminate AD effects from possible age effects (that might be different in DS and control groups) and effects of sex and trisomy 21 (Percy *et al.* 1990a,b). Moreover, interpretation of the biological data must take into account whether the participating subjects are taking drugs or medications, suffering from infections or showing significant nutritional deficiencies, all common factors in DS that may affect test results.

Marker	Reference
Increased mean red cell volume	Eastham and Jancar (1983);
	Welfare and Hewitt (1986);
	Wachtel and Pueschel (1991)
Increased amyloid β-protein deposition in skin	Joachim *et al.* (1989)
Immune complexes with amyloid β-protein	Heinonen *et al.* (1992)
Autoimmune thyroiditis associated with 'subclinical'	Percy *et al.* (1990a)
hypothyroidism	
Increased serum interleukin-6	Mehta *et al.* (1992)
Chromosome 21 loss	Percy *et al.* (1993)

Table 11.5. Peripheral biological markers associated with manifestations of Alzheimer disease and/or ageing in persons with Down syndrome

Some peripheral biological markers that could be associated with clinical features of AD in DS are listed in Table 11.5. Compared with the general population, relatively little investigation in this sphere has been done with DS. This is largely due to difficulties in recruiting persons with DS for such studies and in accurately classifying them as clinically affected or unaffected by AD. Only one of the studies listed in the table (Percy *et al.* 1990a) employed NINCDS-ADRDA criteria to diagnose AD in DS. Two others (Welfare and Hewitt 1986; Mehta *et al.* 1992) have considered the evidence of significant cognitive impairment to reflect the presence of AD, an approach that is controversial. Other studies simply described findings in elderly persons with DS compared with controls of similar age (Joachim *et al.* 1989; Heinonen *et al.* 1992), a procedure that may reflect trisomy 21 rather than specific AD effects. Additional related reports have discussed aberrant biochemical processes underlying the mental retardation in DS, or compared biological markers in adults with DS relative to those in age-matched mentally normal or retarded controls (Hestnes *et al.* 1991).

The mean red cell volume is greater in persons with DS than in those without DS (Eastham and Jancar 1983) and also greater in individuals with DS who show cognitive impairment than in individuals with the syndrome who are cognitively unimpaired (Welfare and Hewitt 1986). Moreover, in demented individuals with DS, measurements of red cell volume showed a significant inverse correlation with measurements of cognitive function. A biological basis for this phenomenon is not known. However, red cell size in DS appears to increase significantly with chronological age (Wachtel and Pueschel 1991). Available data also indicate that red cells in DS contain an unusually high percentage of reticulocytes and that their turn-over time *in vivo* is shorter than normal, suggesting that red cells in persons with DS are particularly young (Wachtel and Pueschel 1991). In further investigations of these findings, it may be possible to develop modified osmotic conditions for measuring red cell volume that provide better discrimination between cells from AD-affected and unaffected individuals with DS.

Increased deposition of amyloid β-protein has been found in the skin of adults with DS. This may be a manifestation of AD, since this phenomenon is a characteristic of persons with AD in general (Joachim *et al.* 1989). On the other hand, it may simply be a consequence of the increased gene dosage of APP in DS. Forsdyke (1992) has pointed out that self-aggregation of a protein usually occurs when the solubility of that protein has been exceeded. Such self-aggregation has been postulated to trigger the amplification of specific populations of T-cells and the induction of specific immune responses to clear the aggregates (Forsdyke 1992). This mechanism may explain the prevalence of immune complexes with amyloid β-protein that have also been found in the skin of elderly persons with DS, as well as in chromosomally normal persons with AD. It may also explain the abundance of autoantibodies with different specificities that have been reported in AD (Table 11.1). These observations unquestionably are of biological interest. However, because the presence of amyloid plaques and tangles in DS brain is not always associated with clinical dementia, analysis of skin biopsies for amyloid β-protein deposition may not aid in confirming a diagnosis of AD in persons with DS.

There has been much interest in the possible roles of superoxide dismutase-1 (SOD-1, a copper- and zinc-requiring enzyme) and glutathione peroxidase (GSHPx, a selenium-requiring enzyme) in the pathogenesis of DS (reviewed by Percy *et al.* 1990a). In addition to almost certainly causing increased oxidation, the increased gene dosage for SOD-1 that occurs in DS may indirectly also cause the subclinical abnormalities in thyroid function that are characteristic of the syndrome—i.e. decreased thyroxine (T4) and triiodothyronine (T3), significantly decreased T3 (rT3), lower thymulin, higher thyrotropin, increased titres of antimicrosomal (AMA) and antithyroglobulin (ATA) autoantibodies, and moderately increased thyroid stimulating hormone (TSH) (Lejeune 1990; Percy *et al.* 1990b; Licastro *et al.* 1992a). The increased gene dosage for SOD-1 induces an increase in the amount of GSHPx. A combination of increased requirements for zinc and selenium coupled with intestinal malabsorption, which has been documented in DS (Abalan *et al.* 1990), may cause deficiencies of both trace elements. Zinc deficiency has been reported in DS. Levels of red cell GSHPx correlate with IQ, suggesting that a deficiency of selenium may contribute to mental retardation in DS. Selenium is also a co-factor of the iodothyronine deiodinase that converts T4 to T3, the metabolically active thyroid hormone. There is now evidence that zinc may likewise be involved in thyroid homeostasis. Licastro *et al.* (1992a) showed that by giving dietary zinc sulphate supplements for four months (treatment that largely restored plasma zinc and thymulin to normal levels) concentrations of thyrotropin and rT3 returned to normal. It has also been speculated that altered levels of O_2^- and H_2O_2, caused by the increased gene dosage of SOD-1, modulate the activity of the 5-deiodase that converts rT3 into T2 (Lejeune 1990) and the peroxidase that catalyzes iodination of thyroglobulin (Percy *et al.* 1990a), respectively. There also may be a connection between decreased zinc and the increased gene dosage for APP in DS, as APP has been shown to bind zinc stoichiometrically (Bush *et al.* 1992b).

Serum parameters of thyroid function have been found to increase in severity with age in DS and to be more exaggerated in persons with DS who probably have AD (M. E. Percy, unpublished data). Thus, thyroid abnormalities might not only contribute to cognitive decline, but also to the development of AD, in DS. Lejeune (1990) pointed out that thyroid function is particularly important in the regulation of monocarbon metabolism (which is disrupted in DS) and tubulin assembly/disassembly (which is disrupted in DS, AD and clinical hypothyroidism). Little is known about the biochemical and physiological role of rT3, but the very low levels in DS may be of particular clinical significance in persons with DS of all ages (Lejeune 1990), and the possibility of therapeutic intervention with dietary zinc (Licastro *et al.* 1992a) and perhaps selenium needs further exploration. It may be that the neuropathological findings in DS brain and the clinical manifestations of AD in DS are the result of increased dosage effects for two different genes: APP and SOD-1, respectively. Although there is some evidence that clinically significant thyroid abnormalities are a risk factor for AD in the general population, this association remains controversial (Kung Sutherland *et al.* 1992).

In a recent pilot study using an enzyme-linked immunoabsorbent assay, levels of serum IL-6, but not IL-1 or B-2 microglobulin, were found to be significantly increased in persons with DS who had severe cognitive impairment compared with those without such impairment (Mehta *et al.* 1992). As there was no significant age effect on the IL-6 levels in this series, it was suggested that this phenomenon may have resulted from an AD-associated inflammatory response in older persons with DS. However, the possibility was not excluded that the elevated IL-6 levels were due to chronic hepatitis B or other infections. These preliminary findings require further investigation in larger studies of persons with DS classified as AD-affected or unaffected by more rigorous diagnostic criteria. As suggested earlier in this chapter, the AD process in normal individuals may also be mediated by IL-6. The fact that AD-affected individuals in the general population do not have significantly elevated serum IL-6 levels is not inconsistent with this hypothesis, as IL-6 production in these individuals may be largely restricted to the CNS (Vandenabeele and Fiers 1991).

A significant age-dependent loss of chromosome 21 was recently described in persons with DS (Percy *et al.* 1993). Although only two to four per cent of PHA-stimulated peripheral blood lymphocytes showed this effect in the most elderly subjects, the observation may have important biological implications. Some of the cells that became diploid through loss of a chromosome 21 may retain two chromosomes 21 from the same parent (i.e. uniparental). If maternally-derived and paternally-derived chromosome 21s are imprinted, they will not be functionally equivalent. There is now evidence that in a high proportion of persons with DS who have trisomy 21 mosaicism, the diploid cell line developed through loss of a chromosome 21 from a trisomic zygote (Dagna Bricarelli *et al.* 1990). It was therefore suggested that bi- or uniparental chromosome 21 mosaicism could account for some phenotypic variability in persons with trisomy 21 and possibly also modulate the rate of development of AD in these cases (Percy *et al.* 1993). Although there currently is no evidence that uniparental disomy mosaicism is associated with AD, this is an avenue of research that requires further exploration. Uniparental chromosome 21 disomy cells of maternal or paternal origin are known to be viable (Niikawa and Kajii 1984). Thus in certain families it may be feasible to determine the parental origin of the chromosome 21s in diploid and trisomic cell lines of DS mosaics with and without dementia. Uniparental disomy is more common than previously believed and is now known to account for a proportion of some clinical disorders of previously unknown aetiology (Hall 1992).

Conclusions

Application of appropriate statistical methodology is necessary to obtain the maximum amount of information from clinical and laboratory data to confirm or predict a diagnosis of AD. Measurements of APP, ADAP and neuropil thread protein in CSF and truncated NGF receptor in urine show particular promise as confirmatory indicators of AD. Measurements of plasma cortisol and α_1-antichymotrypsin are less specific for AD, but may indicate disease progression or regression in longitudinal studies. The role of mutations in mitochondrial or genomic DNA coding for mitochondrial proteins, autoimmunity, and certain mineral (particularly zinc and selenium) and vitamin deficiencies should also be further investigated in the pathogenesis/aetiology of AD in the population at large and in DS. The discovery of mutations in the APP locus of a number of families with early-onset AD has made predictive testing (including prenatally) for AD a prospect for these families. This raises complex ethical, psychological and legal issues for which comprehensive guidelines must be developed.

Finally, genotype–phenotype correlations in unusual cases of DS with partial trisomy 21 may permit an AD region on chromosome 21 to be mapped. A major problem with studies of biological markers for AD in DS is that rigorous diagnostic criteria have not been used consistently to classify subjects as probably affected or unaffected. Also, not all studies have adequately controlled for possible age, sex and trisomy 21 effects. Measurements of serum IL-6 should be further investigated as a possible indicator of AD in DS. Subclinical abnormalities in thyroid function characteristic of DS may contribute to cognitive deterioration and the clinical manifestations of AD in DS. Chromosome 21 loss from trisomic cells has been proposed to contribute to phenotypic variation and variability in the onset of AD in DS.

References

Abalan, F., Jouan, A., Weerts, M. T., Solles, C., Brus, J., and Sauneron, M. F. (1990). A study of digestive absorption in four cases of Down's syndrome. Down's syndrome, malnutrition, malabsorption, and Alzheimer's disease. *Medical Hypotheses* 31, 35-8.

Abraham, C. R., Selkoe, D. J., and Potter, H. (1988). Immunochemical identification of the serine protease inhibitor alpha-1-antichymotrypsin in the brain amyloid deposits of Alzheimer's disease. *Cell*, 52, 487-501.

Abraham, C. R., Shirahama, T., and Potter, H. (1990). Alpha-1-antichymotrypsin is associated solely with amyloid deposits containing the β-protein. Amyloid and cell localization of alpha-1-antichymotrypsin. *Neurobiology of Aging*, 11, 123-9.

Abraham, C. R., Razzaboni, B. L., Papastoitsis, G., and Meckelein, B. (1992). Purification and cloning of APP-processing proteases. *Neurobiology of Aging*, 13, Suppl. 1, Abstract 298, S76.

Adem, A., Nordberg, A., Bucht, G., and Winblad, B. (1986). Extraneural cholinergic markers in Alzheimer's and Parkinson's disease. *Progress in Neuro-Psychopharmacology and Biological Psychiatry*, 10, 247-57.

Adunsky, A., Hershkowitz, M., and Rabinowitz, M. (1989). Alzheimer's dementia and binding to alpha2 adrenoceptors in platelets. *Journal of the American Geriatrics Society*, 37, 741-4.

Alhainen, K., Helkala, E. -L., Reinikainen, K., Hänninen, T., and Riekkinen, P. (1992). The relationship of the CSF monoamine metabolites with clinical response to THA in Alzheimer's disease. *Neurobiology of Aging*, 13, Suppl. 1, Abstract 125, S31-2.

Alom, J., Galard, R., Catalan, R., Castellanos, J. M., Schwartz, S., and Tolosa, E. (1990). Cerebrospinal fluid neuropeptide Y in Alzheimer's disease. *European Journal of Neurology*, 30, 207-10.

Altstiel, L., Lawlor, B., Johannessen, D., Mohs, R., and Davis, K. (1992). Acute phase reactants in Alzheimer's disease. *Neurobiology of Aging*, 13, Suppl. 1, Abstract 107, S27.

Andria-Walkenbaugh, A. M., and Puck, T. T. (1977). Alzheimer disease: further evidence of a microtubular defect. *Journal of Cell Biology*, 75, 279a.

Averback, P. (1983). Two new lesions in Alzheimer's disease. *Lancet*, 2, 1203.

Bahmanyar, S., Moreau-Dubois, M. C., Brown, P., Cathala, F., and Gajdusek, D. C. (1983). Serum antibodies to neurofilament antigens in patients with neurological and other diseases and in healthy controls. *Journal of Neuroimmunology*, 5, 191-6.

Bartha, E., Szelenyi, J., Szilagyi, K., Venter, V., Thu Ha, N. T., Paldi-Harris, P., and Hollan, S. (1987). Altered lymphocyte acetylcholinesterase activity in patients with senile dementia. *Neuroscience Letters*, 79, 190-4.

Basun, H., Forssell, L. G., Wetterberg, L., and Winblad, B. (1991). Metals and trace elements in plasma and cerebrospinal fluid in normal ageing and Alzheimer's disease. *Journal of Neural Transmission*, 4, 231-58.

Besodovsky, H. O., and del Rey, A. (1990). Interactions between immunological cells and the hypothalamus-pituitary-adrenal axis: an example of neuroendocrine immunoregulation. In *Stress and the aging brain: integrative mechanisms,* (ed. G. Nappi, E. Martignani, A. R. Genazzani and F. Petraglia), pp. 163-70. Raven Press, New York.

Blass, J. P., Hanin, I., Barclay, L., Kopp, U., and Reding, M. J. (1985). Red blood cell abnormalities in Alzheimer's disease. *Journal of the American Geriatrics Society*, 33, 401-5.

Blennow, K., and Wallin, A. (1992). A rational approach to the study of biochemical diagnostic markers of Alzheimer's disease. *Neurobiology of Aging*, 13, Suppl. 1, Abstract 109, S28.

Bradford, H. F., Foley, P., Docherty, M., Fillit, H., Luine, V. N., McEwen, B., *et al.* (1989). Antibodies in serum of patients with Alzheimer's disease cause immunolysis of cholinergic nerve terminals from the rat cerebral cortex. *The Canadian Journal of Neurological Sciences*, 16, 528-34.

Brown, P., Preece, M. A., and Will, R. G. (1992). "Friendly fire" in medicine: hormones, homografts and Creutzfeldt-Jakob disease. *Lancet*, 340, 24-7.

Bush, A. I., Whyte, S., Thomas, L. D., Williamson, T. G., Van Tiggelen, C. J., Currie J., *et al.* (1992a). An abnormality of plasma amyloid protein precursor in Alzheimer's disease. *Annals of Neurology*, 32, 57-65.

Bush, A. I., Moir, R. D., Multhaup, G., Williamson, T. G., Rumble, B., Small, D. H., *et al.* (1992b). Specific and saturable binding of the amyloid protein precursor of Alzheimer's disease by zinc(11). *Neurobiology of Aging*, 13, Suppl. 1, Abstract 331, S84.

Butterfield, D. A., Nicholas, M. M., and Markesbery, W. R. (1985). Evidence for an increased rate of choline efflux across erythrocyte membranes in Alzheimer's disease. *Neurochemical Research*, 10, 909-18.

Catania, J., and Fairweather, D. S. (1991). DNA methylation and cellular ageing. *Mutation Research*, 256, 283-93.

Chapman, J., Bachar, O., Korczyn, A. D., Wertman, E., and Michaelson, D. M. (1988). Antibodies to cholinergic neurons in Alzheimer's disease. *Journal of Neurochemistry*, 51, 479-85.

Chipperfield, B., Newman, P. M., and Moyes, I. C. A. (1981). Decreased erythrocyte cholinesterase activity in dementia. *Lancet*, 2, 199.

Christie, J., Hunter, R., Bennie, J., Wilson, N., Carroll, S., and Fink, G. (1990). Reduced plasma oestrogen stimulated neurophysin and delayed response to oestrogen challenge in Alzheimer's disease. *Psychological Medicine*, 20, 773-7.

Cohen, D., and Eisdorfer, C. (1980). Antinuclear antibodies in the cognitively impaired elderly. *Journal of Nervous and Mental Diseases*, 168, 179-80.

Constantinidis, J. (1991). Hypothesis regarding amyloid and zinc in the pathogenesis of Alzheimer disease: potential for preventive intervention. *Alzheimer Disease and Associated Disorders*, 5, 31-5.

Dagna Bricarelli, F., Pierluigi, M., Grasso, M., Strigini, P., and Perroni, L. (1990). Origin of extra chromosome 21 in 343 families: cytogenetic and molecular approaches. *American Journal of Medical Genetics,* Suppl. 7, 129-32.

Dahlström, A., Wigander, A., Lundmark, K., Gottfries, C. G., Carvey, P. M., and McRae, A. (1990). Investigations on auto-antibodies in Alzheimer's and Parkinson's disease using defined neuronal cultures. *Journal of Neural Transmission*, 29, Suppl., 195-206.

Davis, K. L., Davis, B. M., Greenwald, B. S., Mohs, R. C., Mathé, A. A., Johns, C. A., *et al.* (1986). Cortisol and Alzheimer's disease. I. Basal studies. *American Journal of Psychiatry*, 143, 442-6.

De La Monte, S., Growdon, J. H., Volicer, L., Hauser, S. L., and Wands, J. R. (1992). Increased levels of neuronal thread protein in cerebrospinal fluid of patients with probable Alzheimer's disease. *Neurobiology of Aging*, 13, Suppl. 1, Abstract 102, S26.

Deshmukh, V. D., and Deshmukh, S. V. (1990). Stress adaptation failure hypothesis of Alzheimer's disease. *Medical Hypotheses*, 32, 293-5.

Diamond, J. M., Matsuyama, S. S., Meier, K., and Jarvik, L. F. (1983). Elevation of erythrocyte countertransport in Alzheimer's dementia. *New England Journal of Medicine*, 309, 1061-2.

Eastham, R.D., and Jancar, J. (1983). Macrocytosis and Down's syndrome. *British Journal of Psychiatry*, 143, 203-4.

Editorial (1992). Essential trace elements and thyroid hormone. *Lancet*, 1, 1575-6.

Eisdorfer, C., and Cohen, D. (1980). Serum immunoglobulins and cognitive status in the elderly: 2. An immunological behavioural relationship? *British Journal of Psychiatry*, 136, 40-5.

Erickson, D. (1992). Doomsday diagnostic? A precursor protein may predict the risk for Alzheimer's disease. *Scientific American*, 267 (August), 120.

Farlow, M., Ghetti, B., Benson, M. D., Farrow, J. S., van Nostrand, W. E., and Wagner, S. L. (1992). Low cerebrospinal-fluid concentrations of soluble amyloid β-protein precursor in hereditary Alzheimer's disease. *Lancet*, 340, 453-4.

Ferrero, P., Rocca, P., Eva, C., Benna, P., Rebaudengo, N., Ravizza, L., *et al.* (1991). An analysis of lymphocyte ^{3}H-N-methyl-scopolamine binding in neurological patients. *Brain*, 114, 1759-60.

Fillit, H., Luine, V. N., Reisberg, B., Amador, R., McEwen, B., and Zabriskie, J. B. (1985). Studies of the specificity of antibrain antibodies in Alzheimer's disease. In *Senile dementia of the Alzheimer type*, (ed. J. T. Hutton and A. D. Kenny), pp. 307-18. Alan R. Liss, New York.

Fillit, H. M., Kemeny, E., Luine, V., Weksler, M. E., and Zabriskie, J. B. (1987). Antivascular antibodies in the sera of patients with senile dementia of the Alzheimer's type. *Journal of Gerontology*, 42, 180-4.

Fischman, H. K., Reisberg, B., Albu, P., Ferris, S., and Rainer, J. D. (1984). Sister chromatid exchange and cell cycle kinetics in Alzheimer's disease. *Biological Psychiatry*, 19, 319-27.

Forsdyke, D. R. (1992). Two signal model of self/non-self immune discrimination: an update. *Journal of Theoretical Biology*, 154, 109-18.

Franceschi, M., Comola, M., Piattoni, F., Gualandri, W., and Canal, N. (1990). Prevalence of dementia in adult patients with trisomy 21. *American Journal of Medical Genetics*, Suppl. 7, 306-8.

Friedman, E., Sherman, K. A., Ferris, S. H., Reisberg, B., Bartus, R. T., and Schneck, M. K. (1981). Clinical response to choline plus piracetam in senile dementia: relation to red cell choline levels. *New England Journal of Medicine*, 304, 1490-1.

Fudenberg, H. H., Whitten, H. D., Arnaud, P., and Khansari, N. (1984). Is Alzheimer's disease an immunological disorder? Observations and speculations. *Clinical Immunology and Immunopathology*, 32, 127-31.

Fuenfgeld, E. W. (1992). The trace element zinc enhancing brain metabolism. *Neurobiology of Aging*, 13, Suppl. 1, Abstract 389, S98.

Furby, A., Leys, D., Delacourte, A., Buee, L., Soetaert, G., and Petit, H. (1991). Are alpha-1-antichymotrypsin and inter-alpha-trypsin inhibitor peripheral markers of Alzheimer's disease? *Journal of Neurology, Neurosurgery and Psychiatry*, 54, 469.

Gaskin, F., Kingsley, B. S., and Fu, S. M. (1987). Autoantibodies to neurofibrillary tangles and brain tissue in Alzheimer's disease: establishment of Epstein-Barr virus-transformed antibody-producing cell lines. *The Journal of Experimental Medicine*, 165, 245-50 and 937.

Ghanbari, H. A., Miller, B. E., Chong, J. K., Haigler, H. J., and Whetsell, W. O., Jr (1991). Alzheimer's disease associated protein(s) in human brain tissue: detection, measurement, specificity and distribution. In *Alzheimer's disease: basic mechanisms, diagnosis and therapeutic strategies*, (ed. K. Iqbal, D. R. C. McLachlan, B. Winblad and H. M. Wisniewski), pp. 569-76. Wiley, Chichester.

Ghanbari, H., Miller, B., Chong, J., Bennette, D., Azad, N., Wilmer, J., et al. (1992). The road to antemortem biochemical tests for Alzheimer's disease (AD). *Neurobiology of Aging*, 13, Suppl. 1, Abstract 101, S25-6.

Ghiso, J., Jenssen, O., and Frangione, B. (1986). Amyloid fibrils in hereditary cerebral hemorrhage with amyloidosis of Icelandic type is a variant of γ-trace basic protein (cystatin C). *Proceedings of the National Academy of Sciences of the United States of America*, 83, 2974-8.

Giometto, B., Argentiero, V., Sanson, F., Ongaro, G., and Tavolato, B. (1988). Acute-phase proteins in Alzheimer's disease. *European Neurology*, 28, 30-3.

Goate, A., Chartier-Harlin, M. -C., Mullan, M., Brown, J., Crawford, F., Fidani, L., et al. (1991). Segregation of a missense mutation in the amyloid precursor protein gene with familial Alzheimer's disease. *Nature*, 349, 704-6.

Goldstein, S. (1990). Replicative senescence: the human fibroblast comes of age. *Science*, 249, 1129-33.

Greenwald, B. S., Mathé, A. A., Mohs, R. C., Levy, M. I., Johns, C. A., and Davis, K. L. (1986). Cortisol and Alzheimer's disease, II: Dexamethasone suppression, dementia severity and affective symptoms. *American Journal of Psychiatry*, 143, 442-6.

Griffin, W. S. T., Stanley, L. C., Ling, C., White, L., MacLeod, V., Perrot, L. J., et al. (1989). Brain interleukin I and S100 immunoreactivity are elevated in Down syndrome and Alzheimer disease. *Proceedings of the National Academy of Sciences of the United States of America*, 86, 7611-5.

Hall, J. (1992). Genomic imprinting and its clinical implications. *New England Journal of Medicine*, 326, 827-9.

Hardy, J., and Allsop, D. (1991). Amyloid deposition as the central event in the etiology of Alzheimer's disease. *Trends in Pharmacological Sciences*, 12, 383-8.

Harley, C. B. (1991). Telomere loss, mitotic clock or genetic time bomb? *Mutation Research*, 256, 271-82.

Hartmann, T., Rebeck, G. W., Mönning, U., König, G., Masters, C. L., and Beyreuther, K. (1992). Screening of AD-patients for mutations in the APP gene. *Neurobiology of Aging*, 13, Suppl. 1, Abstract 278, S71.

Heinonen, O., Syrjänen, S., Soininen, H., Neittaanmäki, H., Paljärvi, L., Syrjänen, K., et al. (1992). Immune system response in Down's syndrome and Alzheimer's disease. *Neurobiology of Aging*, 13, Suppl. 1, Abstract 126, S32.

Heinrich, P. C., Castell, J. V., and Andus, T. (1990). Interleukin-6 and the acute phase response. *Biochemical Journal*, 265, 621-36.

Henriksson, T., Barbour, R. M., Braa, S., Word, P., Fritz, L. C., Johnson-Wood, K., et al. (1991). Analysis and quantitation of β-amyloid precursor protein in the cerebrospinal fluid of Alzheimer's disease patients with a monoclonal antibody-based immunoassay. *Journal of Neurochemistry*, 56, 1037-42.

Hestnes, A., Stovner, L. J., Husoy, O., Folling, I., Fougner, K. J., and Sjaastad, O. (1991). Hormonal and biochemical disturbances in Down's syndrome. *Journal of Mental Deficiency Research*, 35, 179-93.

Hewitt, K. E., Carter, G., and Jancar, J. (1985). Ageing in Down's syndrome. *British Journal of Psychiatry*, **147**, 58-62.

Huang, H. -M., and Gibson, G. E. (1992). Alterations of β-adrenergic-receptor-stimulated G-protein-mediated cAMP formation in cultured skin fibroblasts from Alzheimer donors. *Neurobiology of Aging*, **13**, Suppl. 1, Abstract 241, S61-2.

Ikeda, T., Yamamoto, K., Takahashi, K., and Yamada, M. (1991a). Immune system-associated antigens on the surface of peripheral blood lymphocytes in patients with Alzheimer's disease. *Acta Psychiatrica Scandinavica*, **83**, 444-8.

Ikeda, T., Yamamoto, K., Takahashi, K., Kaneyuki, H., and Yamada, M. (1991b). Interleukin-2 receptor in peripheral blood lymphocytes of Alzheimer's disease patients. *Acta Psychiatrica Scandinavica*, **84**, 262-5.

Imagawa, M. (1991). Therapy with a combination of coenzyme Q_{10}, vitamin B_6 and iron for Alzheimer's disease and senile dementia of Alzheimer type. In *Alzheimer's disease: basic mechanisms, diagnosis and therapeutic strategies,* (ed. K. Iqbal, D.R.C. McLachlan, B. Winblad and H.M. Wisniewski), pp. 649-51. Wiley, Chichester.

Imagawa, M. (1992). Comparative assessment of regional cerebral blood flow through combination therapy of iron with B_6, C_oQ_{10} for 2 years in Alzheimer's disease and senile dementia of Alzheimer's type. *Neurobiology of Aging*, **13**, Suppl. 1, Abstract 78, S19.

Jamieson, G. A., Maitland, N. J., Wilcock, G. K., Yates, C. M., and Itzhaki, R.F. (1992). Detection by polymerase chain reaction of herpes simplex virus type 1(HSV1) DNA in brain of aged normals and Alzheimer's disease (AD) patients. *Neurobiology of Aging,* **13**, Suppl. 1, Abstract 113, S29.

Jarvik, L. F., Matsuyama, S. S., Kessler, J. O., Fu, T. K., Tsai, S. Y., and Clark, E. D. (1982). Philothermal response of polymorphonuclear leukocytes in dementia of the Alzheimer type. *Neurobiology of Aging*, **3**, 93-9.

Joachim, C. L., Mori, H., and Selkoe, D. J. (1989). Amyloid β-protein deposition in tissues other than brain in Alzheimer's disease. *Nature*, **341**, 226-30.

Jobst, K. A., Navaratnam, D. S., Priddle, J., King, E. M., McDonald, B., Morris, B., *et al.* (1992). A possible antemortem diagnostic test for Alzheimer's disease. *Neurobiology of Aging*, **13**, Suppl. 1, Abstract 105, S26-7.

Kessler, J. A. (1987). Deficiency of a cholinergic differentiating factor in fibroblasts of patients with Alzheimer's disease. *Annals of Neurology*, **21**, 95-8.

Kitaguchi, N., Tokushima, Y., Oishi, K., Takahashi, Y., Shiojiri, S., Nakamura, S., *et al.* (1990). Determination of amyloid β-protein precursors harboring active form of proteinase inhibitor domains in cerebrospinal fluid of Alzheimer's disease patients by trypsin-antibody sandwich ELISA. *Biochemical and Biophysical Research Communications*, **166**, 1453-9.

Korczyn, A. D., and Chapman, J. (1992). The risk of developing Creutzfeld-Jakob disease (CJD) in subjects with the PRNP gene codon 200 point mutation. *Neurobiology of Aging*, **13**, Suppl. 1, Abstract 369, S93.

Korenberg, J. R., Kawashima, H., Pulst, S.- M., Allen, L., Magenis, E., and Epstein, C. J. (1990). Down syndrome: towards a molecular definition of the phenotype. *American Journal of Medical Genetics*, Suppl. 7, 91-7.

Kraus, L. J. (1983). Decreased natural killer cell activity in Alzheimer disease. *Neurosciences Abstracts*, **9**, 115.

Krawczun, M. S., Jenkins, E. C., Lele, K. P., Sersen, E. A., and Wisniewski, H. M. (1991). Study of spindle microtubule reassembly in cells from Alzheimer and Down syndrome patients following exposure to colcemid. *Alzheimer Disease and Associated Disorders*, **4**, 203-16.

Ksiezak-Reding, H., Murphy, C., and Blass, J. P. (1983). Enzyme activities in platelets from patients with Alzheimer's disease. *Age*, **6**, Abstract 30, 137.

Kudo, T., Iqbal, K., Ravid, R., Swaab, D. F., and Grundke-Iqbal, I. (1992). Measurement of ubiquitin immunoreactivity in cerebrospinal fluid (CSF) of Alzheimer disease and control patients. *Neurobiology of Aging*, **13**, Suppl. 1, Abstract 114, S29.

Kumar, M., Cohen, D., and Eisdorfer, C. (1988). Serum IgG brain reactive antibodies in Alzheimer disease and Down syndrome. *Alzheimer Disease and Associated Disorders*, **2**, 50-5.

Kung Sutherland, M., Wong, L., Somerville, M. J., Handley, P., Yoong, L., Bergeron, C., *et al.*

(1992). Reduction of thyroid hormone receptor c-ERB A mRNA levels in the hippocampus of Alzheimer as compared to Huntington brain. *Neurobiology of Aging*, **13**, 301-12.

Lejeune, J. (1990). Pathogenesis of mental deficiency in trisomy 21. *American Journal of Medical Genetics*, Suppl. **7**, 20-30.

Levy, E., Carman, M. D., Fernandez-Madrid, I. J., Power, M. D., Lieberburg, I., van Duinan, S. G., et al. (1990). Mutation of the Alzheimer's disease amyloid gene in hereditary cerebral hemorrhage, Dutch type. *Science*, **248**, 1124-6.

Licastro, F., Mocchenegiani, E., Zannotti, M., Arena, G., Masi, M., and Fabris, N. (1992a). Zinc affects the metabolism of thyroid hormones in Down's syndrome: normalization of thyroid stimulating hormone and of reverse triiodothyronine plasmic levels by dietary zinc supplementation. *International Journal of Neuroscience*, **65**, 259-68.

Licastro, F., Morini, M. C., Malpassi, P., Parente, R., Conte, R., and Savorani, G. (1992b). Altered production of toxic oxygen species in activated peripheral blood neutrophils of patients with Alzheimer's disease. *Neurobiology of Aging*, **13**, Suppl. 1, Abstract 348, S88.

Lindner, M. D., Gordon, D. G., Miller, J. M., Tariot, P. N., McDaniel, K. D., Hamill, R. W., et al. (1992). Urine levels of truncated NGF receptor: increased in mildly demented and decreased in severely demented Alzheimer's patients. *Neurobiology of Aging*, **13**, Suppl. 1, Abstract 245, S62-3.

McGeer, P. L., McGeer, E. G., Kawamata, T., Yamada, T., and Akiyama, H. (1991). Reactions of the immune system in chronic degenerative neurological diseases. *The Canadian Journal of the Neurological Sciences*, **18**, 376-9.

McHarg, A., Naylor, G. J., and Ballinter, B. R. (1983). Erythrocyte ouabain binding in dementia. *Gerontology*, **29**, 140-4.

McKhann, G., Drachman, D., Folstein, M., Katzman, R., Price, D., and Stadlan, E. M. (1984). Clinical diagnosis of Alzheimer's disease: Report of the NINCDS-ADRDA Work Group under the auspices of Department of Health and Human Services Task Force on Alzheimer's disease. *Neurology*, **34**, 939-44.

McRae, A., Ling, E. A., Polinsky, R., Gottfries, C. G., and Dahlström, A. (1991). Antibodies in the cerebral spinal fluid of some Alzheimer's disease patients recognize amoeboid microglial cells in the developing rat central nervous system. *Neuroscience*, **41**, 739-52.

McRae-Deguerce, A., Haglid, K., Rosengren, L., Wallin, A., Blennow, K., Gottfries, C. G., et al. (1988). Antibodies recognizing cholinergic neurons and thyroglobuline are found in the cerebrospinal fluid of a subgroup of patients with Alzheimer's disease. *Drug Development Research*, **15**, 153-63.

Marinani, C., Bresolin, N., Farina, E., Moggio, M., Ferrante, C., Ciafaloni, E., et al. (1991). Muscle biopsy in Alzheimer's disease: morphological and biochemical findings. *Clinical Neuropathology*, **10**, 171-6.

Markesbery, W. R., Leung, P. K., and Butterfield, D. A. (1980). Spin label and biochemical studies of erythrocyte membranes in Alzheimer's disease. *Journal of the Neurological Sciences*, **45**, 323-30.

Matsubara, E., Hirai, S., Amari, M., Shoji, M., Yamaguchi, H., Okamoto, K., et al. (1990). α_1-antichymotrypsin as a possible biochemical marker for Alzheimer-type dementia. *Annals of Neurology*, **28**, 561-7.

Mecocci, P., Ekman, R., Parnetti, L., Cadini, D., Longo, A., Cecchetti, R., et al. (1992). Serum anti-histones and anti-dsDNA autoantibodies in Alzheimer's disease and vascular dementia. *Neurobiology of Aging*, **13**, Suppl. 1, Abstract 112, S28.

Mehta, P. D., Dalton, A. J., Mehta, S. P., Percy, M., and Wisniewski, H. M. (1992). Increased $\beta2$-microglobulin ($\beta2$-M) and interleukin-6 (IL-6) in serum from older persons with Down syndrome (DS). *Neurobiology of Aging*, **13**, Suppl. 1, Abstract 111, S28.

Michaelson, D. M., Faigon, M., Dubovic, V., Chapman, J., and Feldon, J. (1992). Experimental autoimmune dementia (EAD): an immunological model of memory dysfunction and Alzheimer's disease. *Neurobiology of Aging*, **13**, Suppl. 1, Abstract 423, S107.

Miller, A. E., Neighbour, P. A., Katzman, R., Aronson, M., and Lipkowitz, R. (1981). Immunological studies in senile dementia of the Alzheimer type: evidence for enhanced suppressor cell activity. *Annals of Neurology*, **10**, 506-10.

Mirra, S., Gearing, M., Sumi, S. M., Crain, B., Heyman, A., and CERAD neuropathologists. (1992). The neuropathology assessment of Alzheimer's disease and related dementias: the CERAD experience. *Neurobiology of Aging*, 13, Suppl. 1, Abstract 134, S34.

Mönning, U., Schreiter-Gasser, U., Hilbich, C., Bunke, D., Prior, R., Masters, C. L., *et al.* (1991). Alzheimer amyloid β/A4 protein-reactive antibodies in human sera and CSF. In *Alzheimer's disease: basic mechanisms, diagnosis and therapeutic strategies* (ed. K. Iqbal, D. R. C. McLachlan, B. Winblad and H. M. Wisniewski), pp. 557-63. Wiley, Chichester.

Moorhead, P., and Heyman, A. (1983). Chromosome studies of patients with Alzheimer disease. *American Journal of Medical Genetics*, 14, 545-56.

Mowshowitz, S. L., Dawson, G. J., and Elizan, T. S. (1983). Antiviral response of fibroblasts from familial Alzheimer disease and Down's syndrome to human interferon-alpha. *Journal of Neural Transmission*, 57, 121-6.

Nandy, K. (1978). Brain-reactive antibodies. In *Aging and senile dementia: Alzheimer's disease, senile dementia and related disorders,* (ed. R. Katzman, R. D. Terry and K. L. Bick), pp. 506-14. Raven Press, New York.

Nappi G., Martignani, E., Costa, A., Petraglia, F., Blandini, F., Bono, G., *et al.* (1990). Aging and dementia: markers of the hypothalamus-pituitary-adrenal axis and sympathetic activities in biological fluids. In *Stress and the aging brain: integrative mechanisms,* (ed. G. Nappi, E. Martignani, A. R. Genazzani and F. Petraglia), pp. 93-106. Raven Press, New York.

Niikawa, N., and Kajii, T. (1984). The origin of mosaic Down syndrome: four cases with chromosome markers. *American Journal of Human Genetics*, 336, 123-30.

Nijhuis, E., van Duijn, C. M., Wittemen, C., Hofman, A., Rozing, J., and Nagelkerken, L. (1991). T-cell reactivity in patients with Alzheimer's disease. In *Alzheimer's disease, basic mechanisms, diagnosis and therapeutic strategies,* (ed. K. Iqbal, D. R. C. McLachlan, B. Winblad and H. M. Wisniewski), pp. 581-6. Wiley, Chichester.

Nordenson, I., Adolfsson, R., Beckman, G., Bucht, G., and Winblad, B. (1980). Chromosomal abnormality in dementia of the Alzheimer type. *Lancet*, 1, 481-2.

Parker, W. D. Jr, Filley, C. M., and Parks, J. K. (1990). Cytochrome oxidase deficiency in Alzheimer's disease. *Neurology*, 40, 1302-3.

Percy, M. E., Andrews, D. F., and Thompson, M. W. (1981). Duchenne muscular dystrophy: carrier detection using logistic discrimination, serum creatine kinase and hemopexin in combination. *American Journal of Medical Genetics*, 8, 397-409.

Percy, M. E., Rusk, A. C. M., Garvey, M. B., Freedman, J. J., Teitel, J. M., Blake, P., *et al.* (1988). Carrier detection in hemophilia A: ABO blood group, multiple measurements, and application of logistic discrimination. *American Journal of Medical Genetics*, 31, 871-9.

Percy, M. E., Dalton, A. J., Markovic, V. D., Crapper-McLachlan, D. R., Gera, E., Hummel, J. T., *et al.* (1990a). Autoimmune thyroiditis associated with mild 'subclinical' hypothyroidism in adults with Down syndrome: a comparison of patients with and without manifestations of Alzheimer disease. *American Journal of Medical Genetics*, 36, 148-54.

Percy, M. E., Dalton, A. J., Markovic, V. D., McLachlan, D. R., Hummel, J. T., Rusk, A. C., *et al.* (1990b). Red cell superoxide dismutase, glutathione peroxidase and catalase in Down syndrome patients with and without manifestations of Alzheimer disease. *American Journal of Medical Genetics*, 35, 459-67.

Percy, M. E., Markovic, V. D., Dalton, A. J., McLachlan, D. R. C., Berg, J. M., Somerville, M.J., *et al.* (1993). Age-associated chromosome 21 loss in Down syndrome: possible relevance to mosaicism and Alzheimer disease. *American Journal of Medical Genetics*, 45, 584-8.

Petersen, R. B., Tabaton, M., Medori, R., Tritschler, H. J., Berg, L., Schrank, B., *et al.* (1992). Familial thalamic dementia and fatal familial insomnia are prion diseases with the same mutation. *Neurobiology of Aging*, 13, Suppl. 1, Abstract 370, S93.

Peterson, C., Ratan, R. R., Shelanski, M. L., and Goldman, J. E. (1986). Cytosolic free calcium and cell spreading decrease in fibroblasts from aged and Alzheimer's donors. *Proceedings of the National Academy of Sciences of the United States of America* 83, 7999-8001.

Podlisny, M. B., Ostaszewski, B. L., Abrams, E., Tolan, D. R. and Selkoe, D. J. (1992). Search for mutations in the β-amyloid precursor protein (βAPP) by denaturing gradient gel electrophoresis. *Neurobiology of Aging*, 13, Suppl. 1, Abstract 266, S68.

Poduslo, S. E., and Decker, P. (1992). Detection of polymorphisms in candidate genes in Alzheimer's disease. *Neurobiology of Aging*, **13**, Suppl. 1, Abstract 270, S69.

Rapin, J. R., Lespinasse, P., Guard, O., and Yoa, R. -G. (1992). Abnormal erythrocyte deformability and choline transfer in Alzheimer's disease. *Neurobiology of Aging*, **13**, Suppl. 1, Abstract 123, S31.

Read, S., Wagner, S., Vintners, H., and Tomiyasu, U. (1992). Amyloid precursor protein activity in early Alzheimer's disease: diagnostic and prognostic implications. *Neurobiology of Aging*, **13**, Suppl. 1, Abstract 116, S29.

Regland, B., and Gottfries, C. -G. (1992). Hypothesis. The role of amyloid β-protein in Alzheimer's disease. *Lancet*, **340**, 467-9.

Riisoen, H. (1988). Reduced prealbumin (transthyretin) in CSF of severely demented patients with Alzheimer's disease. *Acta Neurologica Scandinavica*, **78**, 455-9.

Robbins, J. H., Otsuka, F., Tarone, R. E., Polinsky, R. J., Brumback, R. A., and Nee, L. E. (1985). Parkinson's disease and Alzheimer's disease: hypersensitivity to Xrays in cultured cell lines. *Journal of Neurology, Neurosurgery and Psychiatry*, **48**, 916-23.

Rogers J., Luber-Narod, J., Styren, S. D., and Civin, W. H. (1988). Expression of immune system-associated antigens by cells of the human central nervous system: relationship to the pathology of Alzheimer's disease. *Neurobiology of Aging*, **9**, 339-49.

Roses, A. D., Alberts, M. J., Saunders, A. M., Gilbert, J. R., Strittmatter, W. J., Schmeckel, D. E., *et al.* (1992). Candidate genes for late-onset Alzheimer's disease. *Neurobiology of Aging*, **13**, Suppl. 1, Abstract 254, S65.

Salbaum, J. M., Weidemann, A., Lemaire, H. -G., Masters, C. L., and Beyreuther, K. (1988). The promoter of Alzheimer's disease A4 precursor gene. *The European Molecular Biology Organization Journal*, **7**, 2807-13.

Schellenberg, G. D., Bird, T. D., Wijsman, E. M., Orr, H. T., Anderson, L., Nemens, E., *et al.* (1992). Genetic linkage evidence for a familial Alzheimer's disease locus on chromosome 14. *Science*, **252**, 668-71.

Shen, Z. X., Ding, Q., Wei, C. Z., Ding, M. C., and Meng, J. M. (1992). CSF cholinesterase in early-onset and late-onset Alzheimer's disease and multi-infarct dementia patients in Chinese kindred. *Neurobiology of Aging*, **13**, Suppl. 1, Abstract 121, S31.

Sheu, K. -F. R., Cooper, A. J. L., Ali, G., Blass, J. P., and Lindsay, J. G. (1992). Mitochondrial α-ketoglutarate dehydrogenase abnormality in Alzheimer disease fibroblasts. *Neurobiology of Aging*, **13**, Suppl. 1, Abstract 250, S64.

Shinoda, T., Kametani, Y., Miyanaga, K., and Iinuma, K. (1992). Preparation of monoclonal antibodies to a β protein subpeptide and screening for immunoreactivity in the sera of patients with Alzheimer's disease. *Neurobiology of Aging*, **13**, Suppl. 1, Abstract 117, S30.

Sims, N. R., Finegan, J. M., and Blass, J. P. (1987). Altered metabolic properties of cultured skin fibroblasts in Alzheimer's disease. *Annals of Neurology*, **21**, 451-7.

Singh, V. K., and Fudenberg, H. H. (1986). Can blood immunoctyes be used to study neuropsychiatric disorders? *The Journal of Clinical Psychiatry*, **47**, 592-4.

Singh, V. K., Fudenberg, H. H., and Brown, F. R. (1987). Immunologic dysfunction: simultaneous study of Alzheimer's and older Down's patients. *Mechanisms of Ageing and Development*, **37**, 257-64.

Skias, D., Bania, M., Reder, A. T., Luchins, D., and Antel, J. P. (1985). Senile dementia of the Alzheimer's type (SDAT): reduced T^{8+} cell-mediated suppressor activity. *Neurology*, **35**, 1635-8.

Smith, R. C., Ho, B. T., Kralik, P., Vroulis, G., Gordon, J., and Wolff, J. (1982). Platelet monoamine oxidase in Alzheimer's disease. *Journal of Gerontology*, **37**, 572-4.

Sorbi, S., and Blass, J. P. (1983). Fibroblast phosphofructokinase in Alzheimer disease and Down's syndrome. *Banbury Report*, **15**, 297-307.

Sorbi, S., Lolli F., Piersanti, P., Sheu, R. -K., Piacentini, S., and Amaducci, L. (1992). Abnormalities of transketolase in Alzheimer's disease fibroblasts. *Neurobiology of Aging*, **13**, Suppl. 1, Abstract 106, S27.

Talamo, B. R., Rudel, R. A., Kosik, K. S., Lee, V. M. -Y., Neff, S., Adelman, L., *et al.* (1989). Pathological changes in olfactory neurons in patients with Alzheimer's disease. *Nature*, **337**, 736-9.

Trounce, I., Byrne, E., and Marzuki, S. (1989). Decline in skeletal muscle mitochondrial respiratory chain function: possible factor in ageing. *Lancet*, 1, 637-9.

Tsuchiya, Y., Hattori, M., Hayashida, K., Ishibashi, H., Okubo, H., and Sakaki, Y. (1987). Sequence analysis of the putative regulatory region of rat alpha-2-macroglobulin gene. *Gene*, 57, 73-80.

Uéda, K., Cole, G., Sundsmo, M., Katzman, R., and Saitoh, T. (1989). Decreased adhesiveness of Alzheimer's disease fibroblasts: is amyloid beta protein precursor involved? *Annals of Neurology*, 25, 246-51.

Vandenabeele, P., and Fiers, W. (1991). Is amyloidogenesis during Alzheimer's disease due to an Il-1/Il-6-mediated "acute phase response" in the brain? *Immunology Today*, 12, 217-9.

Van Nostrand, W. E., Wagner, S. L., Shankle, W. R., Farrow, J. S., Dick, M., Rozemuller, J. M., et al. (1992). Decreased levels of soluble amyloid beta protein precursor in cerebrospinal fluid of live Alzheimer disease patients. *Proceedings of the National Academy of Sciences of the United States of America*, 89, 2551-5.

Wachtel, T. J., and Pueschel, S. M. (1991). Macrocytosis in Down syndrome. *American Journal on Mental Retardation,* 95, 417-20.

Wallace, D. C. (1992). Mitochondrial genetics: a paradigm for ageing and degenerative diseases? *Science*, 256, 628-31.

Wang, G. P., Iqbal, K., Bucht, G., Winblad, B., Wisniewski, H. M., and Grundke-Iqbal, I. (1991). Alzheimer's disease: paired helical filament immunoreactivity in cerebrospinal fluid. *Acta Neuropathologica*, 82, 6-12.

Wasco, W., Hyman, B., and Tanzi, R. (1992). Identification of an amyloid precursor-like protein localized on chromosome 19. *Neurobiology of Aging*, 13, Suppl. 1, Abstract 280, S71.

Watts, H., Kennedy, P. G., and Thomas, M. (1981). The significance of anti-neuronal antibodies in Alzheimer's disease. *Journal of Neuroimmunology*, 1, 107-16.

Welfare, R. W. L., and Hewitt, K. E. (1986). Macrocytosis and cognitive decline in Down's syndrome. *British Journal of Psychiatry*, 148, 482-3.

Whyte, S., Bush, A. I., Currie, J., Williamson, T. G., Van Tiggelen, C. J., Small, D. H., et al. (1992). The abnormality of plasma amyloid protein precursor in Alzheimer's disease. *Neurobiology of Aging*, 13, Suppl. 1, Abstract 115, S29.

Willmer, J., Kisilevsky, R., Guzman, D., Azad, N., and Al-Shamri, S. (1992). Sulphated glycosaminoglycans in the cerebrospinal fluid of individuals with dementia of the Alzheimer type. *Neurobiology of Aging*, 13, Suppl. 1, Abstract 119, S30.

Wolozin, B., Scicutella, A., and Davies, P. (1988). Reexpression of a developmentally regulated antigen in Down syndrome and Alzheimer disease. *Proceedings of the National Academy of Sciences of the United States of America*, 85, 6202-6.

Yankner, B. A., and Mesulam, M. -M. (1991). β-amyloid and the pathogenesis of Alzheimer's disease. *New England Journal of Medicine*, 325, 1849-57.

Zubenko, G. S., Cohen, B. M., Boller, F., Malinakova, I., Keefe, N., and Chojnacki, B. (1987). Platelet membrane abnormality in Alzheimer's disease. *Annals of Neurology*, 22, 237-44.

V

Aetiological aspects

12

The relevance of Down syndrome to aetiological studies of Alzheimer disease

L. J. Whalley

Summary

Alzheimer disease is an age-dependent neurodegenerative disorder in which those cholinergic neurones that project from midbrain to cortex are preferentially lost. Diseased or dying neurones contain abnormal assemblies of cytoskeletal microtubules (neurofibrillary tangles) and typically numerous deposits of proteinaceous cellular debris, including amyloid β-protein and α_1-antichymotrypsin. These changes occur to a lesser extent with ageing in the absence of dementia and to such a severe degree in middle-aged persons with Down syndrome that Down syndrome is widely considered to provide a useful model of Alzheimer disease. Causal studies of Alzheimer disease seek to establish why neurones die and why specific neuronal populations appear to be more vulnerable than others. The availability of the Down syndrome model of Alzheimer disease has been a fertile source of fresh insights into Alzheimer disease.

Numerous possible causes of Alzheimer disease are currently proposed. In this chapter, selected causal models of the disease are outlined and the extent to which each can be supported by studies on Down syndrome is discussed. Epidemiological and experimental neuropathological studies provide much of the data from which these speculative models are derived. It is suggested that Down syndrome may be particularly relevant to two causal models of Alzheimer disease. In one model, amyloid β-protein deposition is considered to be a critical step in the selective loss of neurones. This model places both Alzheimer disease and Down syndrome in the group of cerebral amyloidopathies, some of which are genetically determined whilst others are acquired. The second model considers the selective vulnerability exhibited by some neuronal populations to degenerative changes in Alzheimer disease and Down syndrome. Alterations in dendritic density are observed in both conditions and in ageing in the absence of dementia. Possible mechanisms implicate a role for the neural calcium binding protein S100-β, the gene for which is located on chromosome 21. The association between the neurotrophic actions of S100-β, selective neuronal loss, dendritic proliferation and neurofibrillary tangle formation is relevant to this argument.

The aluminium hypothesis of Alzheimer disease is also reviewed from the standpoint of possible abnormalities in the bioavailability of aluminium in Down syndrome and the relevance this may have to Alzheimer disease in eukaryotic individuals. Finally, the selective vulnerability model is linked to chromosome 21 by (i) Heston's microtubular hypothesis and (ii) the chromosome 21 non-disjunction hypothesis advanced by Potter. Here, data on the proposed familial association between Alzheimer disease and Down syndrome are presented and considered alongside the results of studies on parental age in Alzheimer disease and the source and nature of the genetic contribution to familial Alzheimer disease.

It is concluded that Down syndrome is an important source of causal hypotheses on Alzheimer disease. In addition to the part the model has played in the recognition that genetically determined abnormalities in the synthesis and processing of amyloid precursor protein produce Alzheimer disease changes in Down syndrome, fundamental research on the aetiology of chromosome 21 non-disjunction may yield further understanding of the pathogenesis and role of environmental factors in Alzheimer disease.

Introduction

The typical neuropathological features of Alzheimer disease (AD) are commonly found in the brains of individuals with Down syndrome (DS) who survive into the fifth and sixth decades of life (Whalley 1982). Depositions of abnormal proteins that constitute and define Alzheimer neuropathological changes are found within surviving neurones, in extraneuronal aggregations and in the walls of the cerebral vasculature. These observations are open to wide interpretation.

When first recognized, the association between AD and DS was attributed to abnormal or precocious ageing in DS. Several genetically determined features of ageing occur prematurely in DS (Martin 1977) and, because AD is highly age-dependent (its incidence almost doubles with each successive quinquennium after 64 years) and AD neuropathology may occur with ageing in the absence of dementia (Tomlinson 1992), it was plausible to hypothesize that the association between DS and AD arose because of abnormal ageing in DS. In terms of this hypothesis, pathological ageing in DS might initiate a chain of events completed by formation of the senile plaques and neurofibrillary tangles (NFTs) characteristic of AD.

Once the abnormal karyotype of DS was recognized (Lejeune *et al.* 1959), a second view of the association between DS and AD emerged, most clearly stated by Epstein (see Chapter 13), in which AD neuropathology arose as a consequence of gene dosage effects in DS. Genes assigned to chromosome 21 are summarized in Figure 12.1. Because all individuals with DS have extra genetic material from the long arm of chromosome 21, and because almost all middle-aged persons with DS develop AD, this material was thought sufficient to cause AD. The proposition leads to the hypothesis that in genetically determined AD [frequently referred to as familial AD (FAD)] the same genetic mate-

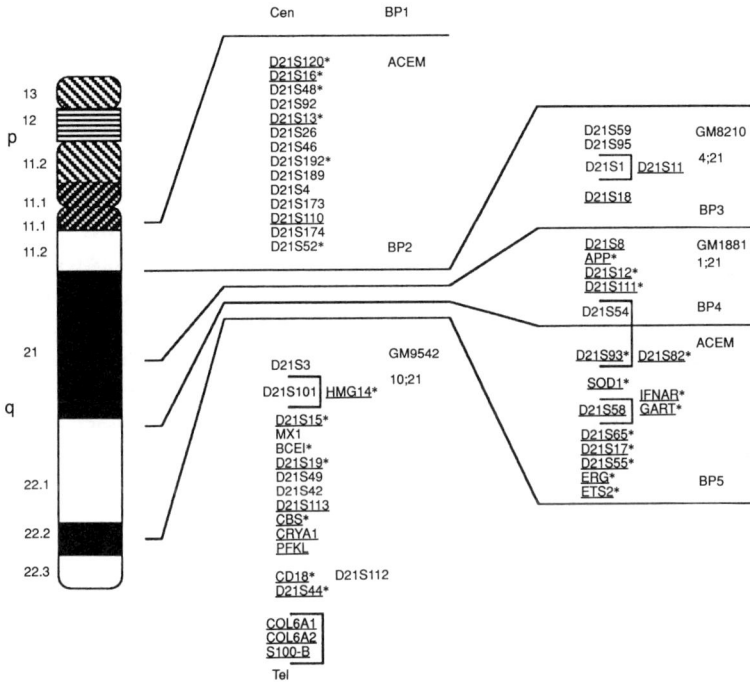

Figure 12.1. Gene map of chromosome 21 (adapted from Cox and Shimizu 1991)

rial has a causal role in neuronal death and the formation of abnormal protein deposits. Although the hypothesis does not contribute directly to discussion on the cause(s) of non-genetically determined AD [sometimes referred to as sporadic AD (SpAD)], it may inform views on the nature of possible gene–environment interactions in SpAD and so help to develop an instructive model of the pathogenesis of AD (Whalley *et al.* 1985). However, no consensus exists on the precise consequences for the individual of the presence of an extra copy of an autosome. Before the components of a pathogenic model of gene–environment interaction of AD can be set out, certain caveats merit consideration. The impact of autosomal trisomy on the complex human genome and its potential for substantial disruption of the phenotype is emphasized by several reviewers (Shapiro 1983; Opitz and Gilbert-Barness 1990). Although it is parsimonious to argue that one or few genes from the 'pathological segment' of chromosome 21 (Figure 12.1) could provide sufficient explanation of the DS phenotype and, by extension, the neuropathology of AD, the evidence to support such a hypothesis is scanty. Shapiro (1983) has argued for an alternative hypothesis that excess genetic material from chromosome 21 is much more likely to produce widespread disruption of developmental and regulatory pathways. If this third view of the association between DS and AD is correct, then the task of disentanglement of genetic and environmental factors when AD develops in DS would be considerable.

The availability of informative animal models of DS may help to resolve these important questions and allow distinction between the widespread disruptive effects of several genes (from the 'pathological segment' of chromosome 21) and the overwhelming effect on the phenotype of duplication of a single gene with a critical function (see Chapter 14). Such considerations have led to a fourth view of the DS-AD association, which accepts the complexity of genetic determinants of nervous system structure and function but chooses to focus on the molecular biology of abnormal protein deposits in AD. This fourth view accepts the DS–AD association as a 'natural experiment' that may assist precise genetic analysis of the pathological mechanisms involved in the formation of amyloid deposits and NFTs (see Chapter 3). Importantly, the molecular pathology of Alzheimer neuropathological changes in eukaryotic individuals [eukaryotic AD (EAD)] may be substantially different from the molecular events that cause Alzheimer changes in DS (AD in DS). It may not be sufficient, therefore, simply to list the possible similarities at a molecular level between EAD and AD in DS. In addition it may be important to consider alternative (largely environmental as opposed to genetic) hypotheses of the causes of AD from the perspective of DS.

In the following two sections, the above views on the association between AD and DS are discussed, first in relation to the molecular neuropathology of AD and then in terms of the abnormal age-related changes present in both AD and DS. The remainder of the chapter reviews four causal hypotheses of AD (selective vulnerability, aluminium, the Heston microtubular hypothesis and the Potter chromosome 21 non-disjunction hypothesis) in the context of the DS–AD association.

Molecular pathology of Alzheimer neuropathological changes in Down syndrome

Diseased or dying neurones in AD occur in both eukaryotic and DS individuals and contain NFTs consisting of abnormal assemblies of cytoskeletal microtubules. The chemical constitution of these assemblies is not well-characterized largely because they are unusually resistant to many techniques of chemical analysis. Instead, most interest has focused on the numerous and relatively more analytically susceptible extracellular deposits of cellular debris, termed the 'senile' or 'neuritic' plaques, which include in their core a mixture of filamentous amyloid protein material shrouded by remnants of deceased neurones. The term 'amyloid' refers to amorphous, eosinophilic debris that is usually extracellular and is made up of protein segments. Studies of amyloid deposition in multiple myeloma and related disorders show that, in myeloma, amyloid proteins comprise primarily the light polypeptide chain of an immunoglobulin. These were produced experimentally by protease digestion of specific proteins and prompted introduction of the term 'amyloidogenic protein'. Subsequently, the amino acid sequences of various amyloid proteins were described and their precursor proteins identified. As indicated below, in AD in DS, hereditary cerebral haemorrhage with amyloidosis of the Dutch type (HCHWA-D) and in certain types of FAD, the amyloid precursor protein (APP) provides the substrate for amyloid deposition. Viewed from this standpoint, AD and AD in DS fall within the proposed molecular classification of cerebral amyloidopathies.

The filamentous material of the plaque core contains two principal proteinaceous components: amyloid β-protein and α_1-antichymotrypsin. In 1984, Glenner and Wong described the amino acid sequence (39–42) of amyloid β-protein. Their very considerable achievement led rapidly to the identification of a much larger precursor protein (APP). Subsequent studies on APP showed it to be a large membrane-spanning protein that perhaps functioned as a receptor molecule and was highly conserved in evolution (Selkoe *et al.* 1987). Its functions in health remain to be established. There are unconfirmed reports of neurotoxic and neurotrophic effects of amyloid β-protein (Yankner *et al.* 1990) and the observation has been made that an inhibitor of platelet coagulation factor XI_a is a truncated form of APP (Smith *et al.* 1990), but the exact relevance of these findings to either EAD or AD in DS is unknown.

Considerable impetus was given to molecular biological studies of AD by the discovery that the APP gene is located on chromosome 21 (Tanzi *et al.* 1987) and that the genetic defect in FAD is also located on chromosome 21 (St George-Hyslop *et al.* 1987). In 1991 Goate *et al.* identified a single point mutation in the APP gene in two of 16 families with early-onset FAD, one English and the other American, and supported the hypothesis that the presence of an abnormal APP gene could cause AD. The hypothesis gained further support from the observation of the same mutation in two of three Japanese FAD families studied (Naruse *et al.* 1991). Identification of molecular pathology specifically associated with AD prompted the proposal (Hardy *et al.* 1991) that a fresh approach be taken to the classification of AD. They recommended that FAD in which the mutation was identified should be reclassified as a β-amyloidopathy followed by the suffix APP_{717} when codon 717 was affected.

Fundamental to any understanding of the development of Alzheimer neuropathological changes in DS, and the relevance this may have to the occurrence of AD in eukaryotic individuals, is the description of the cascade of molecular events that leads to progressive cerebral deposition of the amyloid β-peptide. The precise origins and alternative processing pathways of amyloid β-peptide are currently the focus of much research. If a molecular classification of AD within the cerebral amyloidopathies were accepted, this could have important implications for studies on the relationship between DS and AD. Should it become clear that the molecular pathology of APP and its proteolytic processing are quite distinct between DS and AD, then the relevance of one to the other would require careful qualification.

Amyloid β-peptide is a 4 kDa sequence of the 90 to 130 kDa membrane-bound APP. Three forms of APP contain the amyloid β-peptide sequence and all three are coded by a single gene from which each is derived by alternative mRNA splicing. These three forms are termed APP_{770}, APP_{751} and APP_{695}. Figure 12.2 shows schematically the structure of APP_{770}. Normal secretion and

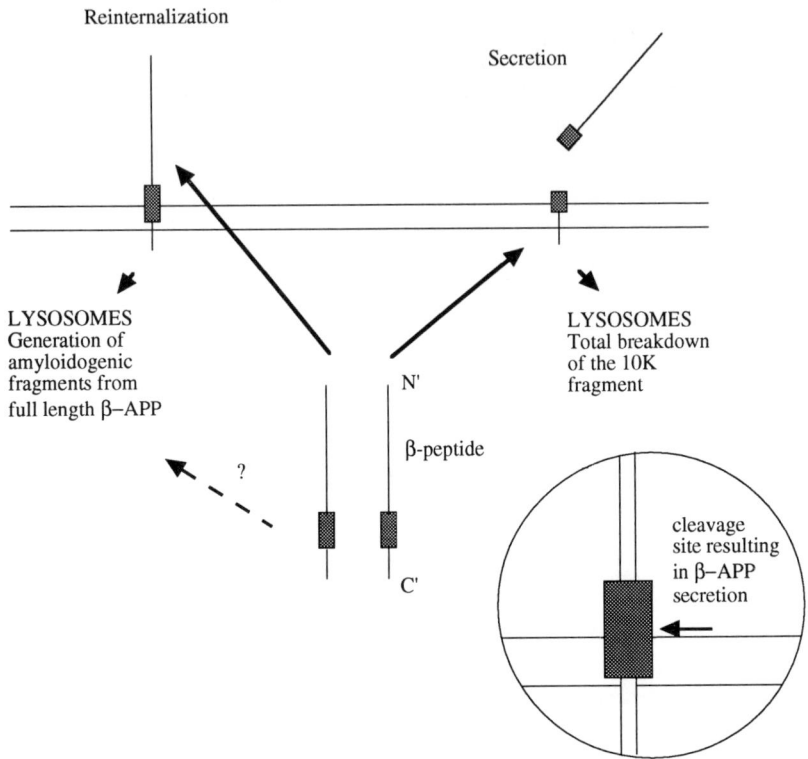

Figure 12.2. Horizontal lines indicate the transmembrane domain. Shaded box indicates the amyloid β-peptide. Schematic representation shows the two normal processing pathways for β-APP. Cleavage and secretion of soluble β-APP at the cell surface or reinternalization and lysosomal targetting of full-length β-APP (adapted from Haass *et al.* 1992)

processing of APP preclude formation of amyloid β-protein because it requires cleavage in the amyloid β region (Esch *et al.* 1990).

If the widespread cerebral deposition of amyloid β-protein in AD and DS is to be satisfactorily explained, alternative proteolytic processing pathways must be available. Haass *et al.* (1992) demonstrated that many β-peptide-containing products can be obtained after incubation of APP with lysosomal proteases. Their findings suggest at least two alternative proteolytic pathways for APP. One yields APP cleaved at the constitutive cleavage site and results in secretion of a soluble fragment. Re-internalization of APP into lysosomes yields amyloid β-peptide-containing fragments. In those rare instances of FAD where mutations of the APP gene exist (probably one per cent of early-onset cases), these mutations may alter the balance between these two pathways and lead in time to the slow accumulation of amyloid β-peptide.

Immunocytochemical studies of brain tissue in DS (Rumble *et al.* 1989) and in ageing in the absence of dementia (Davies *et al.* 1988) show that the deposi-

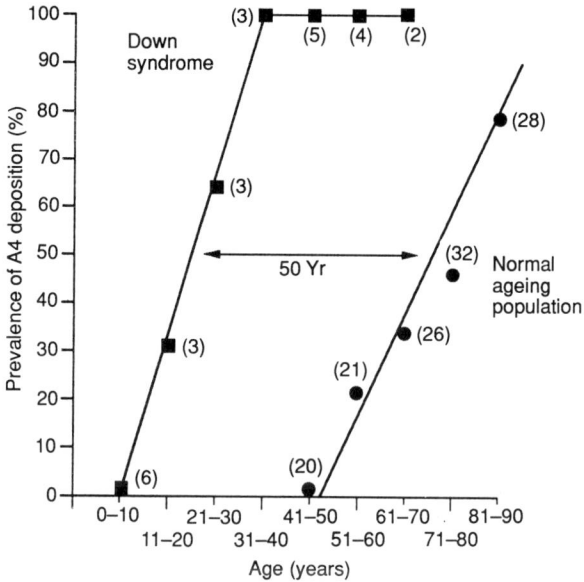

Figure 12.3. The cumulative prevalence of amyloid β-protein deposition in Down syndrome and non-demented eukaryotic individuals (adapted from Rumble *et al.* 1989)

tion of amyloid β-protein differs substantially between the two groups. Figure 12.3 indicates that the cumulative prevalence of amyloid protein deposition (designated as A4 deposition in Figure 12.3) begins before age 30 years in DS but is not detectable until age 50 years in non-demented eukaryotic individuals. Once started, the rate of deposition does not differ between the two groups. Rumble *et al.* (1989) also examined serum concentrations of APP in DS (17 subjects), AD (15 subjects) and elderly controls (33 subjects). Mean serum APP concentrations were significantly higher in the DS (by 1.5 times) than control subjects but did not differ from the subjects with AD. Deposition of APP in brain tissue from DS (two subjects), AD (26 subjects) and elderly controls with neurological disease (17 subjects) was also studied by the same group. APP concentrations were higher (by about 1.5 times) in the subjects with DS than in either the AD or control subjects. Increased APP in DS is probably the result of a gene dosage effect, an inference supported by a report of increased mRNA for APP in fetal DS brain (Tanzi *et al.* 1987). These findings contrast with observations in AD where APP is not increased in brain tissue. However, increased amyloid β-protein production could be associated with increased APP turnover and this could be associated with reduced concentrations of APP in single point measures of a dynamic process. The wide variability of APP concentrations in DS (scattered around a mean value 1.5 times greater than controls) in the study of Rumble *et al.* (1989) supports the latter possibility. A gene dosage effect should be clearcut for all individuals with DS who should each have APP concentrations increased by 1.5 times control values. The results of immunocytochemical studies are consistent with histological studies of the progression of neuropathological lesions in DS (Mann and Esiri 1988; Chapter 5).

HCHWA-D is an autosomal dominant familial cerebral amyloidosis. It features extensive deposition of amyloid in the cerebral vasculature in a manner that resembles early senile plaque formation in AD and DS. It is caused by a mutation in the APP gene that accounts for the abnormal accumulation of amyloid protein in this condition (Levy *et al.* 1990). In contrast, an Icelandic variant of HCHWA-D is caused by a mutation of cystatin C, which, like APP, contains a protease inhibitor sequence (Ghiso *et al.* 1986). In the peripheral nervous system, the amyloidogenic protein is usually transthyretin (prealbumin) with an amino acid substitution. Familial amyloidosis of the Finnish type is not derived from prealbumin but from gelsolin, an actin-modulating protein (Hiltunen *et al.* 1991). Gelsolin binds to actin monomers, nucleates actin filament growth and severs actin filaments. It is regulated by Ca^{2+} and polyphosphoinositides and is encoded by a single gene on chromosome 9.

Senile plaques of the Alzheimer type are also found in familial Creutzfeldt–Jakob disease (CJD) (Adam *et al.* 1982), Gerstmann–Sträussler syndrome (GSS) (Roberts *et al.* 1986) and Kuru (Gajdusek 1977). Cerebral amyloid formation is attributed in familial CJD and GSS to mutations of the prion protein gene located on chromosome 20 (Prusiner 1991). Similarities between the plaques formed in human familial prion diseases and AD have been the occasional source of diagnostic confusion resolved only by careful immunocytochemical investigation. These transmissible dementias complete the context of a current description of the molecular pathologies of the cerebral amyloidopathies. It is within this spectrum of disorders that EAD, SpAD, FAD and AD in DS should be placed.

Abnormal or precocious age-related changes in Alzheimer disease and Down syndrome

The DS–AD association has prompted much intriguing speculation on many aspects of the molecular biology of ageing (Finch 1990). Because there is a profound dysgenesis of brain cytoarchitecture in DS, genetic factors involved in the determination of brain structure and its response to degenerative change are an obvious topic of interest. Neurotrophic regulation of dendritic response to neuronal loss in ageing and a possible association with the formation of neurofibrillary tangles is an example of a well-researched aspect of the DS–AD association. Abnormalities of neuronal plasticity in AD and DS and their relationship to the presence of the gene encoding S100-β, a neurotrophic factor on chromosome 21, are examined below.

Dendritic growth

Neuronal loss is a frequent feature of the ageing brain and brain tissue from old people is less densely packed with neurones, especially in frontal and parietal areas (Terry *et al.* 1987). Presumably in response to neuronal loss with ageing in eukaryotic individuals, surviving neurones make adaptive changes by increase of dendrites (Buell and Coleman 1979, 1981). In normal ageing up to about 80 years, therefore, dendritic proliferation is considered an adaptive response to neuronal loss. By this means, synaptic connectivity (which is the functional unit of central nervous system information processing capacity) is maintained in the face of falling neuronal numbers (Scheibel *et al.* 1976; Buell and Coleman 1981; Flood *et al.* 1985; Flood and Coleman 1986; de Ruiter and Uylings 1987).

In step with neuronal degeneration in AD, surviving neurones also show proliferation and neuritic outgrowth. Scheibel and Tomiyasu (1978) demonstrated pockets of new dendritic growth against a backcloth of dendritic loss. Ferrer *et al.* (1983) described dendritic varicosities and filopodium-like structures in a cortical biopsy specimen in AD. Comparable findings of dendritic growth in DS (Marin-Padilla 1976) and AD (Buell and Coleman 1979) have also been reported (Paula-Barbosa *et al.* 1980; Probst *et al.* 1983; Arendt *et al.* 1986).

New dendritic growth depends on the assembly of new microtubules and these processes are probably regulated by various microtubular assembly proteins (MAPs) including MAP2 and tau (Murphy and Borisy 1975; Drubin *et al.* 1985; Caceres *et al.* 1986; Daniels 1986). In healthy tissue, MAP2 is confined to the somatodendritic compartment of neurones and microtubule-associated tau protein occurs mostly in axons (Kosik *et al.* 1987). However, in AD this distribution is disturbed so that the presence of tau protein is well-established in the somatodendritic compartment among abnormal assemblies of microtubules (NFTs), in 'senile' or 'neuritic' plaques and as dystrophic neurites (Brion *et al.* 1985; Delacourte and Defossez, 1986; Grundke-Iqbal *et al.* 1986; Kosik *et al.* 1986; Nukina and Ihara 1986; Wood *et al.* 1986; Joachim *et al.* 1987; Kowall and Kosik, 1987; Perry *et al.* 1987a; Yen *et al.* 1987). For some neuropathologists, it is the identification of these abnormal assemblies of microtubules that define AD, not the presence of 'senile' or 'neuritic' plaques (Tomlinson *et al.* 1970). Careful electromicroscopic and molecular biological studies demonstrate that NFTs are made up of paired helical filaments in which microtubule-associated tau protein is an integral component. Additionally, ubiquitin is present in intraneuronal NFTs in AD (Mori *et al.* 1987; Perry *et al.* 1987b; Murti *et al.* 1988; Shaw and Chau 1988). Although senile plaques are widespread in the brains of middle-aged individuals with DS, and NFTs also occur, there are reports of significant differences in distribution and density of these changes between DS, AD and non-demented old people (Hirano and Zimmerman 1962; Ball 1976; Hooper and Vogel 1976; Ropper and Williams 1980). It is reasonable, therefore, to be cautious when seeking to draw inferences about similarities between the molecular pathology of NFTs in AD and DS.

Those brain areas containing the highest densities of NFTs in AD display the greatest proliferation of dendrites in a fashion considered typical of older neurones (McKee *et al.* 1991). Comparisons between EAD and AD in DS rarely consider dendritic connectivity. As part of a pioneering series of studies,

Takashima *et al.* (1989) identified dendritic atrophy in children with DS that continued into adulthood. They found a substantial reduction in dendritic branching, dendritic length and spine frequency and related these to the severe learning disabilities in DS and the precocious development of dementia in this condition. Although numbers that could be studied are small, it is now important to compare dendritic spine counts between individuals with DS who do and do not have Alzheimer neuropathological changes. Ferrer and Gullotta (1990) found few differences between two DS subjects (aged 35 and 36 years) without AD and two other DS subjects (aged 47 and 55 years) with AD. However, they considered that dendritic spine counts were lower in DS than in matched EAD individuals and were further reduced in AD in DS. The relevance of these changes in DS to EAD remains unclear. Recently, Flood and Coleman (1990) placed their own and related observations on dendritic spine count with ageing in the absence of dementia and in AD into the wider context of plasticity changes with ageing. They concluded that Alzheimer neuropathological changes were associated in location and extent with the degree of reduction of dendritic spine count. This apparent close relationship merits further study in DS, EAD and AD in DS.

The neurobiological processes that control and implement dendritic sprouting are relevant to the DS–AD association. Failure of the ageing DS brain or that of eukaryotic individuals to adapt successfully to neuronal loss by increased dendritic sprouting could lead to NFT formation, further neuronal loss by disruption of the neuronal cytoskeleton and so, hypothetically, begin a vicious cycle of events leading to AD. The gene coding for the neural calcium binding protein S100-β is firmly located on the long arm of chromosome 21 (Figure 12.1; Allore *et al.* 1988). Because S100-β has neurotrophic actions (Winningham-Major *et al.* 1989; Marshak 1991; Selinfreund *et al.* 1991), duplication of this gene might, by gene dosage effect, provide an alternative explanation of increased abnormal neuritic process growth in DS. S100-β is produced by astroglia and induces growth in cortical neurones. S100-β protein, its specific mRNA and neurotrophic activity are increased 10-20 times in the brain in AD (Marshak *et al.* 1992) and in AD and DS brains (Griffin *et al.* 1989). Speculatively, S100-β could be important in both neural dysgenesis and neurodegeneration in DS and only in neurodegeneration in AD. Some support for up-regulation of S100-β activity in AD that may mimic a gene dosage effect in DS was found by Griffin *et al.* (1989) who examined S100-β immunoreactivity in AD and DS. These authors reported that concentrations of interleukin 1 (IL-1), a component of the immune response that stimulates astrocyte proliferation and reactivity, were increased in the temporal lobe in AD and DS. Elevated IL-1 in AD may, therefore, increase S100-β activity and dendritic sprouting, whereas in DS a gene dosage effect might increase S100-β activity. This hypothesis was weakened by the finding of increased IL-1 concentrations in DS (Griffin *et al.* 1989). The potential complexity of relationships between IL-1, S100-β and APP synthesis and subsequent amyloid β-protein deposition is emphasized by the observation that the promoter of the APP gene contains an IL-1-responsive element (cited by Hardy 1992). In EAD, unspecified neurotoxic

	DS	AD	
	% of control values		
NCAM	100	100	(infers new neuronal membrane formation)
D3 protein	76	86	(infers destruction of neuronal structures)
Glut. syn.	100	100	
GFAP	260	230	
S100-β	690	180	
NCAM/D3 ratio	123	109	

Table 12.1. Neuronal proteins in Down syndrome and Alzheimer disease (adapted from Jorgenson *et al.* 1990)

events could stimulate IL-1 release which in turn could lead to increased APP and S100-β synthesis in a manner that occurs prematurely when AD occurs in DS.

Astrocyte proliferation and reactivity was further examined in DS and AD by Jorgenson *et al.* (1990) in a careful quantitative immunocytochemical study of proteins preferentially located in neurones [neural cell adhesion molecule (NCAM), D3 protein] and in glia [glutamine synthetase (glut. syn.), glial fibrillary acidic protein (GFAP) and S100-β]. Compared with controls, there was marked elevation of S100-β and GFAP in DS and AD (see Table 12.1). Jorgenson *et al.* inferred that their findings were broadly similar in DS and AD (with the exceptional S100-β increase explained by a gene dosage effect in DS) and related to plastic adaptive changes in frontal and/or temporal cortex in response to neuronal degeneration.

The selective vulnerability hypothesis

Nerve cells do not die indiscriminately in ageing in the absence of dementia (Terry *et al.* 1987) or in AD or DS (see Chapter 5). Those neurones that use acetylcholine as a neurotransmitter appear to be more vulnerable to AD degenerative processes than others. Specifically, cholinergic neurones with long axons, whose cell bodies lie mostly in the nucleus basalis and septum, are more susceptible than cholinergic neurones of, for example, the basal ganglia or autonomic nervous system. Similar questions regarding selective vulnerability are posed for Huntington disease, Parkinson disease and motor neurone disease (Calne *et al.* 1986). Phosphorus nuclear magnetic resonance analysis of brain tissue in AD demonstrates a marked increase (about 3-fold) in the ratio of glyceryl-3-phosphorylcholine (GPC) to glycerol phosphorylethamolamine (GPE) (Bárány *et al.* 1985). Decreased GPC phosphodiesterase activity in AD brain may lead to increased GPC content. In turn, increased GPC in AD may point, albeit indirectly, to reduced degradation of GPC to glyceryl phosphate and choline. Since availability of choline limits the synthesis of acetylcholine (Wurtman 1992), this may explain the selective vulnerability of cholinergic neurones in AD. Important new work cited by Wurtman (Nitsch *et al.* 1992) shows that similar increases in GPC concentrations are not observed in DS.

Should these observations be confirmed, they would be of considerable interest and prompt a careful review of the pathogenesis of selective neuronal loss in AD.

Although the above direct estimates of neuronal membrane composition are difficult to obtain *in vivo* in AD or DS, these measurements are fairly straightforward in peripheral cell models. In this respect, the platelet is widely utilized and abnormalities of platelet membrane fluidity (PMF), defined as the inverse of the time taken by labelled molecules to cross the platelet membrane, are reported in ageing in the absence of dementia and AD (Cohen and Zubenko 1985). Zubenko *et al.* (1988) proposed a genetic basis for the PMF abnormality in AD and argued this in some detail. When the ninetieth percentile was used to classify patients with AD, two subgroups emerged. One with increased PMF had earlier onset, a more rapid course and a more frequent family history of dementia. Zubenko and his colleagues in Pittsburgh have extended these studies by taking advantage of the DS–AD association. Earlier, Sinet (1982) had described increased activity of Cu-Zu superoxide dismutase (SOD-1) in DS and linked this via the free radical hypothesis of ageing to the DS–AD association, as the SOD-1 gene is on chromosome 21 (Figure 12.1). Zubenko *et al.* (1989) re-examined patients with AD in whom they had earlier detected increased or normal PMF and measured red blood cell SOD-1 activity, but found this not to differ between groups with increased or normal PMF. This suggests that SOD-1 activity is unchanged in AD and does not explain the altered PMF present in some individuals with AD.

The aluminium hypothesis

The neurotoxic effects of aluminium (Al) are well known. Unfortunate patients undergoing renal dialysis developed an Al encephalopathy before the risk was identified and removed (Candy *et al.* 1992). Also, retrospectively, industrial poisoning by Al dust was thought to cause a dementia-like syndrome (Rifat *et al.* 1990). Some preliminary data have led to the hypothesis that Al plays a causal role in AD. Stated briefly, it is postulated that exposure to Al precipitates a chain of abnormal molecular events that ends in deposition of amyloid β-protein, neuronal death and formation of NFTs. The Al hypothesis of AD remains attractive for a number of reasons. First, many instances of AD are apparently sporadic (SpAD) with no detectable increased family history of dementia and implication of environmental factors, including Al (Henderson 1988). Secondly, some monozygotic twin pairs remain discordant for AD even after lengthy follow-up of the unaffected twin, and thirdly, there are considerable discrepancies between affected monozygotic twins in age at onset of AD (Nee *et al.* 1987). Lastly, some, but not all, studies of Al in the brains of subjects with AD indicate that Al preferentially accumulates at those sites that selectively lose neurones in AD (Candy *et al.* 1988; Morris *et al.* 1989). However, Al is encountered widely and epidemiological evidence points to a weak link, if at all, between Al exposure and AD. A corollary of the Al hypothesis, therefore, would be that in eukaryotic individuals only a minority are vul-

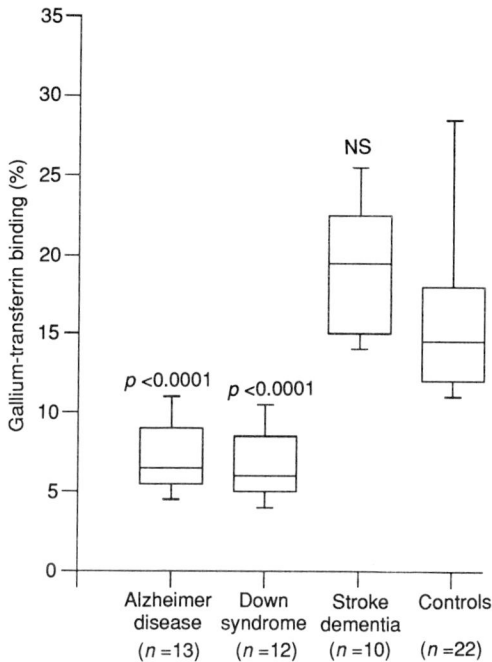

Figure 12.4. [67]Gallium binding to transferrin in plasma from subjects with Alzheimer disease, Down syndrome, stroke dementia and control subjects (adapted from Farrar *et al.* 1990)

nerable to the neurotoxic (i.e. amyloidogenic) effects of Al, but since almost all individuals with DS succumb to AD, the same vulnerability to Al will be present in virtually all of them.

Stated in these terms, the hypothesis has yet to be adequately tested. Farrar *et al.* (1990) examined the distribution of the radioisotope [67]gallium in plasma from subjects with AD, DS, stroke dementia, end-stage renal disease (on haemodialysis) and from control subjects. [67]Gallium was chosen because there are no biologically useful radionuclides of Al and because Al and gallium bind to the same sites on transferrin, a plasma glycoprotein. Figure 12.4 shows that gallium binding was similar in controls and stroke patients. Subjects with AD and those with DS showed significantly lower binding. Transferrin concentrations tended to be lower in subjects with AD and were significantly lower than controls in DS. Transferrin iron saturation was higher in both AD and DS subjects. These findings were interpreted by Farrar *et al.* (1990) as inferring a substantial change in transferrin function (termed the 'gallium-transferrin binding defect'), sufficient in nature and extent to explain susceptibility to the neurotoxic effects of Al in eukaryotic individuals who develop AD and, presumably, also in DS. The exact nature of the defect was not clear but could, it was postulated, be in the structure of the transferrin molecule or determined by the presence of some abnormal substance that modified the binding of gallium to transferrin.

```
┌─────────────────────────┐
│ Aluminium from GIT      │
└─────────────────────────┘
            │
            ▼
┌─────────────────────────┐
│ Aluminium binds to      │
│ serum transferrin       │
└─────────────────────────┘
            │
            ▼
┌─────────────────────────┐
│ Aluminium-transferrin   │
│ enters CNS              │
└─────────────────────────┘
            │
            ▼
┌─────────────────────────┐
│ Aluminium-transferrin   │
│ binds to transferrin    │
│ receptor site in CNS    │
└─────────────────────────┘
            │
            ▼
┌─────────────────────────┐
│ Alteration of APP proteolysis │
│ yielding amyloid-β-protein    │
└─────────────────────────┘
            │
            ▼
┌─────────────────────────┐
│ Extraneuronal amyloid-  │
│ β-protein deposition    │
└─────────────────────────┘
```

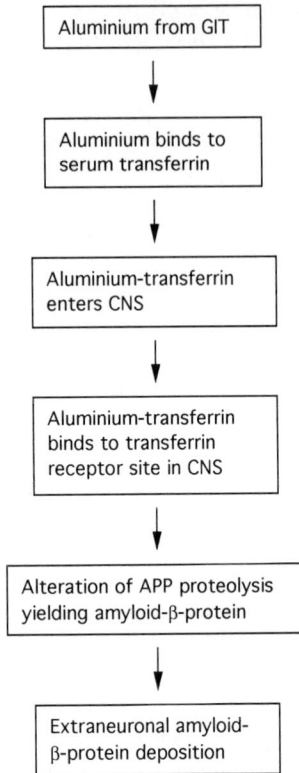

Figure 12.5 Hypothetical cascade of events induced by aluminium and leading to amyloid β-protein deposition

Transferrin-bound Al probably provides the only route of entry for Al into the brain and there is some evidence that Al is intimately associated with senile plaques and NFTs in AD (Candy *et al.* 1986, 1992). The ultimate fate of Al in the brains of subjects without AD is unknown but since the cortex, hippocampus and amygdala contain the highest brain densities of transferrin receptors (Pullen *et al.* 1990) and these are areas where neurones are selectively lost in AD, it is plausible that Al is involved in the pathogenesis of AD. Additionally, amyloid β-protein deposition is increased above expected levels in patients receiving chronic renal dialysis so that about 30 per cent of such patients show evidence of 'immature senile plaques'. In AD and DS, Al absorption from the gastrointestinal tract (GIT) is increased in comparison to age-matched controls and it also increases with age in non-demented individuals (Taylor *et al.* 1992; I. N. Ferrier, personal communication). Although such studies point to an important difference between those susceptible to, or suffering from, AD, nothing is known of the neural consequences of modestly increased Al absorption. At this time is seems reasonable to hypothesize that in DS, and in some instances of AD, there is genetically determined increased Al absorption. In a cumulative or summative fashion, increased intraneuronal Al availability may contribute to the neurodegenerative changes of ageing and accelerate amyloid β-protein deposition, possibly by direct interaction with APP. These events are shown schematically in Figure 12.5.

Familial clustering of Alzheimer disease and Down syndrome: the Heston microtubular hypothesis

Heston and Mastri (1977) in Minnesota systematically scrutinized the medical histories and their interview data obtained from the relatives of 33 patients with neuropathologically validated early-onset AD. In keeping with the already established association of AD with DS, they found an increased incidence of DS among the relatives of AD patients. At that time, the identity and chromosomal location of the APP gene were unknown and Heston and his colleagues sought an explanation within what was then known about genetic factors that could explain their observed association. Their interest understandably focused on the microtubular components of NFTs which had in the previous decade been carefully investigated using electromicroscopic techniques. Heston and Mastri (1977) hypothesized that a genetically determined defect in microtubular organization predisposed to both AD and DS. Noting that microtubular organization provided a critical step in chromosomal alignment during both meiosis and mitosis, they postulated that defective regulation of microtubular protein synthesis may cause chromosomal non-disjunction in DS and NFT formation in AD. The Heston hypothesis stimulated a great deal of research, although somewhat overshadowed of late by substantial recent progress made in molecular biological studies of amyloid β-protein. Nevertheless, recognition that mutation of the APP gene can so far explain but a tiny fraction of cases of AD has caused a renewed interest in Heston's original arguments.

The observations of Heston and Mastri (1977) of an increased incidence of DS in early-onset AD families has proved difficult to replicate, partly because of the considerable care taken by the Minnesota group to obtain comprehensive data from extensive pedigrees to ensure that their AD cases were neuropathologically validated and to supplement medical and public record data with findings from interview. These are each important requirements and no subsequent attempt to replicate the findings of Heston and Mastri has succeeded in meeting all of these conditions. Each is a relevant consideration because the excess of observed over expected numbers of individuals with DS in the original report was small but significant. Diagnostic misclassification in a later series would be likely to reduce the size of this effect. Studies of the incidence of 'dementia' among the relatives of individuals with DS are few and inconsistent. Berr *et al.* (1989) found no excess of AD among relatives of individuals with DS, while Yatham *et al.* (1988) did. In recognition of the importance of Heston's work and acknowledging the low statistical power of some subsequent studies (including the Edinburgh study by Whalley *et al.* 1982), van Duijn (1992) and co-workers in EURODEM re-analysed data from 10 studies of the familial association between AD and DS. They concluded that only three studies (Heston and Mastri 1977; Heyman *et al.* 1984; Broe *et al.* 1990) revealed a significant excess of DS in AD families. In seven studies (including Amaducci *et al.* 1986; Barclay *et al.* 1986; Chandra *et al.* 1987; Fitch *et al.* 1988) there was little support for the association, a finding attributable at least in part to limitations of the study designs. However, when van Duijn (1992) completed a meta-analysis of the pooled data from all 10 studies (Table 12.2), it was concluded that these data supported the Heston hypothesis. There is thus now a consensus that an excess of first-degree relatives of individuals with AD have DS. However, the analysis does not confine this excess to early-onset AD cases but suggests that increased DS incidence was especially marked among those persons with AD who had a family history of dementia. Explanations of familial AD–DS clustering include the possibility that the observation is an artefact caused by the confounding effects of delayed reproduction (attributable to social class) or due to genetically determined reduced reproductive fitness causing a delay in completion of the reproductive cycle in those families predisposed to familial AD–DS. Later reproduction in AD families might be associated with increased parental age in AD and this has been examined in several studies.

Study	Cases	Controls	OR	95% CI
Heyman *et al.* (1984)	7/45	4/91	3.5	1.2–5.7
Amaducci *et al.* (1986)	1/116	0/97	—	—
Chandra *et al.* (1987)	2/64	0/64	—	—
Hofman *et al.* (1989)	5/198	3/198	—	—
Broe *et al.* (1990)	5/165	0/165	—	—
Overall analysis	20/588	7/615	2.7	—
By age at onset				
Before 65 years	9/327	3/348	2.8	1.1–7.5
65 years or over	9/243	4/241	2.6	0.7–10.0

CI= confidence interval
OR= odds ratio adjusted for age, gender, number of siblings and education

Table 12.2. Down syndrome in first-degree relatives and the risk of Alzheimer disease: stratification by gender and age at onset (from van Duijn 1992)

Parental age in Alzheimer disease

Heston and Mastri (1977) reasoned that the same genetic defect predisposed to abnormal assemblies of NFTs as well as chromosomal non-disjunction during meiosis that caused trisomy 21 and thus DS. Consistent with such a hypothesis, maternal age would be expected to be increased in AD as it is in DS. Furthermore, within AD families, reproductive fitness would be reduced. Whalley *et al.* (1982) in Edinburgh examined these and related predictions in a neuropathologically validated series of early-onset AD cases. By direct examination of nineteenth century birth records of 138 parents of AD cases and 414 control parents, data were obtained to support increased parental age in AD, as predicted by the Heston hypothesis. Subsequently, these findings were widely argued. Three later studies (Cohen *et al.* 1982; Amaducci *et al.* 1986; Urakemi *et al.* 1989) supported increased parental age in AD though each study had serious weaknesses in case ascertainment and verification of parental ages. Two studies (English and Cohen 1985; Farrer *et al.* 1991) reported a significant association between AD and decreased paternal age. Commenting on discrepancies between these studies and her own, van Duijn (1992) emphasized the poor quality and incomplete nature of much of the informant-based data on parental age in AD. However, when van Duijn (1992) completed a meta-analysis of four studies (Amaducci *et al.* 1986; Hofman *et al.* 1989; Broe *et al.* 1990; Graves *et al.* 1990), an increased risk of AD was confined to those mothers aged 40 years or over. Farrer *et al.* (1991) found decreased paternal age in 83 late-onset subjects with SpAD.

The mechanism of the association between AD and possible increased parental age remains obscure. Decreased reproductive fitness in females with AD is not implicated nor is decreased birth order in persons with AD (Whalley *et al.* 1982). In these circumstances, the Heston hypothesis requires closer re-examination. Nordenson *et al.* (1980) and Matsuyama and Jarvik (1989) also suggested that a specific microtubular defect could increase NFT formation and increase DS in AD families and thought that age-related mechanisms associated with chromosomal changes in old age may be involved. A provocative and careful review of the Heston hypothesis was recently set out by Potter (1991) who has specifically extended the earlier argument to predict that non-disjunction of chromosome 21 causes both AD and DS.

Potter's chromosome 21 non-disjunction hypothesis

Potter (1991) argues that AD in DS is caused by trisomy 21. Non-disjunction occurs during meiosis (only rarely as 'mosaicism' during later mitosis) and affects all somatic cells. In FAD, genetic mutation occurs and affects centromere division, thus causing mitotic chromosomal non-disjunction and producing a variable degree of somatic chromosomal trisomy 21. Specifically, Potter argues that the genetic mutation in FAD is more likely to be at the centromere than in the APP gene. Environmental factors (which could include aluminium, Ganrot 1986) may induce chromosomal non-disjunction. Although neurones rarely divide, adjacent cerebrovascular endothelial and glial cells undergo varying degrees of chromosomal non-disjunction. However, it may not be necessary to implicate only trisomic central nervous system (CNS) cells as a potential source of abnormal proteins deposited in AD. Selkoe *et al.* (1988) considered that an extra-CNS source of amyloid β-protein could be a plausible alternative.

Demonstration of chromosomal non-disjunction in AD would support the Potter chromosomal non-disjunction hypothesis but may not be straightforward. Jacobs *et al.* (1961) reported increased frequencies of hyperdiploidy with increased age in phenotypically normal individuals. Later studies using chromosome banding techniques identified those chromosomes preferentially lost or gained with age. Galloway and Buckton (1978) established that, in women, hyperdiploidy in cultured lymphocytes is most often attributable to 47, XXX, whereas hypodiploidy (45, X) is more common in men as they age. Of relevance to the proposed association between chromosomal 21 non-disjunction in AD and DS, Nowinski *et al.* (1990) studied aneuploidy in the parents of live-born children with trisomy (usually of chromosome 21) and in the relatives of individuals with DS. As in earlier studies, hypodiploidy was inversely correlated with chromosomal length (i.e. the shorter chromosomes tended to be lost more frequently) but hyperdiploidy was not. Careful analysis by Nowinski *et al.* (1990) using regression techniques showed that once the effects of age were allowed for, sex and relatedness to an individual with trisomy were not associated with the preferential loss or gain of a chromosome. These findings are consistent with earlier chromosomal studies on AD and ageing in the absence of dementia (Finch 1990).

Conclusions

Molecular biological studies of amyloid β-protein deposition in DS provide a useful step in the elucidation of pathogenic mechanisms of AD. Recent research into amyloid fibril deposits in the nervous system has delineated separate conditions, each with its own characteristic clinical features and molecular genetic pathology. Mutations of amyloidogenic protein genes cause the familial amyloidopathies, including all those which feature cerebral amyloid deposition as the principal neuropathological finding. Acquired amyloidotic polyneuropathies are also caused by accumulation of abnormal proteins (as in myeloma). Amyloid β-protein deposition is seen in a closely related spectrum of conditions: AD, the Parkinson dementia complex of Guam, HCHWA-D and DS. The causes of amyloid β-protein deposition in these separate conditions include mutation of the APP gene and age-dependent changes in the processing of APP. In addition, amyloid deposition occurs in both the familial and transmissible spongiform encephalopathies caused by abnormalities of the prion protein gene. Considered together with the APP-derived cerebral amyloidoses, these disorders now comprise a well-understood range of age-related diseases associated with dementia and premature death.

Genetic linkage studies show that AD is not a single homogeneous disorder. Descriptions of specific FAD subpopulations (FAD patients of Volga German extraction or late-onset FAD) show absence of linkage to chromosome 21, whereas earlier studies had revealed linkage between FAD and chromosome 21 at a site distinct from the APP gene locus. These observations led Tanzi *et al.* (1989) to conclude that FAD comprises several diseases. In some instances, FAD is associated with (and probably caused by) genetic abnormalities determined by a gene or genes on chromosome 21. These include mutation of the APP gene and, speculatively, an abnormality of one or more genes affecting chromosome 21 division, perhaps located at or close to the centromere (Potter 1991). In other instances, FAD is caused by genetic defects on chromosomes other than 21.

AD in DS is presently understood to be a single disorder probably caused by an extra copy of the APP gene. So far there is no good evidence that a similar gene dosage effect can account for any other instance of AD, and it is unlikely that AD in DS can provide a single pathogenic model of each putative subtype of EAD. More promising, however, is the use made in several studies of comparisons between AD in DS and EAD. If amyloid β-protein deposition is considered to be the final common pathway shared by AD in DS and the many forms of EAD, then comparative studies of this type should reveal informative differences between EAD and AD in DS that could help disentangle interactions between environmental and non-chromosome 21 genetic factors. Studies of this kind have been undertaken by Blusztajn *et al.* (1990), Nitsch *et al.* (1992) and the Newcastle group (Taylor *et al.* 1992; I. N. Ferrier, personal communication). Observations have been made (Nitsch *et al.* 1992) that some pathological changes thought to be causally important in AD are not found in DS, suggesting that these changes are unlikely to be attributable to the non-specific effects of neurodegeneration. They may prove crucial to understanding the inception of SpAD, assist in the detection of environmental factors and unravel those pathogenic mechanisms that initiate amyloid β-protein deposition in AD.

Studies on the growth and degeneration of the brain in DS may also provide additional insights into the pathogenesis of AD. The brain of the young child with DS differs from that of an age-matched eukaryotic control. The former is reduced in size and has abnormal cytoarchitecture. There are alterations in cortical structure, reduced dendritic densities and synaptic connectivities and these differences become more marked as the child with DS ages. In brains from young children with DS there is increased expression of S100-β but not of APP, which, as in eukaryotic children, disappears in infancy. By the third and fourth decades of life, however, APP is increased in DS, neurodegenerative changes appear and many features of abnormal or precocious senescence are detectable. DS may yet provide additional data relevant to understanding AD. Specifically, genetic factors that determine abnormal or premature ageing in eukaryotic individuals predisposed to AD may first be detected in DS. In these respects DS may provide a prototype for study of the genetic determinants of senescence.

The possibility argued by Potter (1991) that DS is an informative model to understand putative chromosomal changes in AD is supported by some epidemiological evidence. Parental age is increased in AD and there is familial clustering of DS and AD. However, cytogenetic evidence is lacking and may prove elusive (Buckton *et al.* 1983). Fundamental research may further elucidate the pathogenesis of maternal chromosome 21 non-disjunction that causes most instances of trisomy 21 and may prove relevant to AD. Detection of a genetic mutation that predisposes to non-disjunction of chromosome 21 would, as predicted by Heston and Mastri (1977) and Potter (1991), provide a further step in the understanding of AD.

Acknowledgements

I thank Miss Naida Forbes for her assistance in the preparation of this manuscript.

References

Adam, J., Crow, T. J., Duchen, L.W., Scaravilli, F., and Spokes, E. (1982). Familial cerebral amyloidosis and spongiform encephalopathy. *Journal of Neurology, Neurosurgery and Psychiatry*, **45**, 37-45.

Allore, R., O'Hanlon, D., Price, R., Neilson, K., Willard, H. F., Cox, D. R., et al. (1988). Gene encoding the β-subunit of S100 protein is on chromosome 21: implications for Down's syndrome. *Science*, **239**, 1311-2.

Amaducci, L. A., Fratiglioni, L., Rocca, W. A., Fieschi, C., Livrea, P., Pedone, D., et al. (1986). Risk factors for clinically diagnosed Alzheimer's disease: a case-control study of an Italian population. *Neurology*, **36**, 922-31.

Arendt, T., Zvegintseva, H. G., and Leontovich, T. A. (1986). Dendritic changes in the basal nucleus of Meynert and in the diagonal band nucleus in Alzheimer's disease—a quantitative Golgi investigation. *Neuroscience*, **19**, 1265-78.

Ball, M. J. (1976). Neurofibrillary tangles and the pathogenesis of dementia: a quantitative study. *Neuropathology and Applied Neurobiology*, **2**, 395-410.

Bárány, M., Yen-Chung, C., Arus, C., Rustan, T., and Frey, W. (1985). Increased glycerol-3-phosphylcholine in post-mortem Alzheimer brain. *Lancet*, **1**, 517.

Barclay, L. L., Kheyfets, S., Zemcov, A., Blass, J. P., and McDowell, F. H. (1986). Risk factors in Alzheimer's disease. *Advances in Behavioural Biology*, **29**, 141-6.

Berr, C., Borghi, E., Rethoré, M. O., Lejeune, J., and Alperovitch, A. (1989). Absence of familial association between dementia of Alzheimer type and Down syndrome. *American Journal of Medical Genetics*, **33**, 545-50.

Blusztajn, J. K., Gonzalez-Coviella, I. L., Logue, M., Growdon, J. H., and Wurtman, R. J. (1990). Levels of phospholipid catabolic intermediates, glycerophosphocholine and glycerophosphoethanolamine are elevated in brains of Alzheimer's disease but not of Down's syndrome patients. *Brain Research*, **536**, 240-4.

Brion, J. P., van den Bosch de Aguilar, P., and Flament-Durand, J. (1985). Senile dementia of the Alzheimer type: morphological and immunocytochemical studies. In *Advances in applied neurological science: senile dementia of the Alzheimer type*, (ed. J. Traber and J. Gispens), pp. 164-74. Springer, Berlin.

Broe, G. A., Henderson, A. S., Creasey, H., McCusker, E., Korten, A. E., Jorm, A. F., et al. (1990). A case-control study of Alzheimer's disease in Australia. *Neurology*, **40**, 1698-707.

Buckton, K. E., Whalley, L. J., Lee, M., and Christie, J. E. (1983). Chromosomal changes in Alzheimer's presenile dementia. *Journal of Medical Genetics*, **20**, 46-51.

Buell, S. J., and Coleman, P. D. (1979). Dendritic growth in the aged human brain and failure of growth in senile dementia. *Science*, **206**, 854-6.

Buell, S. J., and Coleman, P. D. (1981). Quantitative evidence for selective dendritic growth in normal human aging but not in senile dementia. *Brain Research*, **214**, 23-41.

Caceres, A., Banker, G., and Binder, L. (1986). Immunocytochemical localization of tubulin and microtubule-associated protein tau during development of hippocampal neurons in culture. *Journal of Neuroscience*, **6**, 714-22.

Calne, D. B., Eisen, A., McGeer, E., and Spencer, P. (1986). Alzheimer's disease, Parkinson's disease and motor neurone disease: abiotropic interaction between ageing and environment? *Lancet*, **2**, 1067-70.

Candy, J. M., Oakley, A. E., Klinowski, J., Carpenter, T. A., Perry, R. H., Atack, J. R., et al., (1986). Aluminosilicates and senile plaque formation in Alzheimer's disease. *Lancet*, **1**, 354-7.

Candy, J. M., Oakley, A., and Gauvreau, D. (1988). Association of aluminium and silicon with neuropathological changes in the ageing brain. *Interdisciplinary Topics in Gerontology*, **25**, 140-55.

Candy, J. M., McArthur, F. K., Oakley, A. E., Taylor, G. A., Chen, C. P., Mountfort, S. A., et al. (1992). Aluminium accumulation in relation to senile plaque and neurofibrillary tangle formation in the brains of patients with renal failure. *Journal of the Neurological Sciences*, **107**, 210-8.

Chandra, V., Philipose, V., Bell, P. A., Lazaroff, A., and Schoenberg, B. S. (1987). Case-control study of late onset "probable Alzheimer's disease". *Neurology,* 37, 1295-300.

Cohen, B. M., and Zubenko, G. S. (1985). Aging and the biophysical properties of cell membranes. *Life Sciences,* 37, 1403-9.

Cohen, D., Eisdorfer, C., and Leverentz, J. (1982). Alzheimer's disease and maternal age. *Journal of the American Geriatrics Society,* 30, 656-9.

Cox, D. R., and Shimizu, N. (1991). Report of the committee on the genetic constitution of chromosome 21. *Cytogenetics and Cell Genetics,* 58, 800-26.

Daniels, M. P. (1986). Colchicine inhibition of nerve fiber formation in vitro. *Journal of Cell Biology,* 53, 164-76.

Davies, L., Wolska, B., Hilbich, C., Multhaup, G., Martins, R., Simms, G., *et al.* (1988). A4 amyloid protein deposition and the diagnosis of Alzheimer's disease: prevalence in aged brains determined by immunocytochemistry compared with conventional neuropathologic techniques. *Neurology,* 38, 1688-93.

Delacourte, A., and Defossez, A. (1986). Alzheimer's disease: tau proteins, the promoting factors of microtubule assembly are major components of paired helical filaments. *Journal of the Neurological Sciences,* 76, 173-86.

de Ruiter, J. P., and Uylings, H. B. M. (1987). Morphometric and dendritic analysis of fascia dentata granule cells in human aging and senile dementia. *Brain Research,* 402, 217-29.

Drubin, D. G., Feinstein, S. C., Shooter, E. M., and Kirschner, M. W. (1985). Nerve growth factor-induced neurite outgrowth in PC12 cells involves the coordinated induction of microtubule assembly and assembly-promoting factors. *Journal of Cell Biology,* 101, 1799-807.

English, D., and Cohen, D. (1985). A case-control study of maternal age in Alzheimer's disease. *Journal of the American Geriatrics Society,* 33, 167-9.

Esch, F. S., Keim, P. S., Beattie, E. C., Blacher, R. W., Culwell, A. R., Oltersdorf, T., *et al.* (1990). Cleavage of amyloid β peptide during constitutive processing of its precursor. *Science,* 248, 1122-4.

Farrar, G., Altmann, P., Welch, S., Wychrij, O., Ghose, B., Lejeune, J., *et al.* (1990). Defective gallium-transferrin binding in Alzheimer disease and Down syndrome: possible mechanism for accumulation of aluminium in brain. *Lancet,* 335, 747-50.

Farrar, L. A., Cupples, L. A., Connor, L., Wolf, P. A., and Growdon, J. H. (1991). Association of decreased paternal age and late-onset Alzheimer's disease: an example of genetic imprinting? *Archives of Neurology,* 48, 599-604.

Ferrer, I., and Gullotta, F. (1990). Down's syndrome and Alzheimer's disease: dendritic spine counts in the hippocampus. *Acta Neuropathologica,* 79, 680-5.

Ferrer, I., Aymami, A., Rovira, A., and Grau Veciana, J. M. (1983). Growth of abnormal neurites in atypical Alzheimer's disease. A study with a Golgi method. *Acta Neuropathologica,* 59, 167-70.

Finch, C. E. (1990). *Longevity, senescence and the genome.* University of Chicago Press.

Fitch, N., Becker, R., and Heller, A. (1988). The inheritance of Alzheimer's disease: a new interpretation. *Annals of Neurology,* 23, 14-9.

Flood, D. G., and Coleman, P. D. (1986). Failed compensatory dendritic growth as a pathophysiological process in Alzheimer's disease. *Canadian Journal of Neurological Sciences,* 13, 475-9.

Flood, D. G., and Coleman, P. D. (1990). Hippocampal plasticity in normal aging and decreased plasticity in Alzheimer's disease. *Progress in Brain Research,* 83, 435-43.

Flood, D. G., Buell, S. J., Defiore, C. H., Horwitz, G. J., and Coleman, P. D. (1985). Age-related dendritic growth in dentate gyrus of human brain is followed by regression in the 'oldest old'. *Brain Research,* 345, 366-8.

Gajdusek, D. C. (1977). Unconventional viruses and the origin and disappearance of kuru. *Science,* 197, 943-60.

Galloway, S. M., and Buckton, K. E. (1978). Aneuploidy and ageing: chromosome studies on a random sample of the population using G-banding. *Cytogenetics and Cell Genetics,* 20, 78-95.

Ganrot, P. O. (1986). Metabolism and possible health effects of aluminum. *Environmental Health Perspectives,* 65, 363-41.

Ghiso, J., Jensson, O., and Frangione, B. (1986). Amyloid fibrils in hereditary cerebral hemorrhage with amyloidosis of Icelandic type is a variant of gamma-trace basic protein (cystatin C). *Proceedings of the National Academy of Sciences of the United States of America*, 83, 2974-8.

Glenner, G. G., and Wong, C. W. (1984). Alzheimer's disease and Down's syndrome: sharing of a unique cerebrovascular amyloid fibril protein. *Biochemical and Biophysical Research Communications*, 122, 1131-5.

Goate, A., Chartier-Harlin, M. -C., Mullan, M., Brown, J., Crawford, F., Fidani, L., *et al.* (1991). Segregation of a missense mutation in the amyloid precursor protein gene with familial Alzheimer's disease. *Nature*, 349, 704-6.

Graves, A. B., White, E., Koepsell, T. D., Reifler, B. V., Van Belle, G., Larson, E. B., and Raskind, M. (1990). A case-control study of Alzheimer's disease. *Annals of Neurology*, 28, 766-74.

Griffin, W. S., Stanley, L. C., Ling, C., White, L., MacLeod, V., Perrot, L. J., *et al.* (1989). Brain interleukin I and S-100 immunoreactivity are elevated in Down syndrome and Alzheimer disease. *Proceedings of the National Academy of Sciences of the United States of America*, 86, 7611-5.

Grundke-Iqbal, I., Iqbal, K., Quinlan, M., Tung, Y. C., Zaidi, M. S., and Wisniewski, H. M. (1986). Microtubule-associated protein tau: a component of Alzheimer paired helical filaments. *Journal of Biological Chemistry*, 261, 6084-9.

Haass, C., Koo, E. H., Mellon, A., Hung, A. Y., and Selkoe, D. J. (1992). Targeting of cell surface β-amyloid precursor protein to lysosomes: alternative processing into amyloid bearing fragments. *Nature*, 357, 500-3.

Hardy, J. (1992). An 'anatomical cascade hypothesis' for Alzheimer's disease. *Trends in Neurosciences*, 15, 200-1.

Hardy, J., Mullan, M., Chartier-Harlin, M. -C., Brown, J., Goate, A., Rossor, M., *et al.* (1991). Molecular classification of Alzheimer's disease. *Lancet*, 337, 1342-3.

Henderson, A. S. (1988). The risk factors for Alzheimer's disease: a review and hypothesis. *Acta Psychiatrica Scandinavica*, 78, 267-75.

Heston, L. L., and Mastri, A. R. (1977). The genetics of Alzheimer's disease: associations with hematological malignancy and Down's syndrome. *Archives of General Psychiatry*, 34, 976-81.

Heyman, A., Wilkinson, W. E., Stafford, J. A., Helms, M. J., Sigmon, A. H., and Weinberg, T. (1984). Alzheimer's disease: a study of epidemiological aspects. *Annals of Neurology*, 15, 335-41.

Hiltunen, T., Kiuru, S., Hongell, V., Helio, T., Palo, J., and Peltunen, L. (1991). Finnish type of familial amyloidosis: cosegregation with Asp_{187}-Asn mutation of gelsolin in three large families. *American Journal of Human Genetics*, 49, 522-8.

Hirano, A., and Zimmerman, H. M. (1962). Alzheimer's neurofibrillary changes: a topographic study. *Archives of Neurology*, 7, 227-42.

Hofman, A., Schulte, W., Tanja, T. A., van Duijn, C. M., Haaxma, R., Lameris, A. J., *et al.* (1989). History of dementia and Parkinson's disease in 1st-degree relatives of patients with Alzheimer's disease. *Neurology*, 39, 1589-92.

Hooper, W. M., and Vogel, F. S. (1976). The limbic system in Alzheimer's disease: neuropathological investigations. *American Journal of Pathology*, 85, 1-13.

Jacobs, P. A., Court Brown, W. M., and Doll, R. (1961). Distribution of human chromosome counts in relation to age. *Nature*, 191, 1178-80.

Joachim, C. L., Morris, J. H., Selkoe, D. J., and Kosik, K. S. (1987). Tau epitopes are incorporated into a range of lesions in Alzheimer's disease. *Journal of Neuropathology and Experimental Neurology*, 46, 611-22.

Jorgenson, O. S., Brooksbank, B. W., and Balazs, R. (1990). Neuronal plasticity and astrocytic reaction in Down syndrome and Alzheimer disease. *Journal of the Neurological Sciences*, 98, 63-79.

Kosik, K. S., Joachim, C. L., and Selkoe, D. J. (1986). Microtubule associated protein tau is a major antigenic component of paired helical filaments in Alzheimer's disease. *Proceedings of the National Academy of Sciences of the United States of America*, 83, 4044-8.

Kosik, K. S., Rogers, J., and Kowall, N. W. (1987). Senile plaques are located between apical dendritic clusters. *Journal of Neuropathology and Experimental Neurology*, 46, 1-11.

Kowall, N. W., and Kosik, K. S. (1987). Axonal disruption and aberrant localization of tau protein characterize the neuropil pathology of Alzheimer's disease. *Annals of Neurology*, 22, 639-43.

Lejeune, J., Turpin, R., and Gautier, M. (1959). Le mongolisme, maladie chromosomique (trisomie). *Bulletin de l'Académie Nationale de Médecine (Paris)*, 143, 256-65.

Levy, E., Carman, M. D., Fernandez-Madrid, I. J., Power, M. D., Lieberburg, I., Van Duinen, S. G., *et al.* (1990). Mutation of the Alzheimer's disease amyloid gene in hereditary cerebral hemorrhage, Dutch type. *Science*, 248, 1124-6.

McKee, A. C., Kosik, K. S., and Kowall, N. W. (1991). Neuritic pathology and dementia in Alzheimer's disease. *Annals of Neurology*, 30, 156-65.

Mann, D. M. A., and Esiri, M. M. (1988). The site of the earliest lesions of Alzheimer's disease. *New England Journal of Medicine*, 318, 789-90.

Marin-Padilla, M. (1976). Pyramidal cell abnormalities in the motor cortex of a child with Down's syndrome. A Golgi study. *Journal of Comparative Neurology*, 167, 63-82.

Marshak, D. R. (1991). S100-β as a neurotrophic factor. *Progress in Brain Research*, 86, 169-81.

Marshak, D. R., Pesce, S. A., Stanley, L. C., and Griffin, W. S. (1992). Increased S100-β neurotrophic activity in Alzheimer's disease temporal lobe. *Neurobiology of Aging*, 13, 1-7.

Martin, G. M. (1977). Genetic syndromes in man with potential relevance to the pathobiology of ageing. In *Genetic effects on aging*, (ed. D. Bergsma, D.E. Harrison and N.W. Paul), pp. 5-39. Alan R. Liss, New York.

Matsuyama, S. S., and Jarvik, L. F. (1989). Hypothesis: microtubules, a key to Alzheimer's disease. *Proceedings of the National Academy of Sciences of the United States of America*, 86, 8152-6.

Mori, H., Kondo, J., and Ihara, Y. (1987). Ubiquitin is a component of paired helical filaments in Alzheimer's disease. *Science,* 235, 1641-4.

Morris, C. M., Candy, J. M., Oakley, A. E., Taylor, G. A., Mountfort, S., Bishop, H., *et al.* (1989). Comparison of regional distribution of transferrin receptors and aluminium in the forebrain of chronic renal dialysis patients. *Journal of the Neurological Sciences*, 94, 295-306.

Murphy, B. D., and Borisy, G. G. (1975). Association of high molecular weight proteins with microtubules and their role in microtubule assembly in vitro. *Proceedings of the National Academy of Sciences of the United States of America*, 72, 2696-700.

Murti, K. G., Smith, H. T., and Fried, V. A. (1988). Ubiquitin is a component of the microtubule network. *Proceedings of the National Academy of Sciences of the United States of America*, 85, 3019-23.

Naruse, S., Igarashi, S., Kobayashi, H., Aoki, K., Inuzuka, T., Kaneko, K., Shimizu, T., *et al.* (1991). Mis-sense mutation Val→Ile in exon 17 of the amyloid precursor protein gene in Japanese familial Alzheimer's disease. *Lancet*, 337, 978-9.

Nee, L. E., Eldridge, R., Sunderland, T., Thomas, C. B., Katz, D., Thompson, K. E., *et al.*, (1987). Dementia of the Alzheimer type: clinical and family study of 22 twin pairs. *Neurology*, 37, 359-63.

Nitsch, R. M., Blusztajn, J. K., Pittas, A. G., Slack, B. E., Growdon, J. H., and Wurtman, R. J. (1992). Evidence for a membrane defect in Alzheimer disease brain. *Proceedings of the National Academy of Sciences of the United States of America*, 89, 1671-5.

Nordenson, I., Adolfsson, R., Beckman, G., Bucht, G., and Winblad, B. (1980). Chromosomal abnormality in dementia of Alzheimer type. *Lancet*, 1, 481-2.

Nowinski, G. P., van Dyke, D. L., Tilley, B. C., Jacobsen, G., Babu, V. R., Worsham, M. J., *et al.* (1990). The frequency of aneuploidy in cultured lymphocytes is correlated with age and gender but not with reproductive history. *American Journal of Human Genetics*, 46, 1101-11.

Nukina, N., and Ihara, Y. (1986). One of the antigenic components of paired helical filaments is related to tau protein. *Journal of Biochemistry*, 99, 1541-4.

Opitz, J. M., and Gilbert-Barness, E. F. (1990). Reflections on the pathogenesis of Down syndrome. *American Journal of Medical Genetics*, Suppl. 7, 38-51.

Paula-Barbosa, M. M., Cardoso, R. M., Guimaraes, M. L., and Cruz, C. (1980). Dendritic degeneration and regrowth in the cerebral cortex of patients with Alzheimer's disease. *Journal of the Neurological Sciences*, 45, 129-34.

Perry, G., Friedman, R., Kang, D. H., Manetto, V., Autilio-Gambetti, L., and Gambetti, P. (1987a). Antibodies to the neuronal cytoskeleton are elicited by Alzheimer paired helical filament fractions. *Brain Research*, 420, 233-42.

Perry, G., Friedman, R., Shaw, G., and Chau, V. (1987b). Ubiquitin is detected in neurofibrillary tangles and senile plaque neurites of Alzheimer's disease brains. *Proceedings of the National Academy of Sciences of the United States of America*, **84**, 3033-6.

Potter, H. (1991). Review and hypothesis: Alzheimer disease and Down syndrome—chromosome 21 nondisjunction may underlie both disorders. *American Journal of Human Genetics*, **48**, 1192-200.

Probst, A., Basler, V., Bron, B., and Ulrich, J. (1983). Neuritic plaques in senile dementia of Alzheimer type: a Golgi analysis in the hippocampal region. *Brain Research*, **268**, 249-54.

Prusiner, S. B. (1991). Molecular biology of prion diseases. *Science*, **252**, 1515-22.

Pullen, R. G., Candy, J. M., Morris, C. M., Taylor, G., Keith, A. B., and Edwardson, J. A. (1990). Gallium-67 as a potential marker for aluminium transport in rat brain: implications for Alzheimer's disease. *Journal of Neurochemistry*, **55**, 251-9.

Rifat, S. L., Eastwood, M. R., Crapper-McLachlan, D. R., and Coney, P. N. (1990). Effect of exposure of miners to aluminum powder. *Lancet*, **336**, 1162-5.

Roberts, G. W., Lofthouse, R., Brown, R., Crow, T. J., Barry, R. A., and Prusiner, S. B. (1986). Prion protein immunoreactivity in human transmissible dementias. *New England Journal of Medicine*, **315**, 1231-3.

Ropper, A. H., and Williams, R. S. (1980). Relationship between plaques, tangles and dementia in Down's syndrome. *Neurology*, **30**, 639-44.

Rumble, B., Retallack, R., Hilbich, C., Simms, G., Multhaup, G., Martins, R., *et al.* (1989). Amyloid A4 protein and its precursors in Down's syndrome and Alzheimer's disease. *New England Journal of Medicine*, **320**, 1446-52.

St George-Hyslop, P. H., Tanzi, R., Polinsky, R. J., Haines, J. L., Nee, L., Watkins, P. C., *et al.* (1987). The genetic defect causing familial Alzheimer's disease maps on chromosome 21. *Science*, **235**, 885-9.

Scheibel, A. B., and Tomiyasu, U. (1978). Dendritic sprouting in Alzheimer presenile dementia. *Experimental Neurology*, **60**, 1-8.

Scheibel, M. E., Londsay, R. D., Tomiyasu, U., and Scheibel, A. B. (1976). Progressive dendritic changes in the aging human limbic system. *Experimental Neurology*, **53**, 420-30.

Selinfreund, R. H., Barger, S. W., Pledger, W. J., and Van Eldik, L. J. (1991). Neurotrophic protein S100 β stimulates glial cell proliferation. *Proceedings of the National Academy of Sciences of the United States of America*, **88**, 3554-8.

Selkoe, D. J., Bell, D. S., Podlisny, M. B., Price, D. L., and Cork, L. C. (1987). Conservation of brain amyloid proteins in aged mammals and humans with Alzheimer's disease. *Science*, **235**, 873-7.

Selkoe, D. J., Podlisny, M. B., Joachim, C. L., Vickers, E. A., Lee, G., Fritz, L. C., and Oltersdorf, T. (1988). β-amyloid precursor protein of Alzheimer disease occurs as 110- to 135-kilodalton membrane-associated proteins in neural and nonneural tissues. *Proceedings of the National Academy of Sciences of the United States of America*, **85**, 7341-5.

Shapiro, B. L. (1983). Down syndrome—a disruption of homeostasis. *American Journal of Medical Genetics*, **14**, 241-69.

Shaw, G., and Chau, V. (1988). Ubiquitin and microtubule associated protein tau immunoreactivity each define distinct structures with differing distributions and solubility properties in Alzheimer brain. *Proceedings of the National Academy of Sciences of the United States of America*, **85**, 2854-8.

Sinet, P. M. (1982). Metabolism of oxygen derivatives in Down's syndrome. *Annals of the New York Academy of Sciences*, **396**, 83-94.

Smith, R. P., Higuchi, D. A., and Broze, G. J. (1990). Platelet coagulation factor XIa-inhibitor, a form of Alzheimer precursor protein. *Science*, **248**, 1126-8.

Takashima, S., Ieshima, A., Nakamura, H., and Becker, L. E. (1989). Dendrites, dementia and the Down syndrome. *Brain and Development*, **11**, 131-3.

Tanzi, R. E., Gusella, J. F., Watkins, P. C., Bruns, G. A. P., St George-Hyslop, P., Van Keulem, M. L., *et al.* (1987). Amyloid β-protein gene: cDNA, mRNA distribution, and genetic linkage near the Alzheimer locus. *Science*, **234**, 880-4.

Tanzi, R. E., St George-Hyslop, P. H., and Gusella, J. F. (1989). Molecular genetic approaches to Alzheimer's disease. *Trends in Neurosciences*, 12, 152-8.

Taylor, G. A., Ferrier, I. N., McLouglin, I. J., Fairbairn, A. F., McKeith, I. G., Lett, D., and Edwardson, J. A. (1992). Gastrointestinal absorption of aluminium in Alzheimer's disease: response to aluminium citrate. *Age and Aging*, 21, 81-90.

Terry, R. D., De Terasa, R., and Hansen, L. (1987). Neocortical cell counts in normal human adult aging. *Annals of Neurology*, 21, 530-9.

Tomlinson, B. E. (1992). Ageing and the dementias. In *Greenfield's neuropathology*, (ed. L. W. Duchen and J. H. Adams), (5th edn), pp. 1284-410. Oxford University Press, London.

Tomlinson, B. E., Blessed, G., and Roth, M. (1970). Observations on the brains of demented old people. *Journal of the Neurological Sciences*, 11, 205-42.

Urakemi, K., Adachi, Y., and Takahashi, K. (1989). A community-based study of parental age at the birth of patients with dementia of the Alzheimer type. *Archives of Neurology*, 46, 38-9.

van Duijn, C. M. (1992). *Risk factors for Alzheimer's disease: a genetic-epidemiologic study.* CIP-Data Koniklijke Bibliotheek, Den Haag.

Whalley, L. J. (1982). The dementia of Down's syndrome and its relevance to aetiological studies of Alzheimer's disease. *Annals of the New York Academy of Sciences*, 396, 39-53.

Whalley, L. J., Carothers, S. D., Collyer, S., De Mey, R., and Frackiewicz, A. (1982). A study of familial factors in Alzheimer's disease. *British Journal of Psychiatry*, 140, 249-56.

Whalley, L. J., Wright, A. F., and St Clair, D. M. (1985). Genetic factors in Down's syndrome and their possible role in the pathogenesis of Alzheimer's disease. In *Modern approaches to the dementias. Part 1: Etiology and pathophysiology,* (ed. F. C. Rose), pp. 18-31. Karger, Basel.

Winningham-Major, F. Staecker, J. L., Barger, S. W., Coats, S., and Van Eldik, L. J. (1989). Neurite extension and neuronal survival activities of recombinant S100-β proteins that differ in the content and position of cysteine residues. *Journal of Cell Biology*, 109, 3063-71.

Wood, J. G., Mirra, S. S., Pollock, N. J., and Binder, L. I. (1986). Neurofibrillary tangles of Alzheimer's disease share antigenic determinants with the axonal microtubule-associated protein tau. *Proceedings of the National Academy of Sciences of the United States of America*, 83, 4040-3.

Wurtman, R. J. (1992). Choline metabolism as a basis for the selective vulnerability of cholinergic neurons. *Trends in Neurosciences*, 15, 117-22.

Yankner, B. A., Duffy, L. K., and Kirschner, D. A. (1990). Neurotrophic and neurotoxic effects of amyloid β-protein reversal by tachykinin neuropeptides. *Science*, 250, 279-82.

Yatham, L. N., McHale, P. A., and Kinsella, A. (1988). Down's syndrome and its association with Alzheimer's disease. *Acta Psychiatrica Scandinavica*, 77, 38-41.

Yen, S. -H., Dickson, D. W., Crowe, A., Butler, M., and Shelanski, M. L. (1987). Alzheimer's neurofibrillary tangles contain unique epitopes and epitopes in common with the heat-stable microtubule associated proteins, tau and MAP2. *American Journal of Pathology*, 126, 81-91.

Zubenko, G. S., Huff, F. J., Beyer, J., Auerback, J., and Teply, I. (1988). Familial risk of dementia associated with a biologic subtype of Alzheimer's disease. *Archives of General Psychiatry*, 45, 889-93.

Zubenko, G. S., Moossy, J., Martinez, A. J., Rao, G.R., Kopp, U., and Hanin, I. (1989). A brain regional analysis of morphologic and cholinergic abnormalities in Alzheimer's disease. *Archives of Neurology*, 46, 634-8.

13

Possible causal factors in the development of Alzheimer disease in persons with Down syndrome

D. M. Holtzman and C. J. Epstein

Summary

In humans, the presence of an extra copy of chromosome 21 inevitably leads to Alzheimer disease neuropathology and it is presumed that the increased gene dose of specific HSA-21 gene products is directly or indirectly involved in the development of this condition. Recent data regarding the HSA-21 encoded gene, *APP*, suggest that it may be an interesting candidate for playing a direct role in Alzheimer disease pathogenesis. The roles of other HSA-21 genes, such as *SOD-1* and *S100β*, are under intense investigation. An 'obligate' Down syndrome region of HSA-21 is being narrowed down and further definition of this region may provide detailed information about which genes, when present in three copies, are necessary for the Alzheimer disease phenotype. Studies on the developing nervous system in Down syndrome have suggested that early central nervous system abnormalities may in some way predispose the brain to develop Alzheimer disease neuropathology. Current work with mice transgenic for HSA-21 genes, as well as with trisomy 16 mice (at present the best animal model of Down syndrome), has led to the development of interesting model systems to test specific hypotheses. Further work with these and similar models may provide a unique avenue to pursue both the pathogenesis and treatment of Alzheimer disease.

Introduction

Although the pathological relationship between Down syndrome (DS) and Alzheimer disease (AD) has been recognized for over 60 years (Struwe 1929), the mechanism(s) by which trisomy for human chromosome 21 leads to AD remains to be elucidated. As noted in other chapters in this book, there are numerous similarities between adults with DS and individuals with AD (summarized in Table 13.1). By most criteria, the brain of persons with DS over the age of 40 is indistinguishable from that seen in AD (Wisniewski *et al.* 1985; Holtzman and Mobley 1991). Important changes in the brain include selective

Neuropathology
Similar content, appearance and distribution of β/A4-containing plaques and neurofibrillary tangles. Similar anatomy of neuronal loss.

Neurochemistry
Decreases in cholinergic, noradrenergic, dopaminergic and serotonergic neurotransmitter systems.

Neuropsychology
Dementia in a majority of patients by the sixth decade.

For references, see text.

Table 13.1. Similarities between adults with Down syndrome and individuals with Alzheimer disease

vulnerability of large neurones, loss of specific neurochemical markers and the accumulation of abnormal proteinaceous deposits in extracellular and intracellular spaces. Ultimately, loss of memory and of other intellectual function in AD results from the selective vulnerability of particular neuronal systems. Understanding the relationship, if any, between the neuropathological markers of AD and the pathogenesis of the disease is a focus of current AD research. In DS, AD neuropathology must result in some way from the presence of an extra copy of a specific gene(s) on human chromosome 21 (HSA-21). Thus, an exploration of the effects of an extra copy of HSA-21 genes on the nervous system and of the mechanisms leading to these effects may offer insights into the pathogenesis of AD. In this chapter, we consider two general hypotheses regarding possible causal factors in the development of AD in persons with DS. Recent data regarding gene products on HSA-21, as well as the molecular definition of the DS region of HSA-21, are presented. Lastly, current and future approaches toward understanding the DS and AD phenotype with animal models of DS are discussed.

What is the pathogenic relationship between Down syndrome and the neuropathology of Alzheimer disease?

This question is most easily divided and discussed as two additional subquestions, each of which is considered in turn.

Do genetic abnormalities in Down syndrome directly cause Alzheimer disease?

Central to an understanding of the mechanism by which aneuploidy (in this case trisomy 21) leads to particular phenotypes is the question of how extra copies of normal genes lead to deleterious consequences. The effects of chromosomal imbalance can be divided into two general categories, primary and secondary. Based on the concept of gene dosage effects (Figure 13.1), one would predict that the synthesis and hence the concentration of a gene product is

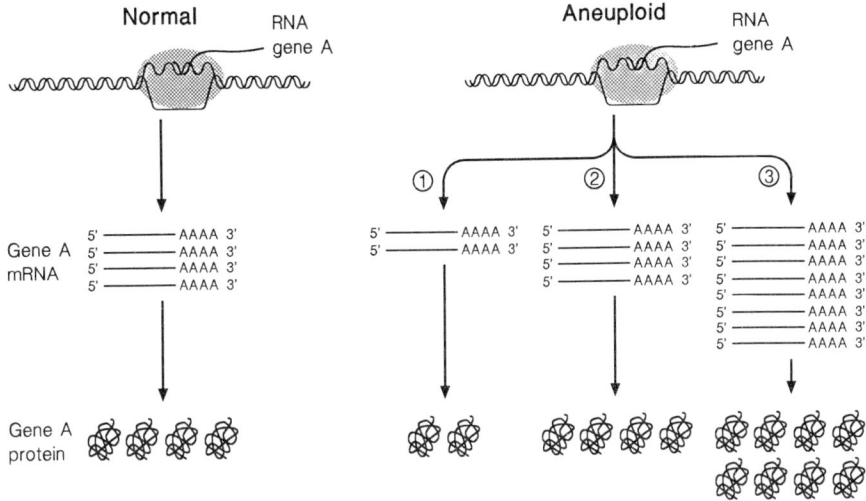

Figure 13.1. There are two general possibilities as to how gene expression is affected by aneuploidy. The first possibility, represented by pathway 2, illustrates the concept of gene dosage effects in which each gene is expressed equally per chromosomal locus, whether in an aneuploid or diploid cell. The second possibility is shown in pathways 1 and 3, in which strict gene dosage effects are not obeyed. Reprinted by permission of the Academic Press from Holtzman and Epstein (1992)

directly proportional to the number of genes coding for its synthesis. In DS, this concept would predict that the three copies of each HSA-21 specific locus is operative and equally expressed. As a result, each HSA-21 gene product in a DS tissue or group of cells would be present at a level 1.5 times that present in appropriate diploid controls. The other general possibility is that the gene dosage effect concept is not operative for all loci in aneuploid cells (Figure 13.1). For example, gene expression may be altered as a secondary effect of a primary gene dosage effect. With a few notable exceptions, a large body of data supports the concept of gene dosage effects in DS and other aneuploid states (Epstein 1986). In either case, it may be that there is a direct traceable relationship between increased gene dose for a specific HSA-21 gene(s) and the neuropathology of AD. Though several HSA-21 genes have now been cloned, three particular genes have received significant attention in regard to the pathogenesis of AD. These include *APP*, *SOD-1* and, more recently, *S100β*. In addition, further analysis of persons with triplication of only a small part of HSA-21 may shed new light on other genes which may be important to the AD phenotype in DS.

APP

One of the hallmarks of AD neuropathology is the extracellular deposition within the brain of a 39–42 amino acid amyloidogenic peptide, termed 'β/A4', within diffuse, neuritic and cerebrovascular plaques. Immunocytochemical studies of brain tissue of persons with DS (Rumble *et al.* 1989) showed that the deposition of β/A4 begins approximately 50 years earlier than in normal ageing populations (Davies *et al.* 1988). Whether β/A4 serves solely as an AD marker or is also involved in the pathogenesis of neurodegeneration remains unclear. *In vitro* and *in vivo* studies (Yankner and Mesulam 1991) suggest that β/A4 may be toxic to neurones. β/A4 is derived from a much larger precursor protein, the amyloid precursor protein (APP) (Kang *et al.* 1987; Holtzman and Mobley 1991). *APP* is expressed in many tissues; however, highest expression occurs in the nervous system where it is expressed predominantly by neurones (Neve *et al.* 1988). APP is an integral membrane protein with three predominant isoforms (APP_{695}, APP_{751} and APP_{770}). It is normally cleaved at a site close to membrane within the β/A4 region (Esch *et al.* 1990; Sisodia *et al.* 1990), a cleavage that would not allow for the eventual deposition of β/A4. Recent studies in PC12 cells and in the human brain (Buxbaum *et al.* 1990; Estes *et al.* 1991), however, suggest that a small fraction of APP may be cleaved N-terminal to the β/A4 region, leading to a fragment that could be further degraded into β/A4.

APP_{751} and APP_{770} contain a protease inhibitor domain and these forms of the molecule are also known as protease nexin II (Oltersdorf *et al.* 1989). Protease inhibitors and proteases have been proposed to be involved in many cellular functions including neurite outgrowth and synaptic transmission (Monard 1988; Rupp *et al.* 1991). In the central nervous system, the presence of APP isoforms with a protease inhibitor domain may be involved in controlling the activity of degratory enzymes during development, regeneration and degeneration (Müller-Hill and Beyreuther 1989). It is thus conceivable that overexpression of APP_{695} as well as the protease inhibitor-containing isoforms of APP (APP_{751} and APP_{770}) as occurs in the DS brain (Neve *et al.* 1988), could alter any one of these processes as well as divert more APP into a pathway leading to β/A4 production.

It has been hypothesized that overexpression of *APP* in DS leads to the premature accumulation of β/A4. Supporting this hypothesis is a recent study in which human APP_{751} (protease inhibitor-containing form of APP) was expressed by neurones in transgenic mice (Quon *et al.* 1991). These mice demonstrated one of the earliest lesions in the adult DS brain, β/A4 deposition in the form of diffuse plaques in the cortex and hippocampus. Whether overexpression of human *APP* in transgenic mice in some way leads to abnormal processing or merely diverts more APP into a normal, though less frequently utilized, pathway remains to be clarified. If other AD neuropathology is indeed caused by β/A4 deposition, one would predict that with time APP_{751} transgenic animals might develop other signs of neurodegeneration such as neurofibrillary pathology, neuronal atrophy, synapse loss and neurochemical alterations. Interestingly, point mutations in *APP* near the β/A4 domain have been recently linked to AD in a small number of families with familial AD (Goate *et al.* 1991; Murrell *et al.* 1991). In these families, it does not appear that the mutations result in overexpression of *APP*, though their presence supports the view that altered processing or stability of APP may be important in AD pathogenesis.

SOD-1

It was postulated several years ago by Sinet and colleagues (Sinet *et al.* 1979; Sinet 1982) that the 1.5-fold elevated SOD-1 (Cu-Zn superoxide dismutase) activity present in the cells of individuals with DS could be injurious to the brain. Although the function of SOD-1 is presumed to be the conversion of the superoxide anion, O_2^-, to H_2O_2 and O_2, it was proposed that the increase in H_2O_2 concentration could lead to lipid peroxidation and cellular dysfunction, either directly or after interaction with O_2^-, to form the highly reactive hydroxyl (OH) radical. Supporting this hypothesis is that PC12 cells transfected with *SOD-1* have been shown to have impaired uptake of the neurotransmitters, dopamine and norepinephrine (Elroy-Stein and Groner 1988). Recently, transgenic mice that overexpress the human *SOD-1* gene have been generated (Epstein *et al.* 1987). Interestingly, these mice share similar pathological abnormalities in the tongue neuromuscular junction as are found in DS and aged mice (Avraham *et al.* 1988). If, however, increased SOD-1 is injurious to the brain, it would be expected that this would manifest itself as an increase in lipid peroxidation that might be worsened by an oxidative stress. In contrast, the brain of *SOD-1* transgenic mice appears to actually be more resistant to ischaemic changes in both a cold-trauma (Chan *et al.* 1991) and vascular occlusion model (Kinouchi *et al.* 1991). Furthermore, there is no evidence of AD-like neuropathology in the brains of these mice. At present, a relationship between increased SOD-1 activity and the pathogenesis of AD in DS remains speculative. This obviously does not rule out an interaction of increased SOD-1 and other alterations in the DS brain.

S100β

Although the exogenous administration of neurotrophic factors, such as nerve growth factor (NGF), may prove to be beneficial for persons with AD (Mobley 1989; Holtzman and Mobley 1991), there are a variety of other growth factors present in the brain that can directly act on both neurones and glia. It has been postulated that in certain pathological conditions, alterations in specific trophic factors may exacerbate rather than improve disease-related processes.

In AD, there is glial proliferation in relation to amyloid-containing plaques (Mancardi *et al.* 1983; Mandybur and Chuirazzi 1990). It is unclear, however, whether gliosis precedes amyloid deposition or whether it is a secondary reaction in response to neuronal dysfunction. Interestingly, in the young DS brain, there is evidence of glial abnormalities as identified by increased size and number of glia immunoreactive for IL-1, GFAP and S100β (Griffin *et al.* 1989). These changes were noted soon after birth and thus may occur before β/A4 deposition is first detected (Rumble *et al.* 1989).

S100β is a protein produced by astrocytes. It has two Ca^{2+} binding regions known as EF hands, characteristics that define a distinct family of S100-related proteins (Marshak 1990). S100β has some *in vitro* activities similar to those found for other factors with neurotrophic activity, such as FGF. Certain neurones in culture respond to S100β by sending out neurites, while antibodies to S100β block this effect (Marshak and Peña 1992). In addition, S100β stimulates the proliferation of glial cell cultures, including primary astrocytes and C6 glioma cells (Selinfreund *et al.* 1991). Recently, elevated levels of S100β have been found in the temporal lobe in AD as measured by several criteria (Marshak *et al.* 1992). To address the significance of increased *S100β* gene dose in isolation, Hilt and co-workers (Yarowsky *et al.* 1991) have constructed transgenic mice that overexpress *S100β*. Increased numbers of glia were noted in three brain regions. Further work with such a model may give new insight into the specific abnormalities resulting from increased gene dose for *S100β* and its relevance to AD neuropathology.

Role of other HSA-21 genes

Although most cases of DS are caused by the presence of an extra copy of HSA-21 in its entirety, molecular analysis of a subset of individuals with many of the phenotypic features of DS reveals that triplication of only a part of HSA-21 is required to produce aspects of this phenotype. Approximately one to five per cent of persons with DS have a partial unbalanced translocation of the long arm of HSA-21, band q22 (Epstein 1986). This band has been called the 'DS region', as defined by the presence of a subset of the major phenotypic features of the syndrome including mental retardation, congenital heart disease, the characteristic facial appearance, hand anomalies and dermatoglyphic changes. Several groups have described rare individuals who have an apparently normal karyotype but harbour a small amount of extra material from HSA-21. Recent molecular analysis of a Japanese family with 'partial' trisomy 21 has been particularly revealing (Korenberg *et al.* 1990). Phenotypic features of affected family members included characteristic DS facies, endocardial cushion defects, hand anomalies and mental retardation. Interestingly, quantitative Southern analysis revealed that the triplicated region excluded sequences mapping in 21q21 (*APP,D21S46*) and in 21q22.1 (*SOD-1, D21S47* and *SF57*), but included DNA sequences mapping in band 21q22.3. In another report, an individual with mental retardation and the DS facial phenotype had a triplication of the region from 21q22.1 to 21qter while *APP* and *SOD-1* were present in only two copies (McCormick *et al.* 1989). At this time, it is unclear as to whether any such persons have developed or will develop AD neuropathology. It may be that increased gene dose for *APP* or another gene linked to AD in the more proximal region of HSA-21 is required. If, however, these persons develop AD, increased gene dose for genes in this 'obligate' region may be candidate molecules, which, when overexpressed, play a direct role in AD pathogenesis.

Are Alzheimer disease changes in Down syndrome an indirect effect of the chromosome imbalance in trisomy 21?

The distinction between the direct and indirect effects of the trisomic state is based on whether or not it is possible to trace an effect directly back to imbalance of a known HSA-21 gene(s). If not, the effect can be considered indirect, keeping in mind the fact that even an indirect effect is ultimately a consequence of the trisomic state and attributable to the imbalance of a specific locus or loci. In regard to indirect effects, it is interesting to look at the phenotypic fea-

Frequency or amount	Features
10- to 18-fold	Increased frequency of leukaemia (also leukaemoid responses)
3- to 15-fold	Enhanced cellular sensitivity to interferon
5- to 9-fold	Exaggerated fibroblast cAMP response to β-adrenergic agonists
3- to 5-fold	Increased adhesiveness of fetal lung and heart fibroblasts
1.4- to 6-fold	Increased cellular sensitivity to viral transformation
≤ 2-fold	Increased cellular sensitivity to radiation
0.4-fold	Decreased platelet serotonin
	Altered immune response, particularly of T-lymphocyte system
	Growth retardation

[a]From Epstein (1987).

Table 13.2. Certain phenotypic features of trisomy 21[a]

tures of DS (Table 13.2). The features in this list, which does not include the major congenital cardiac and gastrointestinal anomalies and the many minor dysmorphic features that characterize DS, point to a variety of abnormalities in cellular responses, membrane function and susceptibility to external infectious agents. It is unknown whether any of these changes are relevant to the pathogenesis of AD; it is also not clear how any of these cellular and functional abnormalities can be implicated on the basis of specific HSA-21 gene products.

One possibility of particular interest in relation to AD is that the nervous system of individuals with DS is intrinsically defective. Somehow, this may predispose the brain to degenerative changes such as those that occur in AD and to a mild degree in normal ageing. In DS, nervous system abnormalities are noted soon after birth in the form of neonatal hypotonia and mental retardation. In addition to a decrease in brain weight and in the size of the cerebellum, there are other more diffuse changes that have been noted at both the morphological and physiological level. Some studies have pointed to abnormalities in the number and structure of dendritic spines (Marin-Padilla 1976; Suetsugu and Mehraein 1980). Detailed studies of neuronal and synaptic density have been carried out by Wisniewski et al. (1986). They studied 73 DS brains from individuals between birth and 14 years of age. Neuronal densities were decreased by 10–50 per cent in several areas, including the parahippocampal and visual cortices. Although synaptic density was not affected at all ages, the mean surface area per synaptic contact zone was reduced by 20–35 per cent over the entire period of development. This is of interest in regard to AD. The work of Terry et al. (1991) has recently shown that an important abnormality in the AD brain that correlates well with dementia is synapse loss. In addition, there appears to be abnormal neuritic sprouting (marked by GAP-43 immunoreactivity) concomitant with synaptic loss (marked by synaptophysin immunoreactivity) (Masliah et al. 1991). It is possible in DS that pre-existing synaptic abnormalities predispose to AD-like synaptic loss associated with other AD neuropathology. For example, based on ultrastructural studies (Probst et al.

1991), β/A4 deposition has been hypothesized to take place in the vicinity of synapses (Beyreuther *et al.* 1992).

A prominent feature in AD brain is that specific brain regions appear to be selectively vulnerable. Selectively vulnerable regions include certain brainstem nuclei (locus coeruleus, raphe), basal forebrain, amygdala, hippocampus, entorhinal cortex and association areas of neocortex (Price 1986). Within the basal forebrain, cholinergic neurones are particularly vulnerable. In DS, two studies have found abnormalities in this system before the occurrence of other significant AD neuropathology. McGeer *et al.* (1985) found a 50 per cent reduction in the number of neurones in the basal forebrain of a 5.5-month-old infant with DS. Casanova *et al.* (1985) found a decrease in the number of cells in the same brain region in young adults with DS. These changes may reflect abnormalities in the genesis of these neurones or in their growth in an underdeveloped brain. In either case, it may predispose these magnocellular neurones to undergo atrophy, lose neurotransmitter markers, develop neurofibrillary tangles and participate in the formation of senile plaques (Price 1986; Vogels *et al.* 1990). Similar changes in a mouse model of DS that may allow study of this and other neurodegenerative phenomena are discussed in the next section on animal models. Despite these observations, it remains unclear as to how deficiency of neuronal number itself would predispose the brain to later losses in neurotransmitters and other changes. It is possible that other abnormalities such as synaptic ones as well as abnormal neuronal electrical properties, previously demonstrated in cultured fetal DS dorsal root ganglion cells (Scott *et al.* 1982), could further compromise neuronal populations that are already decreased in number.

It has been argued that AD in DS may merely reflect a consequence of premature ageing. There is currently no good evidence to support this assertion. AD accounts for 60–70 per cent of cases of late life dementia and its prevalence increases logarithmically with age (Beyreuther *et al.* 1992). It is, however, not a normal accompaniment of ageing. Although there is evidence of premature ageing in some DS organ systems (Sylvester 1974), this is not true for all systems (for more thorough discussion see Epstein 1983). Given recent data that mutations in a specific gene (*APP*) are associated with some cases of AD, it is unlikely that other causes of AD will merely reflect premature ageing as opposed to a true 'disease'.

Another point relating to AD and DS that deserves mention is the observation that the frequency of DS is increased in relatives of persons with AD (Heston *et al.* 1981). This finding led to the suggestion that there may be a 'unitary genetic aetiology' of AD. It has been hypothesized by Potter (1992) that in the brains of persons with sporadic AD, trisomy 21 cells could accumulate over time by unequal chromosome segregation during mitosis. This then could lead either directly or indirectly, as mentioned above, to AD changes as occur in the DS brain, but at a later age due to the modulating effects of mosaicism. Mosaic DS individuals have been described to have dementia and AD neuropathology (Wisniewski *et al.* 1985). In humans, central nervous system neurones do not divide post-natally so that any accumulation of trisomy

21 neurones would have to occur before the final neuronal cell division takes place. Glial cells do divide so that this could occur in these cells post-natally. Although the idea of AD viewed as a late-onset, mosaic form of DS in interesting, there are currently no data from the brains of cases of spontaneous AD to support this hypothesis.

Can animal models of Down syndrome be used to study the relationship of Alzheimer disease to Down syndrome?

A major obstacle in studying possible causal factors in the development of AD in persons with DS is the inability to perform adequate experiments in human subjects. As discussed in detail by Richards *et al.* (Chapter 14), the trisomy 16 (Ts 16) mouse is a genetic model of trisomy 21 (Epstein 1986). Mouse chromosome 16 (MMU-16) contains a cluster of genes and loci also located on the long arm of HSA-21. These include *APP, SOD-1, ETS-2* and markers linked to one form of familial AD (Cheng *et al.* 1988; Coyle *et al.* 1988). Interestingly, Ts 16 mice demonstrate phenotypic features seen in DS including endocardial cushion defects and haematological and immunological abnormalities (Epstein 1986). Although these and other similarities invite further studies, evaluation of Ts 16 mice has been limited to fetal tissues since these animals do not survive birth (Epstein 1986; Coyle *et al.* 1988; Holtzman and Mobley 1991). In recent studies (Holtzman *et al.* 1992a), we have utilized Ts 16 fetal tissues and Ts 16↔2N chimeras (which live into adulthood) to ask the question, 'Are strict gene dosage effects operative in Ts 16?'. In addition, Hohmann *et al.* (1990), Fine *et al.* (1991), Richards *et al.* (1991) and our group at the University of California, San Francisco (Holtzman *et al.* 1992b) have used neurotransplantation strategies to study the effects of Ts 16 on performance of mature neurones. Such models may provide a means to study AD-like alterations that occur in DS. Here, we discuss the relevance of our results in studying the questions raised in this chapter concerning AD. Richards *et al.* (Chapter 14) further discuss the other models mentioned above.

Gene expression studies

Virtually all studies in DS and other aneuploid organisms have shown that gene dose is proportional to gene expression (Epstein 1986). To date, most genes examined have encoded either metabolic enzymes or constitutively expressed products. Since aneuploidy often produces disordered development, it was important to determine whether a departure from strict gene dosage effects may be seen for developmentally regulated genes encoding such products as membrane constituents (receptors and cell adhesion molecules), growth factors, transcription factors and structural proteins. One such protein is APP and the possible relevance of the overexpression of this protein in the pathogenesis of AD in DS has been discussed earlier in this chapter. Overexpression of this gene (beyond that seen on the basis of strict gene dose), as well as dysregulation of other genes, could play a role in creating specific DS phenotypes such as AD. In support of the possibility that gene regulation does not always obey the gene dosage effects concept, two of four anonymous HSA-21 sequences were reported to be markedly overexpressed in the brain of DS fetuses (Stefani *et al.* 1988). In addition, *APP* was overexpressed approximately 4-fold in three fetal DS brains compared with age-matched controls (Neve *et al.* 1988).

Elucidating the molecular basis for changes in gene expression in Ts 16 may make it possible to understand not only how aneuploidy disrupts cellular function but how it is that specific DS phenotypes are produced. We have recently documented marked dysregulation of two developmentally regulated genes, *App* and *Prn-p*, in Ts 16 fetal mice and adult Ts 16↔2N chimeras. As noted, *App* is located on MMU-16 and the gene encoding the prion protein (*Prn-p*) has been localized to MMU-2 (Sparkes *et al.* 1986). On the basis of gene dose, *App* expression would be predicted to be 1.5 times more than that found in appropriate controls and *Prn-p* expression would be predicted to be the same as that found in controls. Our data showed that the expression of *App* and *Prn-p* was increased beyond that predicted by gene dose in several Ts 16 fetal organs and in Ts 16↔2N chimeric adult brain. In embryonic day (ED) 17 Ts 16 skin, *Prn-p* expression was twice that found in normal. At the same age, *App* expression was three to five times greater than normal in placenta and skin. Smaller, but still significant changes (greater than gene dosage effects), were seen in most organs at ED 15. In at least one organ, the lung, there was also an alteration in *App* transcript ratios. In the Ts 16↔2N chimera brain, there was a 2- to 3-fold increase in the level of both APP (3-fold) and Prn-P (2-fold) mRNA. Given the proportion of Ts 16 cells in the brain of these animals, the *APP* mRNA level would be expected to be increased only 1.2- to 1.25-fold compared with controls if *App* expression was proportional to gene dose. In contrast to our findings for *App* and *Prn-p*, the level of *Sod-1* expression was proportional to gene dose in all tissues examined.

One possibility for explaining increased expression in Ts 16 is that the presence of an extra copy of one or more genetic regulatory factors leads to disordered regulation of other genes. Regulatory factors could be acting at the level of gene transcription, RNA processing or mRNA stability. Gene dose can have an important influence on transcriptional regulation. Recent experiments have indicated that the number of copies of the Drosophila gene *Suvar(3)7* is a factor that alters the expression of genes affected by position effect variegation (Reuter *et al.* 1990). The sequence of *Suvar(3)7* suggests that it is a zinc-finger-containing DNA binding protein and that it may act to regulate transcription. Interestingly, the dose of *Suvar(3)7* is not proportional to gene expression (Reuter *et al.* 1990). *Trans*-acting factors are likely to be critical for controlling expression of *App*, *Prn-p* and other developmentally regulated genes. Of particular interest, in light of the abnormal expression described in Ts 16, is that *App* and *Prn-p* have promoters with similar features (Basler *et al.* 1986; Salbaum *et al.* 1989).

The way in which gene dose influences gene expression is central to an understanding of the pathogenesis of disorders resulting from aneuploidy. Our results indicate that aneuploidy due to Ts 16 results in disordered regulation of two developmentally regulated genes. Overexpression beyond that predicted by strict gene dosage effects is likely to disturb cellular function even further. In addition to these effects, prior studies have shown that, for some gene products, a small increase in gene dose and gene expression (i.e. 1.5-fold) may be associated with marked differences in cellular function. For example, the interferon-α (IFN-α) receptor is encoded on HSA-21, and trisomy 21 fibroblasts have 1.5 times the normal amount of receptor. However, when the physiological effects of interferon treatment were examined, it was found that the sensitivity of trisomic cells to the antiviral effect of IFN-α was increased from 3- to 15-fold (Epstein and Epstein 1983). Using the Ts 16 mouse, it should be possible to define molecular events key to gene dysregulation resulting in some of the varied phenotypes in aneuploidy. Such an approach may help to further elucidate the molecular pathogenesis of the AD phenotype in DS.

Neurotransplantation studies

As the Ts 16 mouse is an animal model of DS, it is possible that central nervous system tissue from these animals may provide a means to study important aspects of the pathogenesis of AD in DS. As noted, the major problem in studying these animals is that they do not survive birth. We have recently used an experimental system in which the survival and development of a well-characterized brain region could be studied beyond the fetal period. Basal forebrain cholinergic neurones are one of the vulnerable populations of neurones in adults with DS and AD, and their dysfunction may contribute to dementia (Casanova et al. 1985; Price 1986). In AD, these neurones show reductions in size and cholinergic markers, contain neurofibrillary tangles and participate in the formation of senile plaques (Price 1986; Vogels et al. 1990). Eventually, they undergo atrophy and die. Coyle and colleagues (Coyle et al. 1988; Sweeney et al. 1989) have also shown that in the Ts 16 mouse fetus, hypocellularity of basal forebrain neurones affects cholinergic cells more severely than non-cholinergic cells. The reduction in cholinergic neurones may be responsible for the decreased activity in fetal Ts 16 brain of the cholinergic neurotransmitter synthetic enzyme, choline acetyltransferase (ChAT) (Coyle et al. 1988). It is possible that these changes reflect abnormalities in the genesis of Ts 16 cholinergic neurones or their growth in an underdeveloped brain (Sweeney et al. 1989). Because Ts 16 mice do not survive birth, we transplanted Ts 16 and control basal forebrain cells into the hippocampus of adult mice (Holtzman et al. 1992b). The hippocampus is the normal target zone for these cells. In addition, we used a well-studied paradigm, septohippocampal transection (fimbria–fornix lesion), to examine the effects of host target denervation and trophic influences on the transplants.

At both 1 and 6 months after transplantation into the intact hippocampus, there was evidence of neuronal survival and neurite outgrowth from Ts 16 and control grafts. Cholinergic neurones were consistently identified in grafts from both trisomic and control fetuses. One month after transplantation, Ts 16 and control neurones were similar in appearance, size and number. There was, however, a striking difference between the Ts 16 and control grafts noted at 6 months. In all Ts 16 grafts, cholinergic neurones and their proximal neurites frequently appeared shrunken and atrophic. They were significantly smaller than controls. There was no difference in the size of non-cholinergic neurones, suggesting a selective effect of Ts 16 on cholinergic neurones.

Fimbria–fornix transection denervates the host hippocampus of basal fore-brain afferents and has been shown to stimulate hypertrophy of transplanted fetal cholinergic and sympathetic neurones (Gage *et al.* 1984; Gage and Bjorklund 1986). This may occur through increased availability of trophic factors such as NGF. We asked whether Ts 16 cholinergic neurones would respond to fimbria–fornix transection. As examined 6 months after transplantation, both Ts 16 and control cholinergic neurones were significantly larger in the denervated hippocampus. Ts 16 neuronal area was increased by 49 per cent and controls by 31 per cent. To ask whether Ts 16 grafts demonstrated other evidence of AD neuropathological markers, we looked for evidence of abnormal silver staining in the form of plaques and tangles as well as β/A4 and Alz-50 immunoreactivity. Despite evidence of neuronal dysfunction, there was no evidence of abnormal immunoreactivity after 6 months.

Our current results suggest that Ts 16 basal forebrain transplants provide an animal model of age-related cholinergic neuronal degeneration and raise the possibility that this model can be used to explore the pathogenesis and treatment of cholinergic and other neuronal degeneration in DS and AD. Transplantation of cells into the normal hippocampus was used to provide Ts 16 and control grafts with an equivalent normal target; thus it is probable that a factor(s) intrinsic to Ts 16 cells produced cholinergic atrophy. The nature of this factor is unknown. It may have created its effect by actions within grafted neurones or grafted glial cells or through interactions between grafted cells and host hippocampus. Whatever its cause, its existence must be linked to the presence of an extra copy of one or more genes or regulatory sequences on MMU-16. Future experiments and manipulation with this model should be able to address whether (i) further AD neuropathology will be seen; (ii) neurodegeneration and other neuropathology are directly or indirectly caused by an excess of specific MMU-16 or HSA-21 genes; and (iii) specific neurotrophic factors can reverse spontaneous neurodegeneration as occurs in DS and AD.

Conclusions

Although the relation between DS and AD has now been established for many years, the ultimate events leading to this association still remain unclear. Exciting new information of the genetics and molecular biology of HSA-21 genes and their products are providing novel ways to explore this possibly complex relationship. One of the major unresolved issues is whether the effects of increased gene dosage for specific HSA-21 gene products directly or indirectly leads to accelerated AD neuropathology in DS. Future studies in which the 'obligate' DS region responsible for AD is defined will be forthcoming. In addition, further work in identifying novel HSA-21 gene products and regulatory elements may allow identification of mechanisms by which both developmental and age-related abnormalities in the nervous system can be explained. Current and future animal models of DS may enable one to ask important pathogenetic questions regarding both causation and potential therapy related to the neuronal dysfunction of AD.

Acknowledgements

This work was supported by NIH grants AG00445-02 (D.M.H.) and AG08938 (C.J.E.). D.M.H. is also supported by an American Academy of Neurology research fellowship award and wishes to thank William C. Mobley for his advice and support.

References

Avraham, K. B., Schickler, M., Saponikov, D., Yarom, R., and Groner, Y. (1988). Down's syndrome: abnormal neuromuscular junction in tongue of transgenic mice with elevated levels of human Cu/Zn-superoxide dismutase. *Cell*, **54**, 823-9.

Basler, K., Oesch, B., Scott, M., Westaway, D., Wälchli, M., Groth, D. F., *et al.* (1986). Scrapie and cellular PrP isoforms are encoded by the same chromosomal gene. *Cell*, **46**, 417-28.

Beyreuther, K., Dyrks, T., Hilbich, C., Mönning, U., Konig, G., Multhaup, G., *et al.* (1992). Amyloid precursor protein (APP) and βA4 amyloid in Alzheimer's disease and Down's syndrome. In *Down syndrome and Alzheimer disease,* (ed. C. J. Epstein), pp. 159–82. Academic Press, New York.

Buxbaum, J. D., Gandy, S. E., Ciccheti, P., Ehrlich, M. E., Czernick, A. J., Fracasso, R. P., *et al.* (1990). Processing of Alzheimer's β/A4 amyloid precursor protein: modulation by agents that regulate protein phosphorylation. *Proceedings of the National Academy of Sciences of the United States of America*, **87**, 6003-6.

Casanova, M. F., Walker, L. C., Whitehouse, P. J., and Price, D. (1985). Abnormalities of the nucleus basalis in Down's syndrome. *Annals of Neurology*, **18**, 310-3.

Chan, P. H., Yang, G. Y., Chen, S. F., Carlson, E., and Epstein, C. J. (1991). Cold-induced brain edema and infarction are reduced in transgenic mice overexpressing CuZn superoxide dismutase. *Annals of Neurology*, **29**, 482-6.

Cheng, S. V., Nadeau, J. H., Tanzi, R. E., Watkins, P. C., Jagadesh, J., Taylor, B. A., *et al.* (1988). Comparative mapping of DNA markers from familial Alzheimer's disease and Down syndrome regions of human chromosome 21 to mouse chromosome 16. *Proceedings of the National Academy of Sciences of the United States of America*, **85**, 6032-6.

Coyle, J. T., Oster-Granite, M. L., Reeves, R. H., and Gearhart, J. D. (1988). Down syndrome, Alzheimer's disease, and the trisomy 16 mouse. *Trends in Neurosciences*, **11**, 390-4.

Davies, L., Wolska, B., Hilbich, C., Multhaup, G., Martins, R., Simms, G., *et al.* (1988). A4 amyloid protein deposition and the diagnosis of Alzheimer's disease: prevalence in aged brains determined by immunocytochemistry compared with conventional neuropathologic techniques. *Neurology*, **38**, 1688-93.

Elroy-Stein, O., and Groner, Y. (1988). Impaired neurotransmitter uptake in PC12 cells overexpressing human Cu/Zn-superoxide dismutase—Implication for gene dosage effects in Down syndrome. *Cell*, **52**, 259-67.

Epstein, C. J. (1983). Down's syndrome and Alzheimer's disease: implications and approaches. *Banbury Reports*, **15**, 169-82.

Epstein, C. J. (1986). *Consequences of chromosome imbalance: Principles, mechanisms, and models*. Cambridge University Press, New York.

Epstein, C. J. (1987) Down's syndrome and Alzheimer's disease. What is the relation? In *Advancing frontiers in Alzheimer's disease research,* (ed. G. G. Glenner and R. J. Wurtman), pp. 155-73. University of Texas Press, Austin.

Epstein, C. J., and Epstein, L. B. (1983). In *Lymphokines,* (ed. E. Pick and M. Landy), pp. 277-301. Academic Press, New York.

Epstein, C. J., Avraham, K. B., Lovett, M., Smith, S., Elroy-Stein, O., Rotman, G., *et al.* (1987). Transgenic mice with increased Cu/Zn superoxide dismutase activity: Animal model of dosage effects in Down syndrome. *Proceedings of the National Academy of Sciences of the United States of America,* **84**, 8044-8.

Esch, G. S., Keim, P. S., Beattie, E. C., Blacher, R. W., Culwell, A. R., Oltersdorf, T., *et al.* (1990). Cleavage of amyloid β peptide during constitutive processing of its precursor. *Science,* **248**, 1122-4.

Estes, S., Golde, T. E., Kunishita, T., Blades, D., Lowery, D., Eisen, M., *et al.* (1991). Potentially amyloidogenic, carboxyl-terminal derivatives of the amyloid protein precursor. *Science,* **255**, 726-8.

Fine, A., Ault, B., and Rapoport, S. I. (1991). Mouse trisomy 16 neurons, a model of human trisomy 21 (Down syndrome), can be maintained by intracerebral transplantation. *Neuroscience Letters,* **122**, 4-8.

Gage, F. H., and Bjorklund, A. (1986). Enhanced graft survival in the hippocampus following selective denervation. *Neuroscience,* **17**, 89-98.

Gage, F. H., Bjorklund, A., and Stenevi, U. (1984). Denervation releases a neuronal survival factor in adult rat hippocampus. *Nature,* **308**, 637-9.

Goate, A., Chartier-Harlan, M.-C., Mullan, M., Brown, J., Crawford, F., Fidani, L., *et al.* (1991). Segregation of a missense mutation in the amyloid precursor protein gene with familial Alzheimer's disease. *Nature,* **349**, 704-6.

Griffin, W. S. T., Stanley, L. C., Ling, C., White, L., MacLeod, V., Perrot, L. J., *et al.* (1989). Brain interleukin 1 and S-100 immunoreactivity are elevated in Down syndrome and Alzheimer disease. *Proceedings of the National Academy of Sciences of the United States of America,* **86**, 7611-5.

Heston, L. L., Mastri, A. R., Anderson, A. R., and Anderson, V. E. (1981). The genetics of Alzheimer's disease. Association with hematological malignancy and Down syndrome. *Archives of General Psychiatry,* **34**, 976-81.

Hohmann, C. F., Capone, G., Oster-Granite, M. L., and Coyle, J. T. (1990). Transplantation of brain tissue from murine trisomy 16 into euploid hosts: effects of gene imbalance on brain development. *Progress in Brain Research,* **82**, 203-14.

Holtzman, D. M., and Epstein, C. J. (1992). The molecular genetics of Down syndrome. In *Molecular genetic medicine,* (ed. T. Friedmann), Vol. 2., pp. 105-20. Academic Press, San Diego.

Holtzman, D. M., and Mobley, W. C. (1991). Molecular studies in Alzheimer's disease. *Trends in Biochemical Sciences,* **16**, 140-4.

Holtzman, D. M., Bayney, R. M., Li, Y., Khosrovi, H., Berger, C. N., Epstein, C. J., and Mobley, W. C. (1992a). Dysregulation of gene expression in mouse trisomy 16, an animal model of Down syndrome. *The European Molecular Biology Organization Journal,* **11**, 619-27.

Holtzman, D. M., Li, Y., DeArmond, S. J., McKinley, M. P., Gage, F. H., Epstein, C. J., and Mobley, W. C. (1992b). A mouse model of neurodegeneration: atrophy of basal forebrain neurons in Ts 16 transplants. *Proceedings of the National Academy of Sciences of the United States of America,* **89**, 1383-7.

Kang, J., Lemaire, H. -G., Unterbeck, A., Salbaum, J. M., Masters, C. L., Grzeschik, K. -H., *et al.* (1987). The precursor of Alzheimer's disease amyloid A4 protein resembles a cell surface receptor. *Nature,* **325**, 733-6.

Kinouchi, H., Epstein, C. J., Mizui, T., Carlson, E., Chen, S. F., and Chan, P. H. (1991). Attenuation of focal cerebral ischemia in transgenic mice overexpressing Cu-Zn superoxide dismutase. *Proceedings of the National Academy of Sciences of the United States of America,* **88**, 1158-62.

Korenberg, J. R., Kawashima, H., Pulst, S. M., Ikeuchi, T., Ogasawara, N., Yamamoto, K., *et al.* (1990). Molecular definition of a region of chromosome 21 that causes features of the Down syndrome phenotype. *American Journal of Human Genetics,* **47**, 236-46.

McCormick, M. K., Schinzel, A., Peterson, M. B., Stetten, G., Driscoll, S., Cantu, E. S., *et al.* (1989). Molecular genetic approach to the characterization of the "Down syndrome region" of chromosome 21. *Genomics*, 5, 325-31.

McGeer, E. G., Norman, M., Boyes, B., O'Kusky, J., Suzuki, J., and McGeer, P. L. (1985). Acetylcholine and aromatic amine systems in postmortem brain of an infant with Down's syndrome. *Experimental Neurology*, 87, 557-70.

Mancardi, G. L., Liwnicz, B. H., and Mandybur, T. I. (1983). Fibrous astrocytes in Alzheimer's disease and senile dementia of the Alzheimer's type. *Journal of Neuropathology and Experimental Neurology*, 61, 76-80.

Mandybur, T. I., and Chuirazzi, C. C. (1990). Astrocytes and the plaques of Alzheimer's disease. *Neurology*, 40, 635-9.

Marin-Padilla, M. (1976). Pyramidal cell abnormalities in the motor cortex of a child with Down's syndrome. A Golgi study. *Journal of Comparative Neurology*, 167, 63-71.

Marshak, D. R. (1990). S100β as a neurotrophic factor. In *Molecular and cellular mechanisms of neuronal plasticity in aging and Alzheimer's disease,* (ed. P. D. Coleman, G. A. Higgins, and C. H. Phelps), pp. 169-81. Elsevier Science Publishers, Amsterdam.

Marshak, D. R., and Peña, L. A. (1992). Potential role of S100β in Alzheimer's disease: an hypothesis involving mitotic protein kinases. In *Down syndrome and Alzheimer disease,* (ed. C. J. Epstein), pp. 289–307. Academic Press, New York.

Marshak, D. R., Pesce, S. A., Stanley, L. C., and Griffin, W. S. (1992). Increased S100β neurotrophic activity in Alzheimer's disease temporal lobe. *Neurobiology of Aging*, 13, 1-7.

Masliah, E., Mallory, M., Hansen, L., Alford, M., Albright, T., DeTeresa, R., *et al.* (1991). Patterns of aberrant sprouting in Alzheimer's disease. *Neuron*, 6, 729-39.

Mobley, W. C. (1989). Nerve growth factor in Alzheimer's disease: to treat or not to treat? *Neurobiology of Aging*, 10, 578-80.

Monard, D. (1988). Cell-derived proteases and protease inhibitors as regulators of neurite outgrowth. *Trends in Neurosciences*, 11, 541-4.

Müller-Hill, B., and Beyreuther, K. (1989). Molecular biology of Alzheimer's disease. *Annual Review of Biochemistry*, 58, 287-307.

Murrell, J., Farlow, M., Ghetti, B., and Benson, M. (1991). A mutation in the amyloid precursor protein associated with hereditary Alzheimer's disease. *Science*, 254, 97-9.

Neve, R. L., Finch, E. A., and Dawes, L. R. (1988). Expression of the Alzheimer amyloid precursor gene transcripts in human brain. *Neuron*, 1, 669-77.

Oltersdorf, T., Fritz, L. C., Schenk, D. B., Lieberburg, I., Johnson-Wood, K. L., Beattie, E., *et al.* (1989). The secreted form of the Alzheimer's amyloid precursor protein with the Kunitz domain is protease nexin-II. *Nature*, 341, 144-7.

Potter, H. (1992). Review and hypothesis: Alzheimer disease and Down syndrome—chromosome 21 nondisjunction may underlie both disorders. *American Journal of Human Genetics*, 48, 1192-200.

Price, D. L. (1986). New perspectives on Alzheimer's disease. *Annual Review of Neurosciences*, 9, 489-512.

Probst, A., Langui, D., and Ulrich, J. (1991). Alzheimer's disease: a description of the structural lesions. *Brain Pathology*, 1, 229-39.

Quon, D., Wang, Y., Catalano, R., Scardina, J. M., Murakami, K., and Cordell, B. (1991). Formation of β-amyloid protein deposits in brains of transgenic mice. *Nature*, 352, 239-41.

Reuter, G., Giarre, M., Farah, J., Gausz, J., Spierer, A., and Spierer, P. (1990). Dependence on position-effect variegation in Drosophila on dose of a gene encoding an unusual zinc-finger protein. *Nature*, 344, 219-23.

Richards, S. -J., Waters, J. J., Beyreuther, K., Masters, C. L., Wischik, C. M., Sparkman, D. R., *et al.* (1991). Transplants of mouse trisomy 16 hippocampus provide a model of Alzheimer's disease neuropathology. *The European Molecular Biology Organization Journal*, 10, 297-303.

Rumble, B., Retalack, R., Hilbich, C., Simms, G., Multhaup, G., Martins, R., *et al.* (1989). Amyloid A4 protein and its precursors in Down's syndrome and Alzheimer's disease. *New England Journal of Medicine*, 320, 1446-52.

Rupp, F., Payan, D. G., Magill-Solc, C., Cowan, D. M., and Scheller, R. H. (1991). Structure and expression of a rat agrin. *Neuron*, 6, 811-23.

Salbaum, J. M., Weidemann, A., Lemaire, H. G., Masters, C. L., and Beyreuther, K. (1989). The promoter of Alzheimer's disease amyloid A4 precursor gene. *The European Molecular Biology Organization Journal*, 7, 2807-13.

Scott, B. S., Petit, T. L., Becker, L. E., and Edwards, B. A. V. (1982). Abnormal electric membrane properties of Down's syndrome DRG neurons in cell culture. *Developmental Brain Research*, 2, 257-70.

Selinfreund, R. H., Barger, S. W., Pledger, W. J., and Van Eldik, L. J. (1991). Neurotrophic protein S100β stimulates glial cell proliferation. *Proceedings of the National Academy of Sciences of the United States of America*, 88, 3554-8.

Sinet, P. M. (1982). Metabolism of oxygen derivatives in Down syndrome. *Annals of the New York Academy of Sciences*, 396, 83-94.

Sinet, P. M., Lejeune, J., and Jerome, H. (1979). Trisomy 21 (Down syndrome). Glutathione peroxidase, hexose monophosphate shunt and IQ. *Life Sciences*, 24, 29-34.

Sisodia, S. S., Koo, E. H., Beyreuther, K., Unterbeck, A., and Price, D. L. (1990). Evidence that β-amyloid protein in Alzheimer's disease is not derived by normal processing. *Science*, 248, 492-5.

Sparkes, R. S., Simon, M., Cohn, V. H., Fournier, R. E. K., Lem, J., Klisak, I., *et al.* (1986). Assignment of the human and mouse prion genes to homologous chromosomes. *Proceedings of the National Academy of Sciences of the United States of America*, 83, 7358-62.

Stefani, L., Galt, J., Palmer, A., Affara, N., Ferguson-Smith, M., and Nevin, N. C. (1988). Expression of chromosome 21 specific sequences in normal and Down's syndrome tissues. *Nucleic Acids Research*, 16, 2885-96.

Struwe, F. (1929). Histopathologische Untersuchungen über Entstehung und Wesen der senilen Plaques. *Zentralblatt für die Gesamte Neurologie und Psychiatrie*, 122, 291-307.

Suetsugu, M., and Mehraein, P. (1980). Spine distribution along the apical dendrites of the pyramidal neurons in Down's syndrome. *Acta Neuropathologica*, 50, 207-10.

Sweeney, J. E., Hohmann, C. F., Oster-Granite, M. L., and Coyle, J. T. (1989). Neurogenesis of the basal forebrain in euploid and trisomy 16 mice: an animal model for developmental disorders in Down syndrome. *Neuroscience*, 31, 413-25.

Sylvester, P. E. (1974). Aortic and pulmonary valve fenestrations as aging indices in Down's syndrome. *Journal of Mental Deficiency Research*, 18, 367-76.

Terry, R. D., Masliah, E., Salmon, D. P., Butters, N., DeTeresa, R., Hill, R., *et al.* (1991). Physical basis of cognitive alterations in Alzheimer's disease: synapse loss is the major correlate of cognitive impairment. *Annals of Neurology*, 30, 572-80.

Vogels, O. J. M., Broere, A. J., Ter Laak, H. J., Ten Donkelaar, H. J., Nieuwenhuys, R., and Schulte, B. P. M. (1990). Cell loss and shrinkage in the nucleus basalis of Meynert complex in Alzheimer's disease. *Neurobiology of Aging*, 11, 3-13.

Wisniewski, K. E., Wisniewski, H. M., and Wen, G. Y. (1985). Occurrence of neuropathological changes and dementia of Alzheimer's disease in Down's syndrome. *Annals of Neurology*, 17, 278-82.

Wisniewski, K. E., Laure-Kamionowska, M., Connell, F., and Wen, G. Y. (1986). Neuronal density and synaptogenesis in the postnatal stage of brain maturation in Down syndrome. In *The neurobiology of Down syndrome*, (ed. C. J. Epstein), pp. 29-44. Raven Press, New York.

Yankner, B. A., and Mesulam, M. -M. (1991). β-amyloid and the pathogenesis of Alzheimer's disease. *New England Journal of Medicine*, 325, 1849-55.

Yarowsky, P. J., Krueger, B. K., Michal, T., Gearhart, J. D., Reeves, R. H., and Hilt, D. C. (1991). Astroglial proliferation in vivo in S100β transgenic mouse. *Journal of Cell Biology*, 115 (part 2), 217a.

14

The contribution of mouse models to understanding the neuropathology and functional impairment associated with Alzheimer disease and Down syndrome

S-J. Richards, F. Yamaguchi, and S.B. Dunnett

Summary

Recent research into the neuropathology of Alzheimer disease has re-emphasized the importance of the classical signs of the disease—amyloid plaques and neurofibrillary tangles—with the discovery of their primary molecular constituents, the β/A4 and tau proteins respectively. In other diseases, animal models have provided a powerful means of studying the pathogenesis and treatment of the disease process, but the pathological features of Alzheimer disease do not occur spontaneously in laboratory animals (with the possible exception of rare occurrence in old primates).

Advances in understanding the molecular biology, chromosomal 'linkage' and an association with Down syndrome provide new strategies for inducing the expression of Alzheimer disease-like pathology in experimental mice. The present chapter describes the development of two such strategies. First, since Down syndrome (trisomy 21) in humans is associated with Alzheimer disease-like neuropathology, the development of similar pathologies have been investigated in a murine trisomy of the syntenic chromosome 16, made possible by transplantation of the developing tissues to overcome the inevitable death of the affected mice *in vivo*. Secondly, the role of overexpression of genes located on murine chromosome 16 and human chromosome 21 can be investigated through transgenesis. The first studies of amyloid transgenic mice have yielded variable patterns of pathology and preliminary reports of learning deficits in the genetically engineered mice. It is likely that these preliminary studies will lead to rapid advances in our understanding of the pathological process of Alzheimer disease in conjunction with the development of improved animal models for its study.

Introduction

There are about half a million cases of Alzheimer disease (AD) in the UK and currently this increasingly prevalent disease comes fourth in frequency to heart disease, stroke and cancer. In view of the lack of available information on the current cost of caring for this population, either within the community or in institutional settings, a few approximate calculations have had to be resorted to by the Audit Commission (personal communication), which has conservatively estimated that the direct care costs exceed £3 billion per year in the UK alone. To date, there is no effective treatment or cure for AD and in view of this, together with the extent of the disease and the costliness of caring for people suffering from it, it is surprising that so little is invested into researching the aetiology of AD.

Progress in understanding any disease is invariably characterized by periodic milestones and one such milestone has been the discovery of the similarity between the neuropathology associated with AD and that observed at post-mortem examination of the brains of persons with Down syndrome (DS) (see Chapter 5). This breakthrough inevitably led to investigations of the relative contribution of amyloid, tau and other potentially pathological proteins to the cause of AD dementia. Furthermore, the development of molecular biology techniques has drawn many researchers towards scrutinizing the role of chromosome 21 genes in normal brain development and cognitive function, in addition to the neurodegenerative events underlying AD.

Complete trisomy 21 is the genetic basis of approximately 95 per cent of DS cases (see Chapter 2). These individuals are known to be at risk of developing AD in middle life (Burger and Vogel 1973; Heston *et al.* 1982; Price *et al.* 1982; Oliver and Holland 1986). This has been attributed to the presence of gene(s) on chromosome 21, which, in excess, lead to the neuropathological changes observed in AD. It is not yet known whether the relatively rare unbalanced translocation DS cases, which have only a small region of chromosome 21, telomeric of the amyloid precursor protein (APP) gene, in excess, are any more vulnerable to AD than anyone else in the general population.

The main neuropathological features of AD are neuritic amyloid plaques and neurofibrillary tangles (NFT), which are found in their highest density in the entorhinal cortex and hippocampus implicating the perforant pathway (Kidd 1963; Terry 1963; Wisniewski *et al.* 1985). Risk considerations for AD include (i) changes in the regulation of the APP gene as a function of normal ageing, (ii) a few early age-of-onset AD families providing 'linkage' to chromosome 21 and (iii) individuals with trisomy 21 demonstrating the pathological changes associated with AD. Indeed, in trisomy 21, amyloid is deposited some 50 years before that observed in the normal ageing population, with the amyloid depositions predating the neurofibrillary pathology (Mann and Esiri 1989; Mann *et al.* 1989; Rumble *et al.* 1989).

APP is a 35 kDa protein that is anterogradely and retrogradely transported along axons and co-exists with synatophysin at pre- and post-synaptic dendritic membranes (Koo *et al.* 1990). The protein comprises an extracellular domain, a transmembrane region and an intracellular C-terminus domain. The APP is localized in the central nervous system within all neurones in the hippocampus, cortex, olfactory bulb and cerebellum, and may be induced within astrocytes, macrophages and microglia. It is also found in the peripheral nervous system

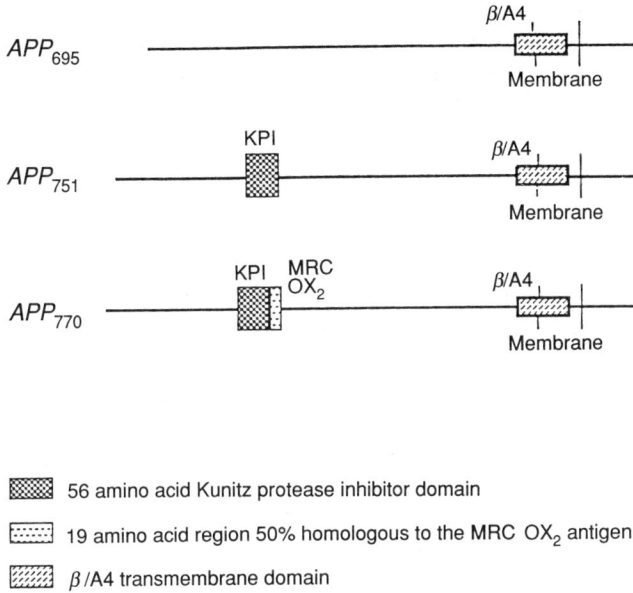

Figure 14.1. The major isoforms of the amyloid precursor protein

specifically within sertoli cells, lymphocytes, stimulated T-cells and polymy-ocytes (Schubert *et al.* 1991).

Five APP isoforms have been identified with three major APP transcripts appearing within the central nervous system: APP_{695}, $_{751}$ and $_{770}$ (Kitaguchi *et al.* 1988; Ponte *et al.* 1988; Tanzi *et al.* 1988). APP_{695} is the most abundant form in the brain; it is highly expressed during development and is localized within neurones (Shivers *et al.* 1988). By contrast, the APP_{751} and $_{770}$ isoforms are localized within astrocytes. The differences between these three isoforms may be accounted for by the presence of a 56 amino acid region homologous to the Kunitz protease inhibitor (KPI) region and a 19 amino acid region that has 50 per cent homology to the MRC Ox2 antigen (Figure 14.1). Tentative evidence suggests that the balance of these various isoforms changes during ageing (Johnson *et al.* 1989, 1990). This is not surprising since the neuronal content of the brain diminishes with ageing and concomitantly the number of astrocytes increases in response to neurodegenerative processes associated with normal ageing. This apparent increase in isoforms containing the KPI region may have implications for the generation of Alzheimer-type neuropathology within the ageing brain. It is the subject of much research interest and is currently being evaluated in human AD, DS and animal models.

Normal cleavage of the APP occurs at residue 616 (or residue 16 of the β/A4 region) and involves the proteolytic enzyme secretase to yield a secreted form of amyloid and a residual membrane inserted fragment that subsequently under-goes a form of intracellular processing (Kang *et al.* 1987; Esch *et al.* 1990;

Sisodia *et al.* 1990). The β/A4 protein is generated by cleavage from the larger transmembrane glycoprotein APP (Kang *et al.* 1987). The toxic β/A4 fragment comprises an 8.4 kDa β-pleated sheet of 42–43 amino acids in length (Glenner and Wong 1984; Masters *et al.* 1985a,b). It is generated by abnormal cleavage with a protease, as yet undefined, with residues 10–39 being required to generate the β/A4 aggregate. This is the fragment considered to be central to amyloid pathogenesis.

The β/A4 amyloid protein has been isolated from the amyloid plaque and its amino acid composition is homologous to the amyloid protein isolated from AD cerebral vasculature (Glenner and Wong 1984; Masters *et al.* 1985a,b; Roher *et al.* 1986). Although the APP gene encoding the β/A4 protein has been mapped to chromosome 21 (Reeves *et al.* 1987), genetic 'linkage' studies have shown a recombination event occurring between the APP gene and the disease locus, thus ruling out the APP as the gene causative of familial AD in the majority of 'late' and 'early' age-of-onset cases (Van Broeckhoven *et al.* 1987).

Rare, single base mutations at position 717 of the APP gene have been identified in a few AD pedigrees (Chartier-Harlin *et al.* 1991; Goate *et al.* 1991; Murrell *et al.* 1991; Naruse *et al.* 1991). These mutations comprise substitutions of valine for isoleucine, glycine or phenylalanine. However, the contribution of these single base substitutions, in the transmembrane region of the APP gene, to accelerated production of amyloid pathogenesis is unclear. Certainly these mutations occur exterior to the neurotoxic β/A4 region and following normal secretase cleavage would be retained within the membrane inserted intracellular fragment. Thus, any influence of these base changes on the pathogenesis of AD are likely to be exerted during the intracellular processing of the membrane inserted fragment, post-cleavage. A further mutation at the N-terminus of the β/A4 gene close to the endosomal–lysomal cleavage site (Estus *et al.* 1992; Golde *et al.* 1992) has also been reported (Mullan *et al.* 1992).

NFT have been identified immunocytochemically with antibodies to microtubule-associated protein tau (Brion *et al.* 1985; Kosik *et al.* 1986), ubiquitin (Mori *et al.* 1987; Perry *et al.* 1987) and β/A4 amyloid (Masters *et al.* 1985b). However, the major components of the NFT are paired helical filaments (PHF), of which tau protein has been demonstrated at the ultrastructural level to be a constituent of the protease-resistant core (Wischik *et al.* 1988). The possibility of abnormal post-translational modification of tau is being investigated by several groups as accounting for the abnormal tau observed in AD, which furthermore may give rise to the PHF.

Individuals with DS until the last decade have been susceptible to premature death often because of a congenital heart defect and altered immune response causing a level of immune deficiency. Such death over a broad spectrum of ages led to an availability of post-mortem tissue for assessment of the ontogeny of AD-associated neuropathological changes. With improved health care through antibiotics, corrective heart surgery and other treatments, many individuals with DS are now living to their fifth, sixth and seventh decades (see Chapter 2). Today one of the fundamental difficulties of studying the consequences of increased gene dosage upon development, is the markedly reduced availability of fetal/neonatal DS and age-matched control material for post-mortem examination. Thus, for both technical and ethical reasons there is a distinct advantage in developing *in vivo* and *in vitro* models of DS within which controlled and systematic studies of the developmental processes may be undertaken. The development of such models may also help to elucidate the pathogenesis of AD itself.

The search for animal alternatives, within which the processes leading to amyloid depositions could be observed, has led to aged wild and domestic animals being scrutinized. Amyloid accumulations have subsequently been detected in the dog, the non-human primate and polar bear (Johnstone *et al.* 1991). However, the duration, in years, for the depositions to occur within these animals has created technical, experimental and ethical difficulties. Thus the search for simpler animal models continues.

The trisomy 16 mouse model of Alzheimer disease pathology

In 1975, Gropp *et al.* pioneered the development of trisomic mouse models by generating trisomies of specific autosomes, using a breeding regime based on matings from heterozygous, balanced translocation strains. The contribution of these trisomic strains to animal models of human disease was heightened by the identification of homologues to several human chromosome 21 loci on mouse chromosome 16 (Cox and Epstein 1985; Reeves *et al.* 1987, 1989; Coyle *et al.* 1988). This strongly implies that a sizable region of human chromosome 21 has been evolutionarily conserved between mouse and man (Figure 14.2).

The efficacy of this model has been severely hampered by the aneuploidy being lethal to the developing fetus at around 16–20 days of gestation. DS fetuses are vulnerable to being miscarried during the first trimester of pregnancy, often due to severity of congenital heart defects. Similarly, such defects can result in a failure of the embryonic trisomic mice to survive full-term.

To overcome the limitations presented by death *in utero*, several techniques have been employed to increase survival times of specific trisomic tissues or populations of cells, including: chimeras prepared from trisomy 16 and euploid embryos (Cox *et al.* 1984; Gearhart *et al.* 1986), neural transplantation (Hohmann *et al.* 1990; Richards *et al.* 1990, 1991; Fine *et al.* 1991), and long-term tissue culture (Orozoco *et al.* 1987; Bredesen *et al.* 1991; Plioplys 1991). The latter two techniques have permitted successful comparisons between euploid and trisomic central nervous system tissues during post-natal development.

While these methodologies represent innovative ways in which to investigate critical developmental questions, they too are not without their limitations. While tissue culture studies are able to monitor cell-to-cell contact in the short-term, they lack the ability to recreate normal neuroanatomical networks, and so are deprived of important regulatory inputs. To some extent a solid graft of embryonic tissue transplanted to its normal site of development may overcome some of these difficulties. One of the critical parameters for grafted cell survival is the rapid vascularization of the transplanted tissue. This is possible to

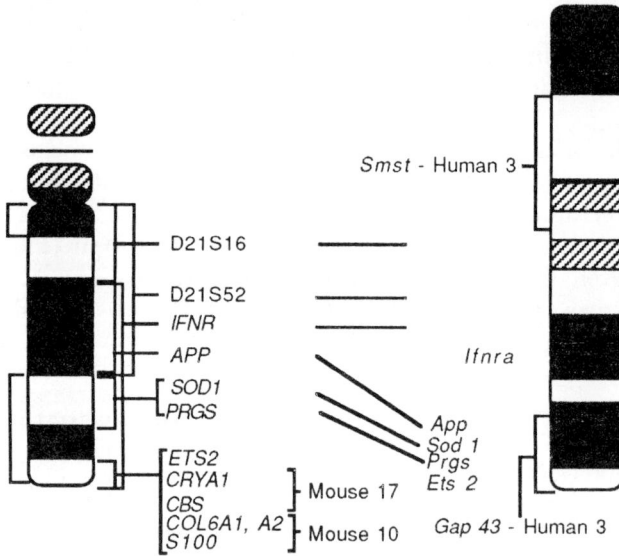

Figure 14.2. Synteny between human chromosome 21 and mouse chromosome 16 (adapted from Epstein 1988)

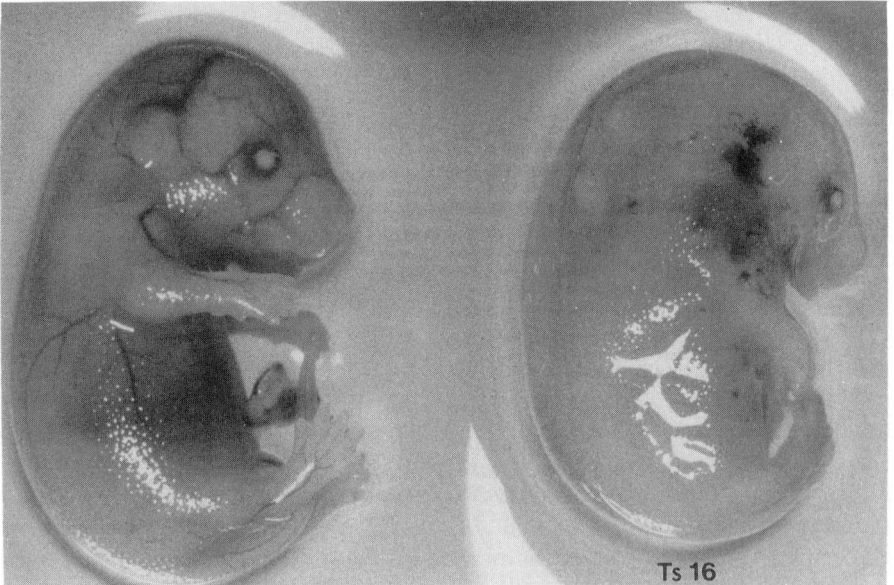

Figure 14.3. A trisomy 16 mouse embryo (Ts.16) and its normal littermate

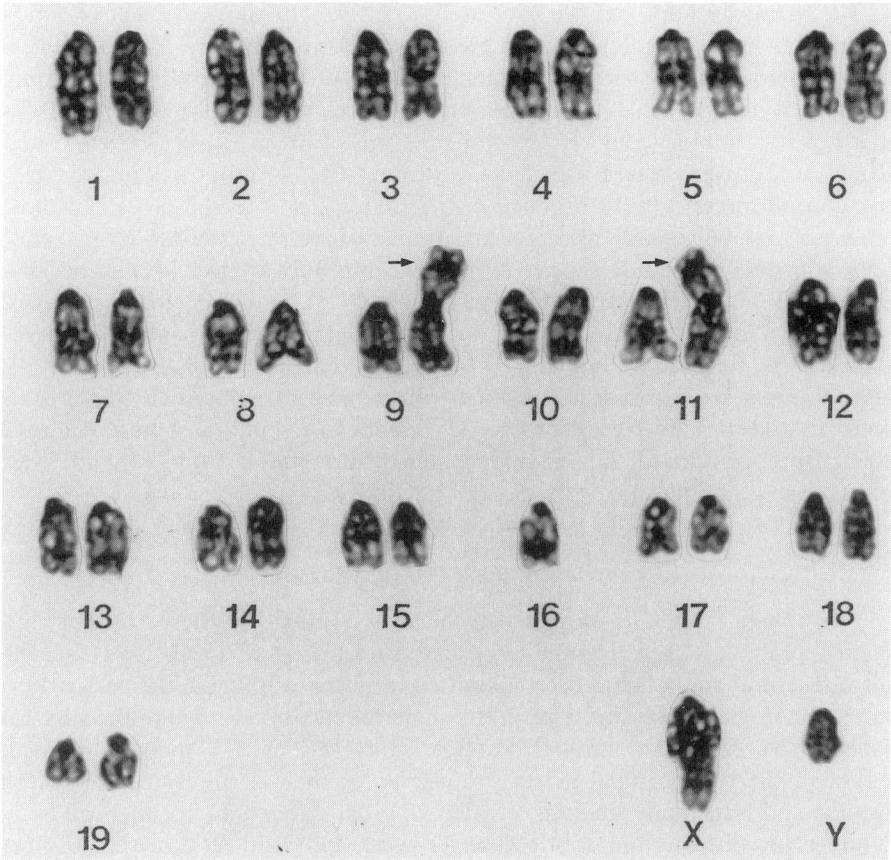

Figure 14.4. Karyotype confirmation of a trisomy 16 embryo. The two translocation chromosomes are indicated with an asterisk

achieve by preparing a pial surface in the host transplantation site using a preparatory procedure prior to grafting. The association between the endothelial cells and the amyloidosis observed within human AD has yet to be determined.

Another problem posed by the exploitation of the synteny between these two chromosomes is that murine chromosome 16 is a larger autosome than human chromosome 21. Therefore, a gene dosage effect associated with genes located on mouse chromosome 16 but not located on human chromosome 21 will be observed during developmental analyses of trisomic tissues. Hohmann *et al.* (1990) have specifically examined the effects of growth-associated protein 43 (GAP43) and pre-prosomatostatin (ppSmst) upon the development of trisomy 16 cortical tissues transplanted into neocortical sites within neonatal mice. Both of these genes map to murine chromosome 3 and not to murine chromosome 16 or human chromosome 21.

The particular approach we have used, to investigate the possibility of reproducing AD neuropathology within trisomy 16 cortical tissues, involves neural transplantation as a method of extending the life of these tissues. Trisomy 16 mice were generated by a breeding regime in which male offspring from matings of homozygous Robertsonian translocations Rb(9:16)9Rma and Rb(11:16)2H were mated with females of the CFLP strain possessing acrocentric chromosomes only to produce a Rb9Rma/Rb2H X CFLP cross. Trisomic embryos may be visually distinguished from their normal siblings by morphological criteria, such as a shorter crown to rump measurement, severe oedema of the neck and flattened nasal bridge (Figure 14.3). However, for purposes of scientific validation, all tissues used for transplantation were confirmed karyotypically as trisomic or euploid (Waters and Bartlett 1988) (Figure 14.4). Hippocampal tissues dissected from trisomic mice of embryonic days 16–18 were grafted as solid pieces into cortical cavities in young host mice. Control transplants comprised similar hippocampal dissections from euploid litter mates, also grafted into cortical cavities in CFLP hosts.

Histological and immunocytochemical analysis

The histological and immunocytochemical assessment of these trisomic and control grafts has been reported elsewhere (Richards *et al.* 1990, 1991) and the ultrastructural analysis has been submitted for publication. To summarize these data, histological assessment of grafts undertaken prior to 4 months survival failed to reveal any abnormal features. However, at 4–6 months survival, although the hippocampal grafts irrespective of their donor source appeared healthy and Nissl body staining revealed no gross differences in either cell body numbers or distribution, subtle changes were apparent. Within the trisomic grafts, some cell bodies appeared atrophied and with a tendency to aggregate. Palmgren's silver staining showed fibrillary accumulations within the dendritic processes of pyramidal neurones within the trisomic grafts. Thioflavin-S, a histological marker for amyloid, revealed positive staining of a few cells and around the blood vessels within these trisomic tissues. By contrast, no abnormal histopathology was observed in either the surrounding host parenchyma or in the euploid grafts.

Euploid and trisomic mouse transplants were assessed at 4–6 months using antibodies currently employed for post-mortem confirmation of human AD neuropathology. Polyclonal antibodies were raised against (i) APP (Masters *et al.* 1985a), (ii) the β/A4 protein (Masters *et al.* 1985a), (iii) α_1-antichymotrypsin (Abraham *et al.* 1988) and (iv) purified PHF preparations (antibody A128) (Sparkman and White 1989). Monoclonal antibodies were raised recognizing (i) a specific form of tau bound within the protease-resistant core of the PHF (Tau 6.423) (Wischik *et al.* 1988) and (ii) ubiquitin (gift of Prof. B. Anderton).

These antibodies were used to assess the pathological properties of the trisomic and euploid grafts. All of these antibodies were initially tested for their specificity in human AD tissue prior to being applied to trisomic and euploid

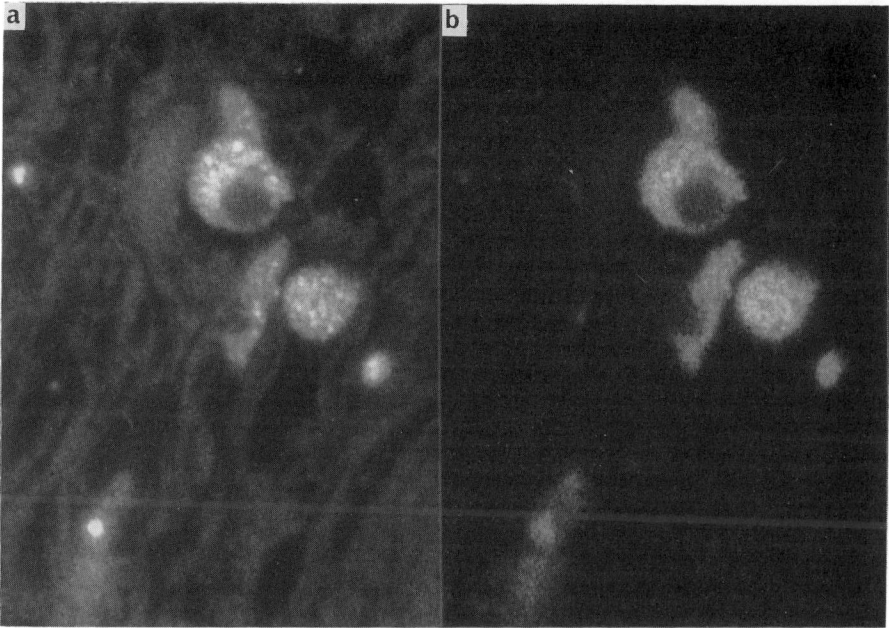

Figure 14.5. Intracytoplasmic immunoreactivity with (a) β/A4 amyloid and (b) Tau 6.423 antibodies demonstrate co-existence of these two proteins within the same cell body

tissue grafts. Within formalin fixed, wax embedded human AD tissue immunoreactivity for plaques was observed with the APP, β/A4, and α_1-antichymotrypsin antisera and intracellular immunoreactivity with the A128, Tau 6.423 and β/A4. Some intracellular co-localization of β/A4 with Tau 6.423 within intracytoplasmic granular inclusions was observed within these human AD post-mortem tissues.

Results of the immunocytochemical analysis of the mouse grafts revealed immunoreactive cells, amounting to approximately one to five per cent of total cells stained by Nissl histology. These immunoreactive cells were either isolated individual cells or aggregated cell masses in which some of them appeared to be losing their normal cell morphology. No immunoreactivity was observed within euploid grafts or host parenchyma. Co-localization of some of these AD-associated proteins were observed in the trisomic grafts. Co-localization of APP with α_1-antichymotrypsin was predominantly intracellular, and β/A4 with Tau 6.423, exclusively intracellular (Figure 14.5). In areas where cells appeared as aggregates, extracellular staining was observed as a fine filamentous product in close proximity to the cell soma. These extracellular filaments were APP and α_1-antichymotrypsin immunoreactive (Figure 14.6). β/A4 and α_1-antichy-motrypsin were observed in the walls of small blood vessels exclusively within the trisomic grafts (Richards et al. 1991).

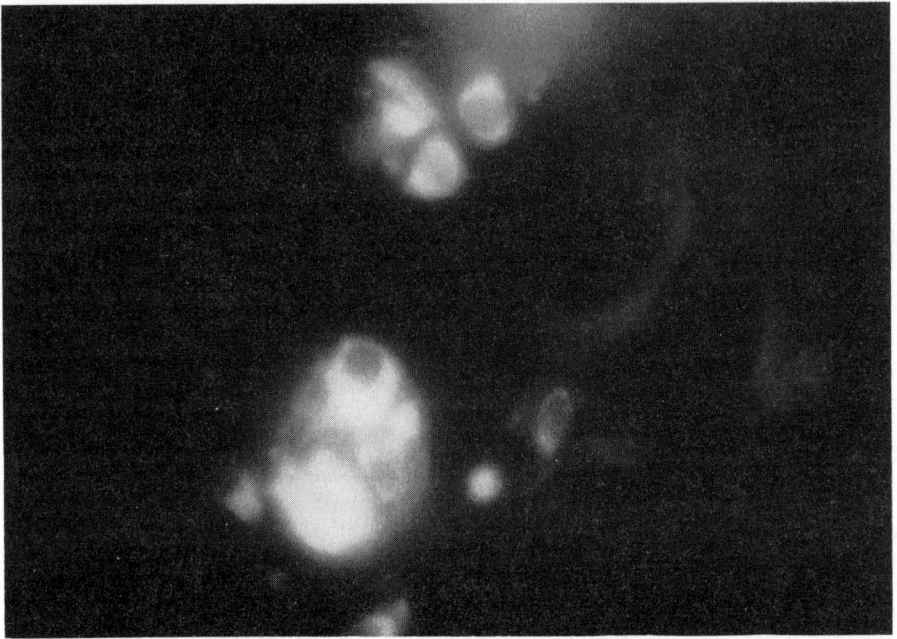

Figure 14.6. A trisomy 16 graft demonstrating intra- and extracellular immunoreactivity for (a) APP and (b) α,-antichymotrypsin. An immunoreactive blood vessel is depicted

Ultrastructural analysis

Tissues from three trisomy 16 hippocampal grafts at one year post-transplantation, three control grafts of the same age and one normal hippocampus from an 18-month-old mouse were prepared by one per cent formaldehyde and one per cent glutaraldehyde perfusion in a 0.12 M phosphate buffer. Dehydration and subsequent embedding were undertaken according to standard EM procedures.

Analysis of trisomic tissues has failed to reveal any amyloid plaque or PHF formations at either the light or ultrastructural level. However, in the trisomy 16 grafts and in the normal aged hippocampal tissue, vast amounts of electron dense deposits have been observed within the cell cytoplasm of large neurones (Figure 14.7). Such depositions are generally absent from similar cells in the age-matched control grafts. In human AD such electron dense, intracellular depositions are associated with lipofuscin. These deposits are reported to be of lysosomal origin and to contain fragments of the β/A4 domain normally retained intracellularly after synaptic cleavage of the secreted portion of the larger APP (Bancher *et al.* 1989; Benovitz *et al.* 1989).

In addition to these massive, electron dense depositions of lipofuscin, we have also observed other abnormal features within the trisomy 16 grafts that are generally absent from the age-matched control tissues. Intracellularly the cytoplasmic changes are associated with (i) the Golgi apparatus frequently

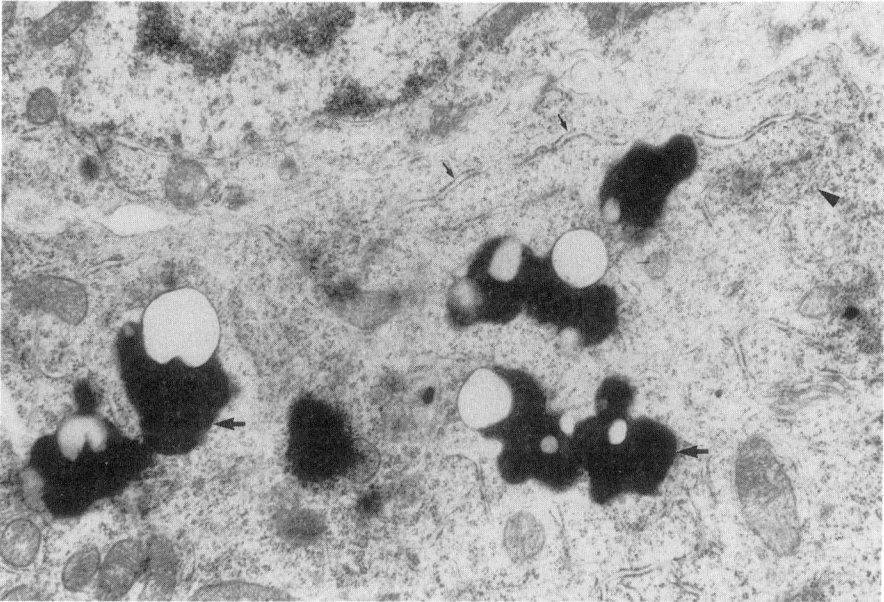

Figure 14.7. Ultrastructural visualization of a neurone within a >1-year-old trisomy 16 graft containing cytoplasmic electron dense deposits of lipofuscin (arrows). Note the large numbers of free ribosomes (arrowhead) and the elongated endoplasmic reticulum (small arrows)

being distended, with large numbers of free ribosomes, (ii) endoplasmic reticulum being abnormally extended and (iii) membrane changes in the nuclear envelope and outer cell membrane being apparent. Extracellular changes concern axonal degeneration and degenerating profiles within axons in populations of neurones and glia. These profiles appear to be mitochondrial and lysosomal in origin and are observed in varying stages of degeneration within the parenchyma, axon and at the synapse. The neuropil of trisomic grafts is less compact than age-matched normal controls—indicative of axonal loss.

The degenerative profiles observed within the trisomy 16 grafts have also been observed within normal hippocampal tissue obtained from laboratory raised mice over two years old at the time of sacrifice. In fact, it would be difficult to distinguish between these two sources of tissues on the basis of their ultrastructural analysis alone. Conversely, the aged-matched control grafts have yielded only isolated incidences of intracellular, electron dense depositions, axonal degeneration and degenerating profiles within axons and at synapses.

Amyloid transgenic mice

The classical pathological features of AD, upon which post-mortem confirmation of the disease is made, is the β/A4 amyloid plaque and the NFT within specific brain regions. As previously mentioned, studies of DS brains at post-

Reference	Gene construct	Promoter	Abnormality	Location	Status of model
Quon et al. (1991)	Full-length APP$_{751}$	Rat neurone-specific enolase	Extracellular amyloid depositions	Hippocampus and cortex	Unclear
Kawabata et al. (1991)	100 aa C-terminal fragment	Human Thy-1	Plaques, tangles, degeneration	Hippocampus and cortex	Retracted
Wirak et al. (1991)	β/A4 protein	Human APP regulatory region	Intracellular amyloid depositions	Hippocampus	Retracted
Beer et al. (1991)	Full-length APP$_{695}$	Human metallothionen IIA	Increased APP expression, but no amyloidosis	Brain	Details not published
Neve/ Kammesheidt	104 aa C-terminal fragment	?	Intracellular amyloid in lysosomes	Hippocampus and cortex	Details not published

aa, amino acids

Table 14.1. Review of amyloid precursor protein transgenic mice

mortem have revealed that amyloid plaque pathology predates the occurrence of the NFT. In elucidating the contribution of the various isoforms of the APP gene to the development of this amyloid plaque pathology, it is possible to generate transgenic mice with the different APP isoforms inserted into their genome. Furthermore, it is possible through transgene technology to determine the contribution of the different APP gene fragments (in particular the neurotoxic C-terminal region and the different APP mutations) to plaque formation, as well as the contribution of these APP regions to memory.

The success of a transgenic mouse may be judged upon the expression of the inserted gene and translated protein product by specific cell types and within appropriate brain regions. In an attempt to achieve appropriate gene expression within the host system, a variety of different promoters have been used in combination with the gene construct. Results of preliminary APP transgenic experiments have been variable and a review of current data can be seen in Table 14.1. However, the most successful to date has been the transgenic work of Quon et al. (1991) who used the rat neurone-specific enolase (NSE) promoter in conjunction with the human APP$_{751}$ cDNA to construct their transgenic mice. The construct was microinjected into the JU strain of mice with the NSE promoter directing the neural-specific expression of β-galactosidase. By exploiting this visualization system it has been possible for Quon and colleagues to detect increased neuronal expression of the APP$_{751}$ isoform. Apparent extracellular β/A4 amyloid depositions have been observed as early as two months old and which increased with age.

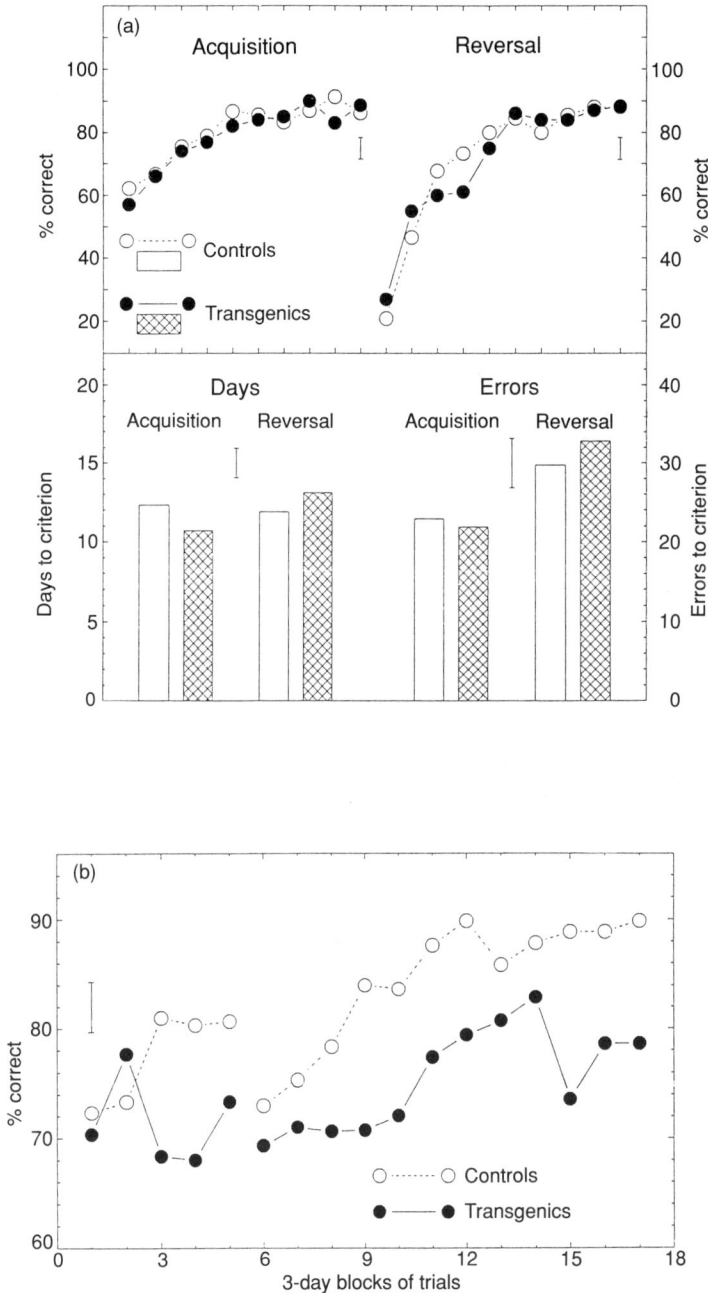

Figure 14.8. (a) Results from the spatial position habit and reversal T-maze task demonstrate no statistically significant difference in performance between transgenic and control groups. (b) The delayed alternation T-maze task demonstrated impaired performance within the transgenic group

In addition to the ability of transgenic mice to produce some of the neu-ropathological features associated with AD, it is also possible to assess the functional implication of extra copies of a particular gene construct upon behaviour. We have assessed a transgenic mouse inserted for the APP_{695} construct under regulation of the metallothionine IIA promoter (Beer *et al.* 1991). Although we have confirmed the presence of the construct inserted within the transgenic mouse genome, we have failed to observe overexpression of the APP_{695} gene or any neuropathological features up to one year survival. The same transgenic line maintained within the McLaughlin Research Institute, Great Falls, has similarly failed to demonstrate any neuropathology at one to three years of age.

To evaluate the spatial learning abilities of these APP_{695} transfected mice, 43 transgenic and 43 control mice were tested at eight weeks old in the Morris water maze (Morris 1981). Data collected over 12 weeks of testing indicated that the transgenic mice had a small, but statistically significant, task acquisi-tion deficit (Yamaguchi *et al.* 1991). When tested on a simple T-maze spatial position habit and reversal task there was little difference in performance be-tween control and transgenic groups (Figure 14.8a). However, when the T-maze task became more complex and required a greater demand on memory, as repre-sented by the delayed spatial alternation task, the differences in performance between the two groups increased with the transgenics performing less well (Figure 14.8b) (F. Yamaguchi, S-J. Richards and S.B. Dunnett, unpublished data).

These data pose a number of difficulties for interpretation, as a small but nevertheless significant memory impairment on the more complex spatial mem-ory tasks has been observed within the transgenic mice when compared with the control strain non-transgenics. However, even up to three years of age, the APP_{695} transgenic strain fails to produce any of the neuropathological features associated with human AD. This raises questions about the direct relationship between amyloid pathology and dementia. Furthermore, it raises the question of whether increasing expression of any genomic DNA fragment might influ-ence cognitive function. Certainly, trisomy and unbalanced translocations in the human invariably result in mental retardation. Thus, the observations of a memory impairment in transgenic mice may reflect no more than a non-specific mental impairment induced by incorporating extra amounts of DNA into the genome.

Discussion

Amyloid plaques are one of the pathological features upon which the post-mortem confirmation of a diagnosis of AD is made. Indeed, the β/A4 protein was initially isolated from the neuropathological plaque core and was subse-quently found to share amino acid homology with, and to be antigenically related to, the amyloid protein isolated from AD cerebral vasculature. However, the presence of APP and β/A4 amyloid does not alone confirm AD neuropathology. Amyloid neuropathology may be observed within other neurodegenerative disorders such as Lewy Body dementia, dementia pugilistica and Creutzfeldt–Jakob disease. By contrast, the serine protease inhibitor, α_1-antichymotrypsin, which is predominantly associated with amyloid deposits in neuritic plaques and around blood vessels in the AD brain, is absent from the amyloid pathology observed in Creutzfeldt–Jakob disease (Abraham *et al.* 1988). Thus this protein has become a distinctive marker for distinguishing at least these two disorders. Similarly Tau 6.423 enables distinction to be made between AD and other neurodegenerative pathologies.

The strongest evidence for the trisomy 16 mouse neural transplantation model capturing the primary defect causative of AD is the presence of (i) the α_1-antichymotrypsin antigen and (ii) Tau 6.423, a protein that selectively recognizes isolated stripped PHFs and the tau released from PHFs. Tau 6.423 does not cross-react with other tau proteins in control brain tissues. Further evidence supporting the similarity of neurodegenerative events occurring within the trisomic mouse and human AD comes from the observations of β/A4 and Tau 6.423 co-localization in intracellular granular inclusions within cortical tissues. There is an absence of comparable staining in control mouse grafts and also in control human cortical tissues.

The intracellular inclusions of β/A4 and Tau 6.423, seen at the light level of microscopy, appear at the ultrastructural level to be associated with cytoplasmic lysosomes. The electron dense depositions observed within these neuronal lysosomes have been reported as being responsible for the abnormal proteolytic events that give rise to the insoluble and toxic amyloid peptide (Benovitz *et al.* 1989). The intracytoplasmic abundance of ribosomes and abnormalities of the Golgi apparatus and endoplasmic reticulum observed within these trisomic cortical tissues more than one year old, to some extent mimic those observed in the ultrastructural observations made by Wisniewski (1992) and colleagues. Our data suggest that the neuronal processing and packaging of the amyloid protein product is impaired within trisomy 16 mice and that this impairment is occurring at the same time, and may be related to, the aberrant processing of amyloid within the lysosomal degradative pathway. Synaptic membrane damage is required as one of the primary degenerative events, thus exposing the normally membrane-embedded cleavage site of the β/A4 fragment for abnormal proteolytic cleavage. The subsequent aberrant β/A4 amyloid fragment has been demonstrated to comprise the major constituent of the extracellular amyloid plaque. A failure to degrade and reprocess the residual membrane inserted and intracellular APP following normal secretase cleavage could be causative of this membrane damage. In this respect, post-cleavage APP may resemble the IgA receptor, which also has a membrane-inserted region and an intracellular truncated C-terminus and which is processed by lysosomal proteolytic enzymes. Furthermore, it is likely that the APP mutations are hampering this lysosomal degradative pathway and imposing additional difficulties upon intracellular processing. Based on our ultrastructural observations of aged trisomy 16 cortical tissue, we propose the series of events leading to the pathological features observed in DS and AD shown in Figure 14.9.

Degeneration of certain mitochondria at synapses and in the processes of glia within the normal aged and trisomy 16 hippocampal tissue grafts, together

Failure of the lysosomal degradative pathway or other mechanisms to retrieve efficiently the post-cleavage, C-terminus fragment of the APP from synaptic membranes.

Breakdown of the synaptic membranes leading to exposure of APP transmembrane sites for abnormal proteolytic cleavage and the production of the β/A4 amyloid plaque protein.	Abnormal or partial reprocessing of the C-terminus APP fragment due to impaired lysosomal proteolysis or rare APP$_{717}$ mutations leading to the formation of large depositions of lipofuscin and possible lysosome death. Lipofuscin accumulation impairs normal cell functioning.

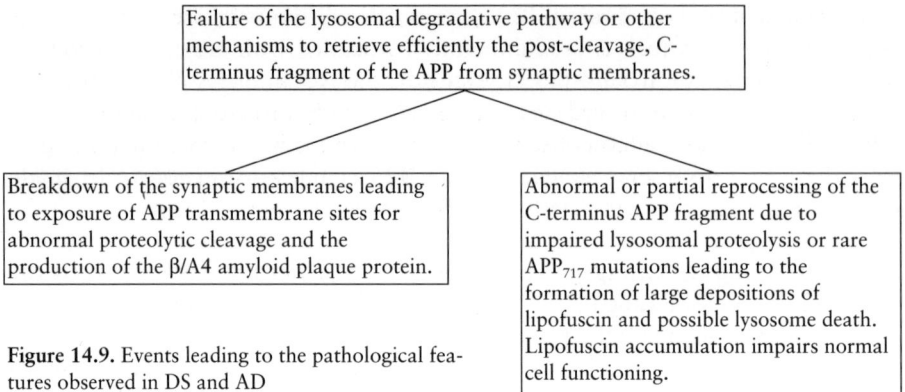

Figure 14.9. Events leading to the pathological features observed in DS and AD

with the absence of the abnormal profiles in control tissues, lead to speculation about the role of mitochondria in synaptic maintenance. The association of amyloid with these mitochondria is of particular interest especially if certain fragments of the β/A4 demonstrate the ability to bind to NFT (Yamaguchi *et al.* 1990; Caputo *et al.* 1991).

Animal models in which the ontogeny of AD may be studied are now realized as being central to further developments in understanding the disease. While the trisomy 16 mouse relies on the synteny that exists between human chromosome 21 and murine chromosome 16, and that human trisomy 21 (DS) individuals represent the largest population 'at risk' for AD, experiments exploiting this aneuploidy are hampered by its being lethal *in utero*. Innovative experimental procedures have been successful in rescuing cortical tissues from this aneuploidy for sufficient periods of time to permit the pathological changes associated with AD to be observed. However, the inability to preserve complete neuroanatomical circuitry, or to monitor the behavioural consequences of this trisomy throughout ageing, impose essential limitations on these transplantation and tissue culturing techniques.

In contrast, mouse transgenesis offers the potential of a more powerful experimental system of AD pathology and cognitive impairment analysis. Although initial reports of transgenic mice derived from genomes inserted with constructs containing the different amyloid isoforms and fragments of the APP gene have not always been substantiated (Table 14.1), these reflect the need for the parameters of transgenesis to be more clearly understood. Different promoters, variable copy numbers and lack of control over the insertion site within the genome introduce variability and continue to make it difficult to draw comparisons between the various APP regions contained within the construct. Once these variables have been controlled and reliable embryo recipient strains identified, transgenesis is likely to provide a powerful system to evaluate the contribution of these individual fragments to AD pathology.

The prospect of assessing the relationship between the neuropathology and the memory impairment associated with AD through transgenesis is exciting but much work is still required in first establishing good transgenic strains.

The effect of incorporating extra fragments of DNA into the recipient genome, the influence of the insertion site and the effect of the inserted gene on host gene expression have yet to be clarified. Certainly, technological developments in the field of transgenesis have led to the production of transgenic mice derived by homologous recombination in which the insertion site is predetermined. Now that the limitations of present transgenic strains are clearly identified, rapid progress in the development of improved strains can be expected.

In the meantime, transgenesis is even now offering the research scientist a possibility of investigating some of the questions which have eluded those neuropathologists dependent on human post-mortem samples. For example, whether the primary pathology is initiated within the cell or extracellularly, to determine the relationship, if any, between amyloid and fibrillary changes and whether amyloid is central to the dementia. Certainly, with little fine tuning, transgenesis will be an exciting and influential *in vivo* system in which to study human disease processes and disorders.

References

Abraham, C. R., Selkoe, D. J., and Potter, H. (1988). Immunochemical identification of the serine protease inhibitor α_1-antichymotrypsin in the brain amyloid deposits of Alzheimer's disease. *Cell*, 52, 487-501.

Bancher, C., Grundke-Iqbal, I., Iqbal, K., Kim, K. S., and Wisniewski, H. (1989). Immunoreactivity of neuronal lipofuscin with monoclonal antibodies of the amyloid β-protein. *Neurobiology of Aging*, 10, 125-32.

Beer, J., Salbaum, J. M., Schlichtmann, E., Hoppe, P., Early, S., Carlson, G., *et al* (1991). Transgenic mice and Alzheimer's disease. In *Alzheimer's disease: basic mechanisms, diagnosis and therapeutic strategies*, (ed. K. Iqbal, D. R. C. McLachlan, B. Winblad and H. M. Wisniewski), pp. 473-8. Wiley, Chichester.

Benowitz, L., Rodriguez, W., Paskevich, P., Mufson, E. J., Schenk, D., and Neve, R. L. (1989). The amyloid precursor protein is concentrated in neuronal lysosomes in normal and Alzheimer's disease subjects. *Experimental Neurology*, 106, 237-50.

Bredesen, D. E., Kane, D. J., Holtzman, D. M., and Epstein, C. J. (1991). Reaggregating cultures of mouse trisomy 16 brain. *Neuroscience Abstracts*, 421.1.

Brion, J. P., Couck, A. M., Passareiro, E., and Flament-Durand, J. (1985). Neurofibrillary tangles of Alzheimer's disease: an immunohistochemical study. *Journal of Submiscroscopy and Cytology*, 17, 89-96.

Burger, P. C., and Vogel, F. S. (1973). The development of the pathologic changes of Alzheimer's disease and senile dementia in patients with Down syndrome. *American Journal of Pathology*, 73, 457-68.

Caputo, C. B., Wischik, C., Sobel, I. R. E., Kirschener, D. A., Frazer, P. E., and Brunner, W. F. (1991). Possible contribution of the C-terminus of β-amyloid protein precursor (APP) to Alzheimer's paired helical filaments (PHFs). *Neuroscience Abstracts*, 573. 6.

Chartier-Harlin, M. -C., Crawford, F., Houlden H., Warren, A., Hughes, D., Fidani, L., *et al.* (1991). Early-onset Alzheimer's disease caused by mutations at codon 717 of the β-amyloid precursor protein. *Nature*, 353, 844-6.

Cox, D. R., and Epstein, C. J. (1985). Comparative gene mapping of human chromosome 21 and mouse chomosome 16. *Annals of the New York Academy of Science*, 450, 169-77.

Cox, D. R., Smith, A., Epstein, L. B., and Epstein, C. J. (1984). Mouse trisomy 16 as an animal model of human trisomy 21 (Down syndrome): production of viable 16↔diploid mouse chimeras. *Developmental Biology*, 101, 416-24.

Coyle, J. T., Oster-Granite, M. -L., Reeves, R. H., and Gearhart, J. D. (1988). Down syndrome, Alzheimer's disease and trisomy 16 mouse. *Trends in Neuroscience*, 11, 390-4.

Epstein, C. J. (1988). Mouse models for Down syndrome and Alzheimer's disease. *Discussions in Neuroscience*, 5, 127-34.

Esch, F. S., Keim, P. S., Beattie, E. C., Blacher, R. W., Culwell, A. R., Oltersdorf, T., *et al.* (1990). Cleavage of amyloid β peptide during constitutive processing of its precursor. *Science*, 248, 1122-4.

Estus, S., Golde, T. E., Kunishita, T., Blades, D., Lowery, D., Eisen, M., *et al.* (1992). Potentially amyloidogenic, carboxyl-terminal derivatives of the amyloid protein precursor. *Science*, 255, 726-8.

Fine, A., Ault, B., and Rapoport, S. I. (1991). Mouse trisomy 16 neurons, a model of human trisomy 21 (Down syndrome), can be maintained by intracerebral grafting. *Neuroscience Letters*, 122, 4-8.

Gearhart, J. D., Singer, H. S., Moran, T. H., Tiemeyer, M., Oster-Granite, M. -L., and Coyle, J. T. (1986). Mouse chimeras composed of trisomy 16 and normal (2N) cells: preliminary studies. *Brain Research Bulletin*, 16, 815-25.

Glenner, G. G., and Wong, C. W. (1984). Alzheimer's disease: an initial report on the purification and characterisation of a novel cerebrovascular amyloid protein. *Biochemical and Biophysical Research Communications*, 120, 885-90.

Goate, A., Chartier-Harlin, M. -C., Mullan, M., Brown, J., Crawford, F., Fidani, L., *et al.* (1991). Segregation of a missense mutation in the amyloid precursor protein gene with familial Alzheimer's disease. *Nature*, 349, 704-6.

Golde, T. E., Estus, S., Younkin, L. H., Selkoe, D. J., and Younkin, S. G. (1992). Processing of the amyloid protein precursor to potentially amyloidogenic derivatives. *Science*, 255, 728-30.

Gropp, A., Kolbus, U., and Giers, D. (1975). Systematic approach to the study of trisomy in the mouse II. *Cytogenetics and Cell Genetics*, 14, 42-62.

Heston, L. L., Mastri, A. R., Anderson, V. E., and White, J. (1982). Dementia of the Alzheimer type: clinical genetics, natural history and associated conditions. *Archives of General Psychiatry*, 38, 1085-90.

Hohmann, C. F., Capone, G., Oster-Granite, M. -L., and Coyle, J. T. (1990). Transplantation of brain tissue from murine trisomy 16 into euploid hosts: Effects of gene imbalance on brain development. *Progress in Brain Research*, 82, 203-14.

Johnson, S. A., Rogers, J., and Finch, C. E. (1989). APP-695 transcript prevalence is selectively reduced during Alzheimer's disease in cortex and hippocampus but not in cerebellum. *Neurobiology of Aging*, 10, 267-72.

Johnson, S. A., McNeill, T., Cordell, B., and Finch, C. E. (1990). Relation of neuronal APP_{751}/APP_{695} mRNA ratio and neuritic plaque density in Alzheimer's disease. *Science*, 248, 854-7.

Johnstone, E. M., Chaney, M. O., Norris, F. H., Pascual, R., and Little, S. P. (1991). Conservation of the sequence of the Alzheimer's disease amyloid peptide in dog, polar bear and five other mammals by cross-species polymerase chain reaction. *Molecular Brain Research*, 10, 299-305.

Kang, J., Lemaire, H. -G., Unterbeck, A., Salbaum, J. M., Masters, C. L., Grzeschik, K. -H., *et al.* (1987). The precursor of Alzheimer's disease amyloid A4 protein resembles a cell surface receptor. *Nature*, 325, 733-6.

Kawabata, S., Higgins, G., and Gordon, J. W. (1991). Amyloid plaques, neurofibrillary tangles and neuronal loss in brains of transgenic mice overexpressing a C-terminal fragment of human amyloid precursor protein. *Nature*, 354, 476-8.

Kidd, M. (1963). Paired helical filaments in electron microscopy of Alzheimer's disease. *Nature*, 197, 192-3.

Kitaguchi, N., Takahashi, Y., Tokushima, Y., Shiojiri, S., and Ito, H. (1988). Novel precursor of Alzheimer's disease amyloid protein shows protease inhibitory activity. *Nature*, 331, 530-2.

Koo, E. H., Sisodia, S. S., Archer, D. R., Martin, L. J., Weidemann, A., Beyreuther, K., *et al.* (1990). Precursor of amyloid protein in Alzheimer's disease undergoes fast anterograde axonal transport. *Proceedings of the National Academy of Sciences of the United States of America*, 87, 1561-5.

Kosik, K.S ., Joachim, C. L., and Selkoe, D. J. (1986). Microtubule-associated protein tau is a major antigenic component of paired helical filaments in Alzheimer's disease. *Proceedings of the National Academy of Sciences of the United States of America*, 83, 4044-8.

Mann, D.M.A., and Esiri, M.M. (1989). The pattern of acquisition of plaques and tangles in the brains of patients under 50 years of age with Down's syndrome. *Journal of the Neurological Sciences*, 89, 169-79.

Mann, D. M. A., Prinja, D., Davies, C. A., Ihara, Y., Delacourte, A., Defossez, A., *et al.* (1989). Immunocytochemical profile of neurofibrillary tangles in Down syndrome patients of different ages. *Journal of the Neurological Sciences*, 92, 247-60.

Masters, C. L. Simms, G., Weinman, N. A., Multhaup, G., McDonald, B. L., and Beyreuther, K. (1985a). Amyloid plaque core protein in Alzheimer's disease and Down syndrome. *Proceedings of the National Academy of Sciences of the United States of America*, 82, 4245-9.

Masters, C. M., Multhaup, G., Simms, G., Pottgeisser, J., Martins, R. N., and Beyreuther, K. (1985b). Neuronal origin of cerebral amyloid: neurofibrillary tangles of Alzheimer's disease contain the same protein as the amyloid of plaque core and blood vessels. *The European Molecular Biology Organization Journal*, 4, 2757-63.

Mori, H., Kando, J., and Ihara, Y. (1987). Ubiquitin is a component of paired helical filaments in Alzheimer's disease. *Science*, 235, 1641-4.

Morris, R. G. M. (1981). Spatial localisation does not require the presence of local cues. *Learning and Motivation*. 12, 239-49.

Mullan, M., Crawford, F., Axelman, K., Houlden, H., Lilius, L., Winblad, B., and Lannfelt, L. (1992). A pathogenic mutation for probable Alzheimer's disease in the APP gene at the N-terminus of β-amyloid. *Nature Genetics*, 1, 345-7.

Murrell, J., Farlow, M., Ghetti, B., and Benson, M. D. (1991). A mutation in the amyloid precursor protein associated with hereditary Alzheimer's disease. *Science*, 254, 97-9.

Naruse, S., Igarashi, S., Aoki, K., Kaneko, K., Iihara, K., Miyatake, T., *et al.* (1991). Mis-sense mutation Val↔Ile in exon 17 of the amyloid precursor protein gene in Japanese familial Alzheimer's disease. *Lancet*, 337, 978-9.

Oliver, C., and Holland, A. (1986). Down syndrome and Alzheimer's disease: a review. *Psychological Medicine*, 16, 307-22.

Orozoco, C. B., Smith, S. A., Epstein, C. J., and Rapoport, S. I. (1987). Electrophysiological properties of cultured dorsal root ganglion and spinal cord neurons of normal and trisomy 16 fetal mice. *Brain Research*, 429, 111-22.

Perry, G., Mulvihill, G., Manetto, V., Autilio-Gambetti, L., and Gambetti, P. (1987). Immuno-cytochemical properties of Alzheimer straight filaments. *Journal of Neuroscience*, 7, 3736-8.

Plioplys, A. (1991). Trisomy 16 mouse model of Alzheimer's disease. In *Alzheimer's disease: basic mechanisms, diagnosis and therapeutic strategies*, (ed. K. Iqbal, D. R. C. McLachlan, B. Winblad and H. M. Wisniewski), pp. 479-85. Wiley, Chichester.

Ponte, P., Gonzalez-DeWhitt, P., Schilling, J., Miller, J., Hsu, D., Greenberg, B., *et al.* (1988). A new A4 amyloid mRNA contains a domain homologous to serine proteinase inhibitors. *Nature*, 331, 525-7.

Price, D. L., Whitehouse, P. T., Struble, R. G., Coyle, J. T., Clarke, A. W., DeLong, M. R., *et al.* (1982). Alzheimer's disease and Down syndrome. *Annals of the New York Academy of Science*, 396, 145-6.

Quon, D., Wang, Y., Catalano, R., Scardina, J. M., Murakami, K., and Cordell, B. (1991). Formation of β amyloid protein deposits in brains of transgenic mice. *Nature*, 352, 239-41.

Reeves, R. H., Robakis, N. K., Oster-Granite, M. -L., Wisniewski, H. M., Coyle, J. T., and Gearhart, J.D. (1987). Genetic linkage in the mouse of genes involved in Down syndrome and Alzheimer's disease in man. *Molecular Brain Research*, 2, 215-21.

Reeves, R. H., Crowley, M. R., Lorenzon, N., Pavan, W. J., Smeyne, R. J., and Goldowitz, D. (1989). The mouse neurological mutant Weaver maps within the region of chromosome 16 that is homologous to human chromosome 21. *Genomics*, 5, 522-6.

Richards, S-J., Waters, J. J., Rogers, D., Martel, F., Sparkman, D. R., White, C. L., *et al.* (1990). Hippocampal grafts derived from embryonic trisomy 16 mice exhibit amyloid (A4) and neurofibrillary pathology. *Progress in Brain Research*, 82, 215-23.

Richards, S -J., Waters, J. J., Sparkman, D. R., White, C. L., III, Beyreuther, K., Masters, C. L., *et al.* (1991). Transplants of mouse Trisomy 16 hippocampus provide a model of Alzheimer's disease neuropathology. *The European Molecular Biology Organization Journal*, 10, 297-303.

Roher, A., Wolfe, D., Palutke, M., and KuKuruga, D. (1986). Purification, ultrastructure and chemical analysis of Alzheimer disease amyloid plaque core protein. *Proceedings of the National Academy of Sciences of the United States of America*, 83, 2662-6.

Rumble, C. M., Retallack, R., Hilbich, C., Simms, G., Multhaup, G., Martins, R., *et al.* (1989). Amyloid A4 protein of Alzheimer's disease and its precursors in Down syndrome. *New England Journal of Medicine*, 320, 1446-52.

Schubert, W., Prior, R., Weidemann, A., Dircksen, H., Multhaup, G., Masters, C. L., and Beyreuther, K. (1991). Localisation of Alzheimer's β/A4 amyloid precursor protein at central and peripheral synaptic sites. *Brain Research*, 563, 184-94.

Shivers, B., Hilbich, C., Multhaup, G., Salbaum, M., Beyreuther, K., and Seeburg, P. H. (1988). Alzheimer's disease amyloidogenic glycoprotein: expression pattern in rat brain suggests a role in cell contact. *The European Molecular Biology Organization Journal*, 7, 1365-70.

Sisodia, S. S., Koo, E. H., Beyreuther, K., Unterbeck, A., and Price, D. (1990). Evidence that β-amyloid protein in Alzheimer's disease is not derived by normal processing. *Science*, 248, 492-5.

Sparkman, D., and White, C. L., III (1989). Clinical and research applications of monospecific antibodies to the abnormal proteins of Alzheimer's disease. In *Familial Alzheimer's disease–molecular genetics and clinical perspectives*, (ed. G. D. Miner, R. W. Richter, J. P. Blass, J. L. Valentine and L. A. Winters-Miner), pp. 269-85. Marcel Dekker, New York.

Tanzi, R. E., McClatchey, A. I., Lamperti, E. D., Villa-Komaroff, L. T., Gusella, J. F., and Neve, R. L. (1988). Protease inhibitor domain encoded by an amyloid protein precursor mRNA associated with Alzheimer's disease. *Nature*, 331, 528-30.

Terry, R. D. (1963). The fine structure of neurofibrillary tangles in Alzheimer's disease. *Journal of Neuropathology and Experimental Neurology*, 22, 629-42.

Van Broeckhoven, C., Genthe, A. M., Vandenburghe, A., Horstemke, B., Backhovens, H., Raeymaekers, P., *et al.* (1987). Failure of familial Alzheimer's disease to segregate with the A4 amyloid gene in several European families. *Nature*, 329, 153-5.

Waters, J. J., and Bartlett, D. J. (1988). Direct trophoblast processing and disease. *Clinical Cytogenetics Bulletin*, 2, 30-1.

Wirak, D. O., Bayney, R., Ramabhadran, T. V., Fracasso, R. P., Hart, J. T., Hauer, P. E., *et al.* (1991). Deposits of amyloid β protein in the central nervous system of transgenic mice. *Science*, 253, 323-5.

Wischik, C. M., Novak, M., Edwards, P., Klug, A., Tichelaar, W., and Croucher, R. A. (1988). Structural characterisation of the core of the paired helical filament of Alzheimer disease. *Proceedings of the National Academy of Sciences of the United States of America*, 85, 4506-10.

Wisniewski, H. M. (1992). Data presented at the Keystone Symposia on Molecular and Cellular Biology—Advances in Understanding Neurodegenerative Disorders. Montana, USA.

Wisniewski, K. E., Wisniewski, H. M., and Wen, G. Y. (1985). Occurrence of neuropathological changes and dementia of Alzheimer's disease in Down syndrome. *Annals of Neurology*, 17, 282-7.

Yamaguchi, F., Richards, S -J., Beyreuther, K., Salbaum, M., Carlson, G. A., and Dunnett, S. B. (1991). Transgenic mice for the amyloid precursor protein 695 isoform have impaired spatial memory. *NeuroReport*, 2, 781-4.

Yamaguchi, H., Ishiguro, K., Shoji, M., Yamazaki, T., Nakazato, Y., Ihara, Y., and Hirai, S. (1990). Amyloid β/A4 protein precursor is bound to the neurofibrillary tangles in Alzheimer-type dementia. *Brain Research*, 537, 318-22.

Index

Abbreviations: AD, Alzheimer disease; DS, Down syndrome; NFT, neurofibrillary tangle; SP, senile plaque.

Note that the use in subentries of the term 'AD alone' or 'DS alone' pertains to references about either AD or DS exclusively.